D1083140

Advances in
INORGANIC CHEMISTRY
AND
RADIOCHEMISTRY

Volume 21

CONTRIBUTORS TO THIS VOLUME

Frank J. Berry

Jeremy K. Burdett

R. D. Cannon

J. Stephen Hartman

Maria de Sousa Healy

S. S. Krishnamurthy

Gleb Mamantov

Jack M. Miller

Antony J. Rest

A. C. Sau

N. R. Smyrl

M. Woods

Advances in

INORGANIC CHEMISTRY

AND

RADIOCHEMISTRY

EDITORS

H. J. EMELÉUS

A. G. SHARPE

University Chemical Laboratory
Cambridge, England

VOLUME 21

1978

ACADEMIC PRESS  New York San Francisco London

A Subsidiary of Harcourt Brace Jovanovich, Publishers

ACADEMIC PRESS, INC.
111 Fifth Avenue, New York, New York 10003

United Kingdom Edition published by
ACADEMIC PRESS, INC. (LONDON) LTD.
24/28 Oval Road, London NW1 7DX

LIBRARY OF CONGRESS CATALOG CARD NUMBER: 59–7692

ISBN 0–12–023621–4

PRINTED IN THE UNITED STATES OF AMERICA

CONTENTS

LIST OF CONTRIBUTORS ix

Template Reactions

MARIA DE SOUSA HEALY AND ANTONY J. REST

I. Introduction 1
II. Types of Template Effect 4
III. Template Synthesis of Macrocycles 7
IV. Choice of Templates for Specific Syntheses 33
V. Physical Studies of Template Reactions 34
VI. Applications of Template Reactions 36
References 37

Cyclophosphazenes

S. S. KRISHNAMURTHY, A. C. SAU, AND M. WOODS

I. Introduction 41
II. Synthetic Routes to Cyclophosphazenes 43
III. Halogen Replacement Reactions of Cyclophosphazenes 46
IV. Other Reactions of Cyclophosphazenes 66
V. Physical Methods 75
VI. Bonding and Electronic Structure 94
VII. Potential Applications 96
References 97

A New Look at Structure and Bonding in Transition Metal Complexes

JEREMY K. BURDETT

I. Introduction 113
II. Applications of the Angular Overlap Method 114
III. Conclusions 143
References 143

Adducts of the Mixed Trihalides of Boron

J. Stephen Hartman and Jack M. Miller

I. Introduction 147
II. Preparation, Detection, and Properties of Mixed Boron Trihalide Adducts 149
III. Mechanisms of Halogen Redistribution 158
IV. Equilibria in Halogen Redistribution 162
V. Donor-for-Halogen Exchange: Difluoroboron Cations 166
VI. NMR Applied to Adducts: Advantages and Pitfalls 167
VII. Conclusion 172
References 172

Reorganization Energies of Optical Electron Transfer Processes

R. D. Cannon

I. Introduction 179
II. Review of Data 185
III. Theory 211
References 225

Vibrational Spectra of the Binary Fluorides of the Main Group Elements

N. R. Smyrl and Gleb Mamantov

I. Introduction 231
II. Inorganic Binary Fluorides 232
III. Matrix Isolation Studies of Transient, Inorganic, Binary Fluoride Species 246
References 250

The Mössbauer Effect in Supported Microcrystallites

Frank J. Berry

I. Introduction 255
II. Iron and Iron Oxides 259

III. Tin 280

IV. Gold, Europium, and Ruthenium 281

References 282

SUBJECT INDEX 287

CONTENTS OF PREVIOUS VOLUMES 297

LIST OF CONTRIBUTORS

Numbers in parentheses indicate the pages on which the authors' contributions begin.

FRANK J. BERRY (255), *University of Cambridge Chemical Laboratory, Cambridge CB2 1EW, England*

JEREMY K. BURDETT (113), *Department of Inorganic Chemistry, The University, Newcastle upon Tyne NE1 7RU, England*

R. D. CANNON (179), *School of Chemical Sciences, University of East Anglia, Norwich NR4 7TJ, England*

J. STEPHEN HARTMAN (147), *Department of Chemistry, Brock University, St. Catharines, Ontario, Canada L2S 3A1*

MARIA DE SOUSA HEALY (1), *Department of Chemistry, The University, Southampton S09 5NH, England*

S. S. KRISHNAMURTHY (41), *Department of Inorganic and Physical Chemistry, Indian Institute of Science, Bangalore-560012, India*

GLEB MAMANTOV (231), *Department of Chemistry, University of Tennessee, Knoxville, Tennessee 37916*

JACK M. MILLER (147), *Department of Chemistry, Brock University, St. Catharines, Ontario, Canada L2S 3A1*

ANTONY J. REST (1), *Department of Chemistry, The University, Southampton S09 5NH, England*

A. C. SAU (41), *Department of Inorganic and Physical Chemistry, Indian Institute of Science, Bangalore-560012, India*

N. R. SMYRL* (231), *Department of Chemistry, University of Tennessee, Knoxville, Tennessee 37916*

M. WOODS (41), *Department of Chemistry, Birkbeck College, University of London, London WC1E 7HX, England*

* Present address: Union Carbide Corporation, Nuclear Division, P. O. Box Y, Oak Ridge, Tennessee 37830.

Advances in
INORGANIC CHEMISTRY
AND
RADIOCHEMISTRY

———

Volume 21

TEMPLATE REACTIONS

MARIA DE SOUSA HEALY and ANTONY J. REST

Department of Chemistry, The University, Southampton, England

I. Introduction 1
II. Types of Template Effect 4
  A. Kinetic 4
  B. Thermodynamic 5
  C. Equilibrium 6
III. Template Synthesis of Macrocycles 7
  A. Nitrogen-Donor Macrocycles 7
  B. Oxygen-Donor Macrocycles 22
  C. Sulfur-Donor Macrocycles 26
  D. Mixed Donor Macrocycles 27
IV. Choice of Templates for Specific Syntheses 33
V. Physical Studies of Template Reactions 34
VI. Applications of Template Reactions 36
  References 37

I. Introduction

During the past decade there has been a growing interest in the synthesis of multidentate ligands and in the complexes such ligands form with metal ions. The principal types of multidentate ligand are illustrated for quadridentate ligands, e.g., the tripod (**I**), the open chain (**II**), and the closed chain or macrocycle (**III**). Macrocycles may be single-ring systems, e.g., structures **IV–VIII**, or multiring systems, e.g., the cryptates (**IX** and **X**). The obvious resemblance between a

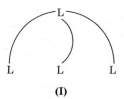

(I)

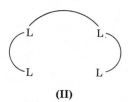

(II)

1

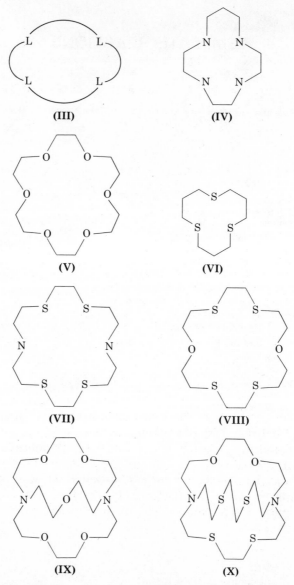

(III)   (IV)

(V)   (VI)

(VII)   (VIII)

(IX)   (X)

planar metal complex of structure (**IV**) and the prosthetic groups in hemoglobin, chlorophyll, and vitamin B_{12} has stimulated research on the synthesis of macrocycles and on the structure, bonding, and reactions of macrocycle–metal complexes.

For many years the synthesis of macrocycles has been a largely unsuccessful and wasteful endeavor because of the low yields, the many side products of the reactions, and the large volumes of solvents that were required to give sufficient dilution to minimize polymerization and encourage cyclization. Two exceptions were compounds **XI** and **XII**. Von Baeyer first reported the formation of **XI** from the

(XI)   (XII)

violent reaction of pyrrole with dry acetone on the addition of *one* drop of hydrochloric acid (*142*). Subsequently, the reaction was made less violent (*44–46*) and eventually compound **XI** was obtained with an 88% yield (*121*). Ackman, Brown, and Wright condensed furan with acetone in the presence of concentrated hydrochloric acid and produced an 18–20% yield of **XII** (*1*), but attempts to broaden the scope of the reaction, by using other carbonyl compounds and a variety of furans (*8, 22*), or to obtain mixed donor macrocycles from pyrrole and furan (*21*), gave low yields of macrocycles.

One of the first examples of a metal or metal salts facilitating the formation of a macrocycle was the self-condensation of *o*-phthalonitrile to give metal phthalocyanine complexes from which the free ligand was easily displaced (*89*):

$$4 \quad \xrightarrow{M/M^+/M^{2+}/M^{3+}} \quad \tag{1}$$

(M = Na, Mg, Cu, Ni, and Sb
M^+ = Na$^+$, K$^+$ and Cu$^+$;
M^{2+} = Ca^{2+}, Mg^{2+}, Ni^{2+}, and Cu^{2+}
M^{3+} = Fe^{3+})

free macrocycle

The role of the metal ions in promoting cyclization was not understood until much later when Hurley *et al.* (*74*) isolated a series of intermediates in the reaction between 1,3-diiminoisoindoline with nickel chloride. The widespread utilization of metal ions in the synthesis of macrocycles was developed largely through the work of the group led by Busch. The formation of macrocycles using Ni(II) ions (*136*) and complexes (*138*), e.g.,

$$(2)$$

$$Ni(H_2NCH_2CH_2S)_2 + 2CH_3I \rightarrow Ni(H_2NCH_2CH_2SCH_2)I_2 \qquad (3)$$

led Busch to recognize that the coordination sphere of the metal ion would hold the reacting groups in the correct positions for cyclization reactions, i.e., the metal ion acts as a "template." Since the time of these observations, many more examples of template reactions have been discovered; these are quoted in reviews on macrocycle synthesis and the properties of macrocycle–metal complexes (*9, 12, 23, 24, 30, 39, 85, 86, 88, 94, 110*).

II. Types of Template Effect

A. KINETIC

A reaction is described as proceeding by a kinetic template effect if it provides a route to a product that would not be formed in the absence of the metal ion and where the metal ion acts by coordinating the reactants. An alternative description for this process is the coordination template effect that more aptly describes how the stereochemistry imposed by the metal ion, through coordination, promotes a series of controlled steps in a multistep reaction, e.g., Scheme 1 (*17, 136, 137*).

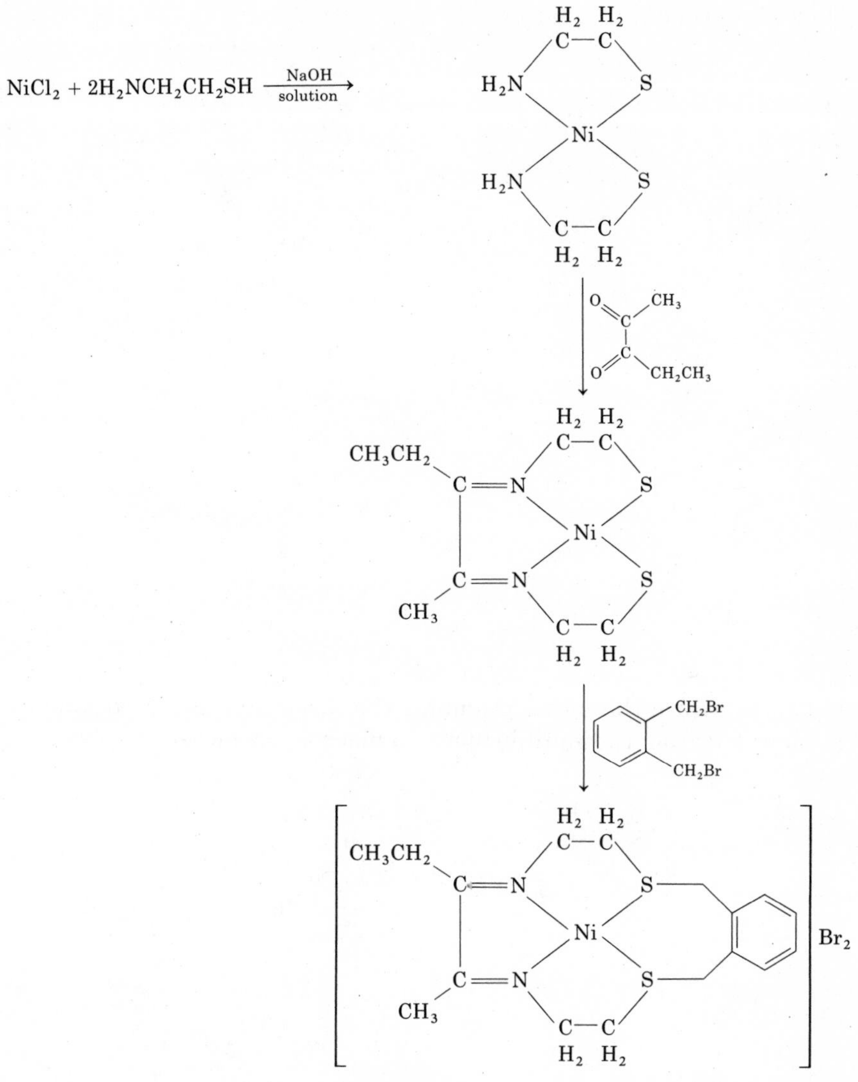

$$NiCl_2 + 2H_2NCH_2CH_2SH \xrightarrow[\text{solution}]{\text{NaOH}}$$

SCHEME 1

B. THERMODYNAMIC

Macrocycles formed by reactions that are described as proceeding by the thermodynamic template effect can take place in the absence

of metal ions (40),

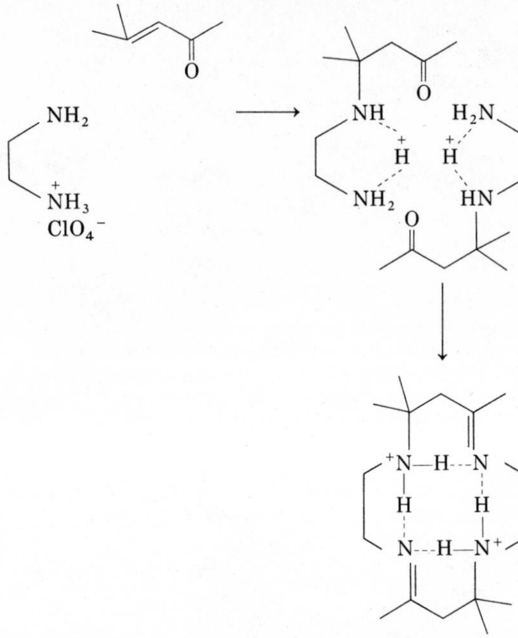

(4)

but in this case the metal promotes the formation of macrocycle by removing it from the equilibrium as a macrocycle-metal complex (39), e.g.,

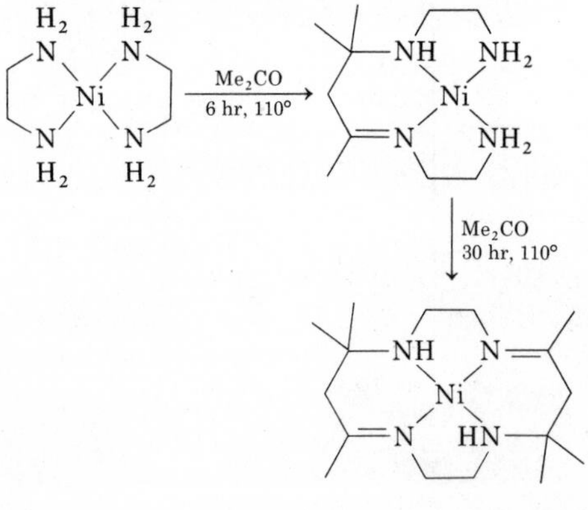

(5)

C. EQUILIBRIUM

The equilibrium template effect, so named by Thompson and Busch (137), is a combination of the two previous effects. In this case the reactants react reversibly to give an intermediate that forms a stable complex with the metal so that all the reactants proceed to a macrocycle–metal complex. The distinctive feature of the equilibrium effect is the formation of *different* products in the metal-assisted and metal-free reactions (*137*), e.g.,

(6)

tetradentate Schiff's base complex

Mixture of products including thiazoles, thiazolines, and mercaptals

whereas in the thermodynamic effect the two reactions give the *same* products.

III. Template Synthesis of Macrocycles

A requirement for a metal ion template has been established for the majority of reactions described in this section. Other reactions described are those in which metal ions, particularly Na^+, are part of the stoichiometry of the reactions but for which a template function has not been identified.

A. NITROGEN-DONOR MACROCYCLES

1. Macrocycles with Benzenoid Units

The self-condensation of *o*-aminobenzaldehyde has been the most studied reaction under this heading. In the absence of metal ions, self-condensation was found (*6, 127*) to be a very slow reaction that gave a mixture of products including a bisanhydrotrimer and a trisanhydrotetramer. The same reaction was repeated by McGeachin (*95*) and by Taylor *et al.* (*135*) and a structure (**XIII**) was assigned (*135*) to the bisanhydrotrimer. When compound **XIII** is heated with an equimolar amount of $[Ni(H_2O)_6](NO_3)_2$ in absolute ethanol for 3 hr, a complex of macrocycle **XIV** is formed. Eichhorn and Latif (*50*)

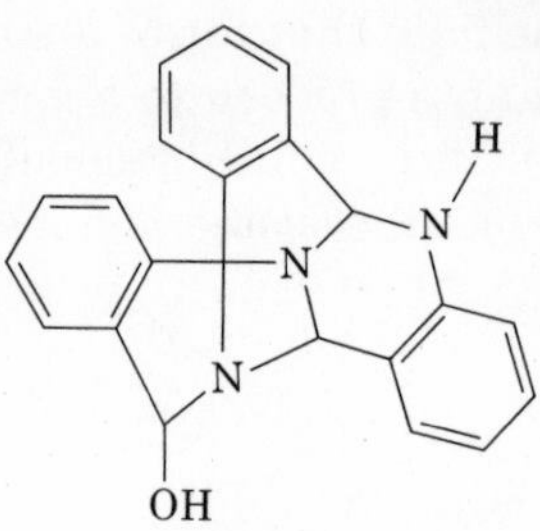

(XIII)

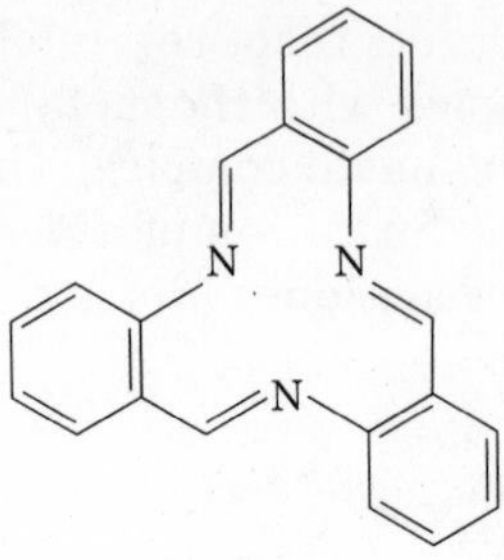

(XIV)

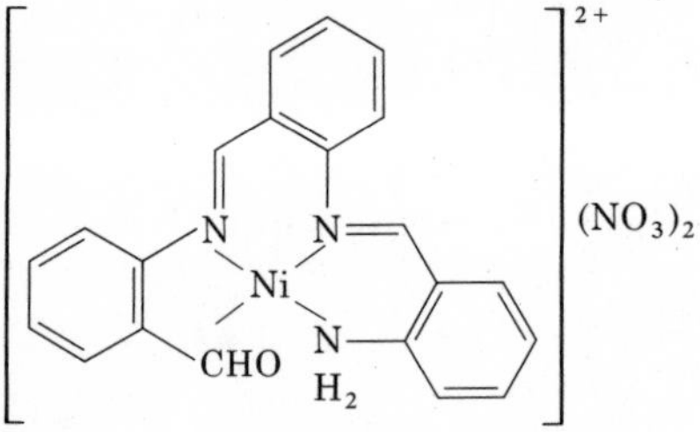

(XV)

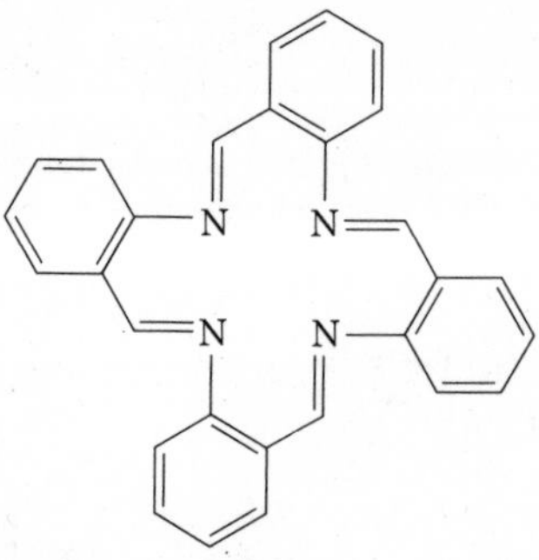

(XVI)

carried out the self-condensation of o-aminobenzaldehyde in the presence of divalent metal nitrates and found that Ni and Co formed M(II) complexes of the trisanhydrotrimer (**XIII**), which were assigned as having structure **XV**; Cu gave Cu(I) complex of **XIII**, and Mn gave uncomplexed **XIII**. Later investigation (*54, 96, 97, 135*) of the self-condensation of o-aminobenzaldehyde in the presence of Ni(II) salts showed that a mixture of complexes containing trimeric (**XIV**) and tetrameric (**XVI**) macrocycles {tribenzo[*b,f,j*][1,5,9]triazacycloduo-decine (TRI) and tribenzo[*b,f,j,n*][1,5,9,13]tetraazacyclohexadecine (TAAB), respectively} was formed. Nickel(II) complexes of **XIV**, i.e., [Ni(TRI)(H$_2$O)X$_2$], with pseudo-octahedral geometry around nickel, and of **XVI**, i.e. [Ni(TAAB)]X$_2$, with square planar geometry around nickel, have been obtained with a variety of counteranions (X$^-$ or Y^{2-}). In general, the nature of the complexes depends on the coordination geometry preference of the metal ion, e.g., self-condensation in the presence of Cu(II) ions gives only the square planar complex [Cu(TAAB)]$^{2+}$, whereas the same reaction with Co(II) ions [from CoBr$_2$ and with subsequent oxidation of Co(II) to Co(III) by concentrated HBr] has led to the isolation of both complexes **XIV** and **XVI** and also to the formation of octahedral complexes $[Co(TRI)_2]^{3+}$ (**XVII**) and $[Co(TAAB)X_2]^+$ (**XVIII**) (*34, 35*). Chemical reactions can be carried out on a macrocycle while it is bound to the metal, e.g., $[Ni(TAAB)]^{2+}$ reacts (*134*) with alkoxide ions to give compound **XIX**, and nucleophilic attack on $M(TAAB)^{2+}$ by bis(2-hydroxyethyl)methyl-amine or bis(2-hydroxyethyl)sulfide has led (*80*) to the formation of square pyramidal complexes (**XX**).

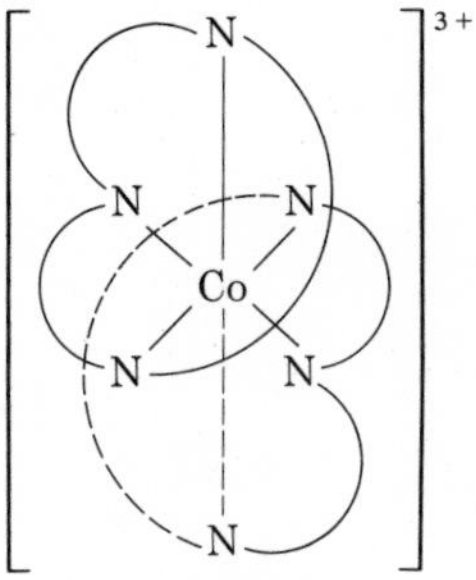

(**XVII**) 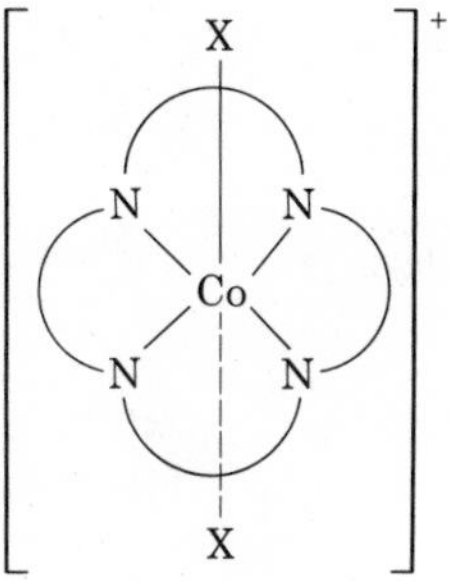 (**XVIII**)

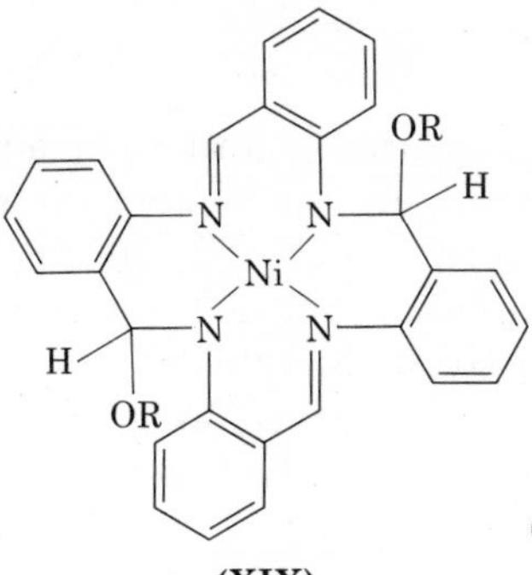

$(R = CH_3—, CH_3CH_2—)$

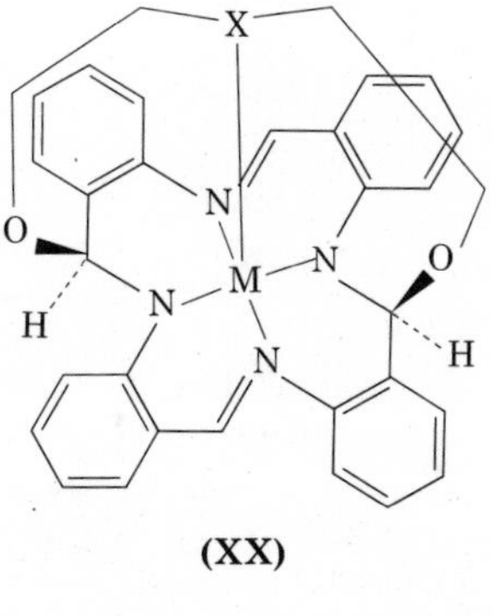

(**XIX**)

$(M = Ni, Cu$
$X = S, NCH_3)$

(**XX**)

Condensation of a series of diaminodialdehydes with ethylene-diamine, o-phenylenediamine, or 1,8-diaminonaphthalene in the presence of Ni(II), Co(II), and Cu(II) acetates has afforded a series of macrocyclic quadridentate complexes containing 14-, 15-, or 16-membered rings (10, 11, 16):

$$n = 2,3 \tag{7}$$

A template synthesis (28) has been used to form the macrocycle **XXI** by heating 1,1,3,3-tetramethoxypropane with concentrated HCl in ethanol, adding the templating agent (CuCl$_2$·2H$_2$O), and refluxing for 8 hr. A yellow-green solid is obtained from this solution and refluxing this with an ethanolic solution of o-phenylenediamine affords the copper complex of the macrocyclic ligand. Bromomalondialdehyde has been found to react rapidly with the diacetate metal complexes of o-phenylenediamine (72) to give the macrocycle **XXII**. The yield of

R = H   (XXI)
R = Br   (XXII)

complex **XXII** is 60% when Co(II) ions are used as the template agent, and almost quantitative when Cu(II) ions are used. This is a vast improvement in the yield of macrocycle and in the simplicity of the reaction over the alternative synthesis starting from propynal and using no metal ions (*68*). Macrocycles **XXIII–XXVI** can all be prepared by similar template reactions using bromomalonaldehyde with

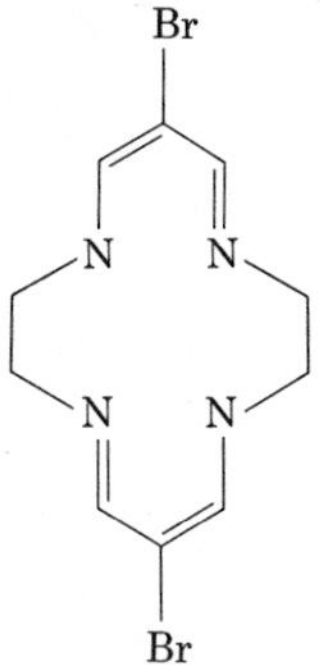

(XXIII)

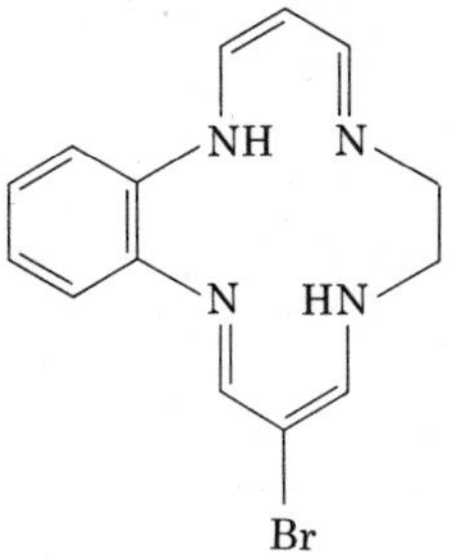

(XXIV)

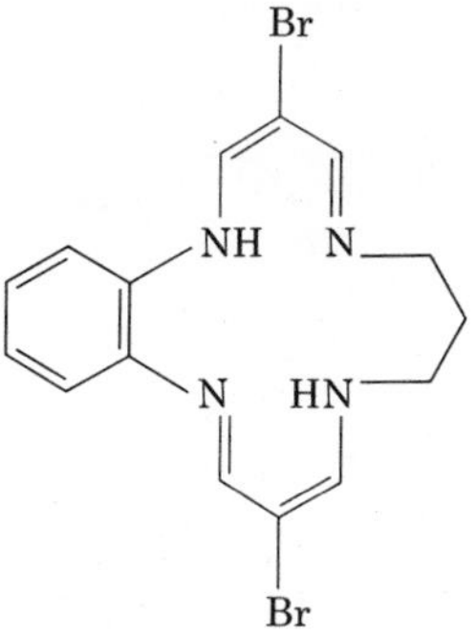

(XXV)

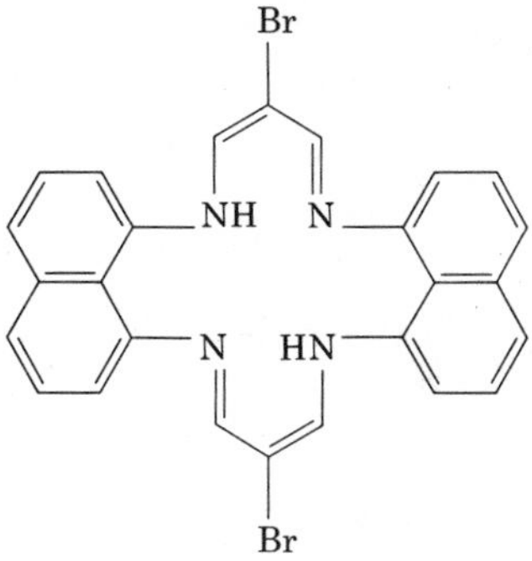

(XXVI)

ethylenediamine, namely, with a mixture of ethylenediamine and *o*-phenylenediamine, with a mixture of 1,3-diaminopropane and *o*-phenylenediamine, and with bis(1,8-diaminonaphthalene)copper(II) acetate, respectively. The bromine atoms provide centers in which side chains can be substituted to yield macrocycle complexes that resemble metalloporphyrins.

The free ligand (**XXVII**) was first prepared by a template condensation of *o*-phenylenediamine with pentane-2,4-dione in the presence of Ni(II) ions (*56*), with removal of the metal ion by subsequent reaction with anhydrous HCl in ethanol, and then isolated as the hydrochloride

salt. A slight modification (**XXVIII**) can be achieved as a complex of
Fe(II) or Fe(III) (*57*), and a Zn(II) complex of **XXIX** is also known
(*99*). The original analog (**XXX**) was prepared by Jager (*75*). At high
dilution and in the presence of a trace of H_2SO_4, *o*-phenylenediamine
reacts with 2,6-diacetylpyridine to yield a hexadentate macrocycle
(**XXXI**) which is known to form a binuclear Cu(II) complex (*130*).

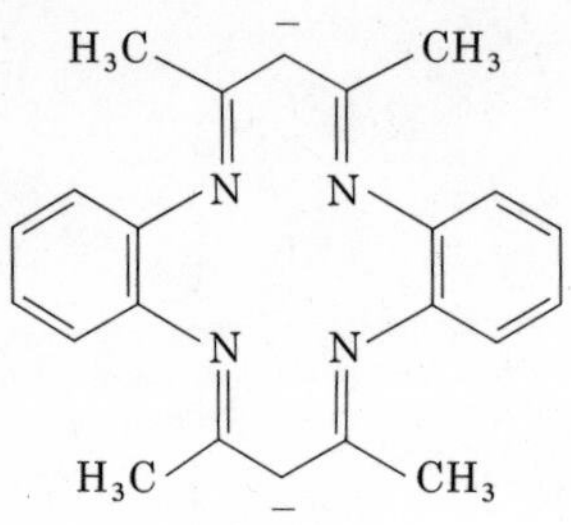

(**XXVII**)

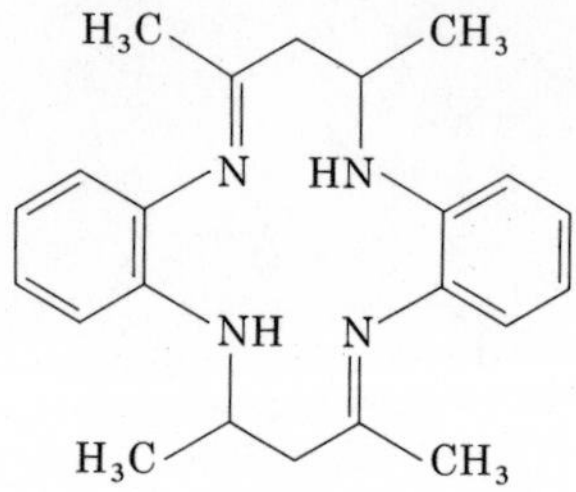

(**XXVIII**)

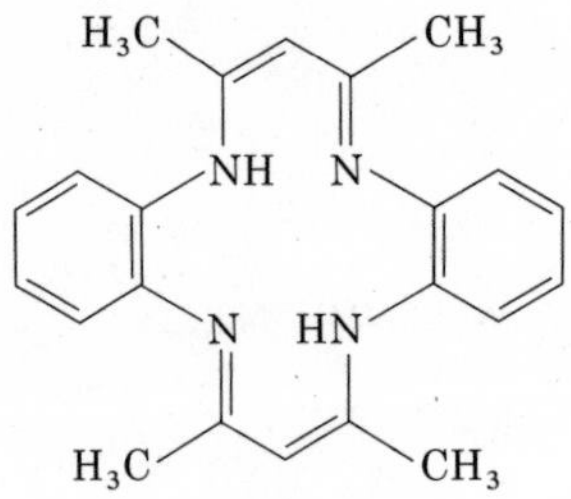

(**XXIX**)

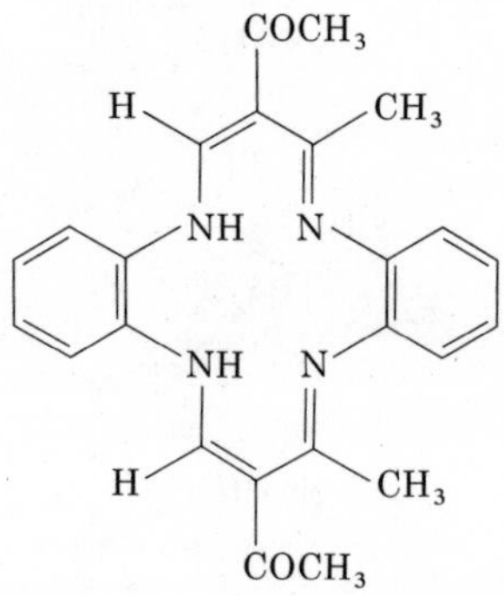

(**XXX**)

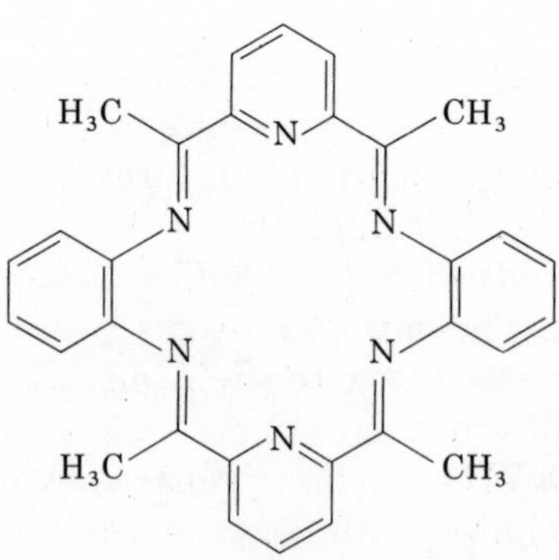

(**XXXI**)

Ochai and Busch (*101*) briefly mention the synthesis of a macro-cycle containing a pyridyl unit (**XXXII**) from the condensation of 2,6-diacetylpyridine with bis(3-aminopropyl)imine in the presence of Ni(II) ions. A similar condensation (*38*), using the same pyridyl source, with tetraethylaminetetramine in the presence of Fe(II) ions produces macrocycles **XXXIII** and **XXXIV**, containing five and six donor

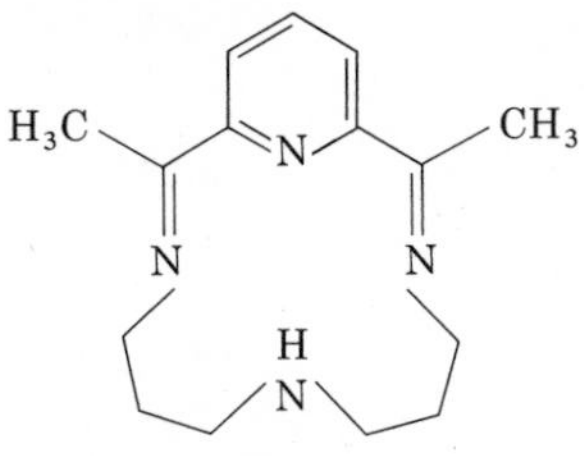

H_3C CH_3

(**XXXII**)

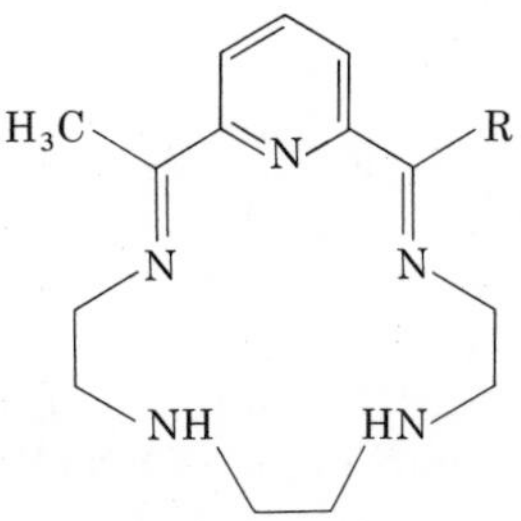

H_3C R

(**XXXIIIa**) R = Me
(**XXXIIIb**) R = H

H_3C CH_3

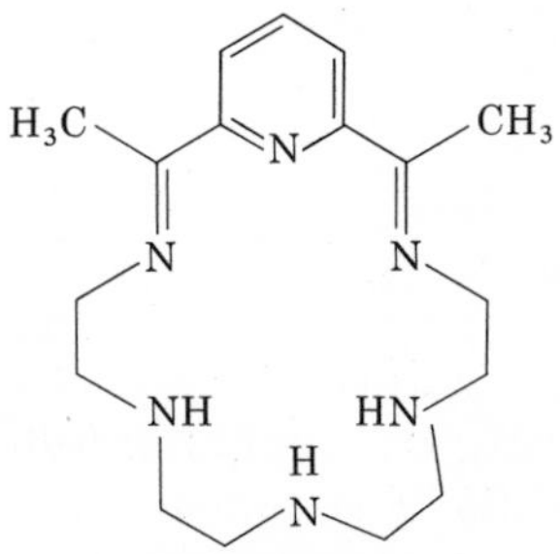

(**XXXIV**)

atoms, respectively. Macrocycle **XXXII** acts as a ligand with a number of metals, e.g. Cu(II) (*116*), Ni(II) complexes of the reduced ligand that give rise to meso and racemic forms (*79*), and Co(II) (*91*). The interesting complexes [RCo(**XXXII**)X]Y [R = alkyl, X = halide, Y = $B(C_6H_5)_4^-$ or PF_6^-] have been isolated; they can be considered as model compounds for vitamin B_{12} because reduction affords neutral Co(I) complexes from which Co(III) complexes can be obtained by oxidative alkylation (*52*). Macrocycle **XXXIII** forms unusual seven-coordinate Fe(III) complexes (*53, 98*). A comprehensive study of the formation of ligands of the types **XXII–XXXIV** gives independent confirmatory evidence for the operation of the template effect (*113*).

Other macrocycles containing pyridyl units (**XXXV–XXXVIII**) have been synthesized on Co(II) and Ni(II) templates by Lewis and Wainwright (*84*). In the case of Ni(II) the free macrocycle **XXXVIII** can be generated; it reacts almost quantitatively with Fe(II) ions to give an air-stable high-spin complex. However, most macrocycles containing pyridyl units are derived from 2,6-diacetylpyridine (*58*), e.g.,

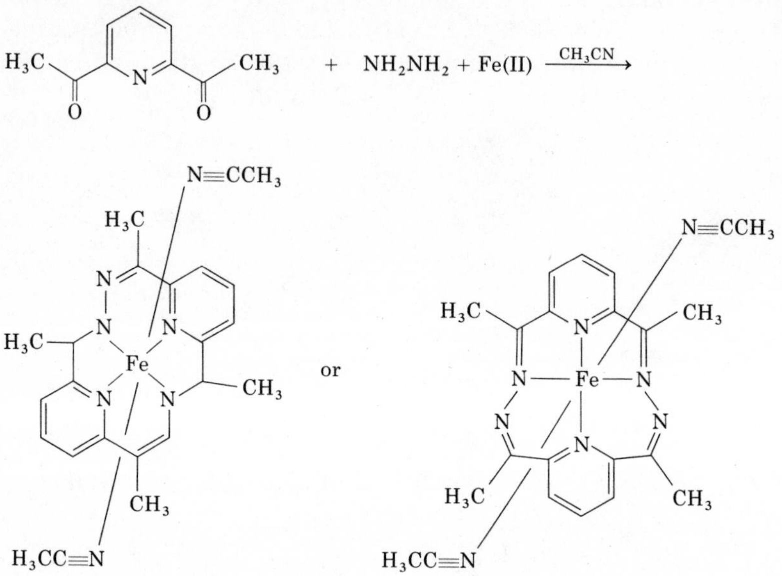

$$\text{(8)}$$

In the synthesis of **XXXIX** when using an Mg(II) template (*49*), the key reagent is 2,6-diacetylpyridine. The formation of **XXXIX** is a particularly significant result because in nature Mg(II) is bound to a porphyrin in chlorophyll, and in this reaction Mg(II) has been used successfully to cyclize a nitrogen macrocycle, thereby giving a possible indication of the conditions required for the biological synthesis of chlorophyll. The macrocycles were produced in 40–60% yield in the presence of Mg(II) ions, but in the absence of metal ions the products were of a polymeric nature.

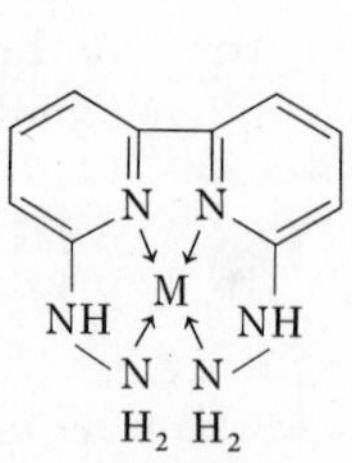

(M = Ni, Co)

(XXXV)

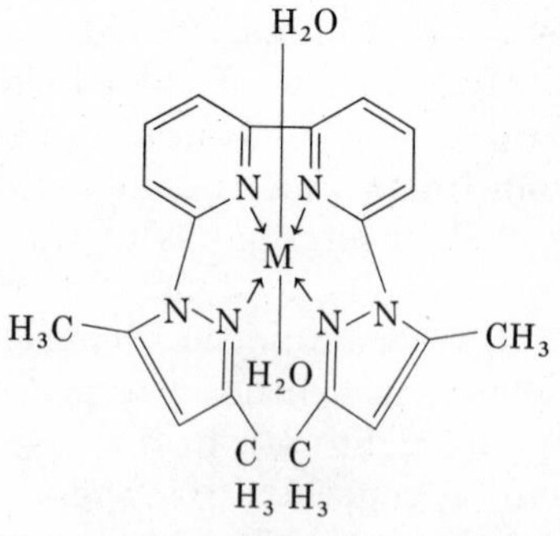

(XXXVI)

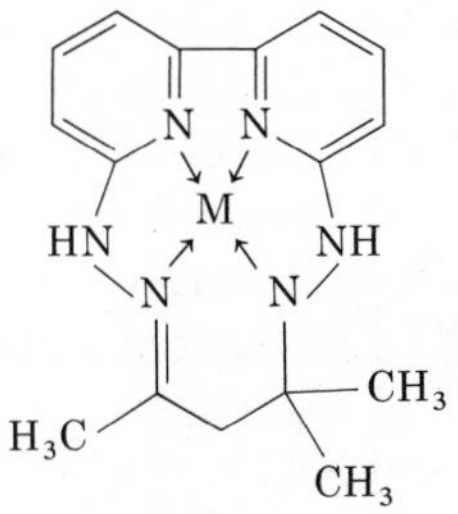

(XXXVII)

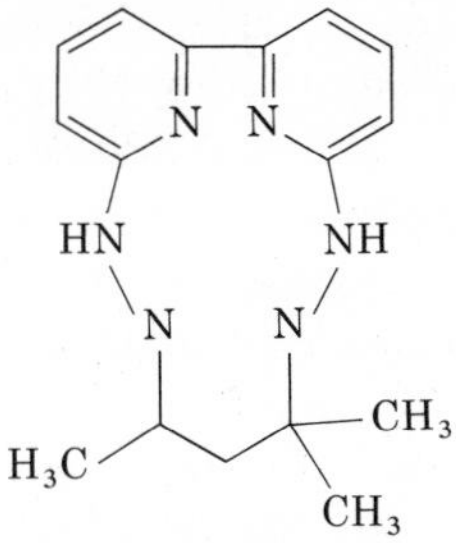

(XXXVIII)

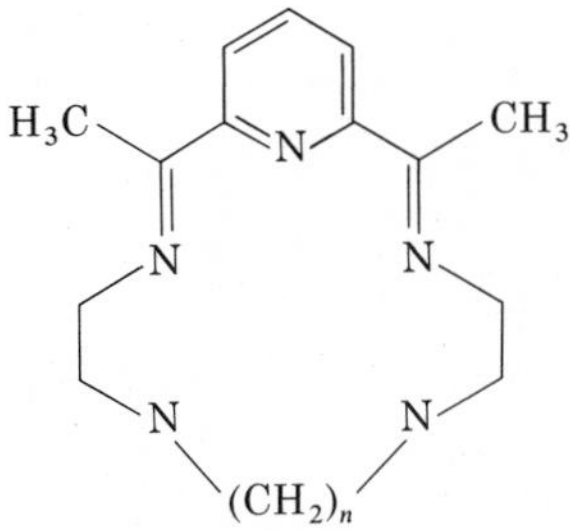

(XXXIX)

Condensation of the symmetric dialdehyde XL with diamines affords quadridentate complexes (XLI) when M(II) (M = Co, Ni, Cu) acetates are used as template agents, and hexadentate complexes (XLII) in the

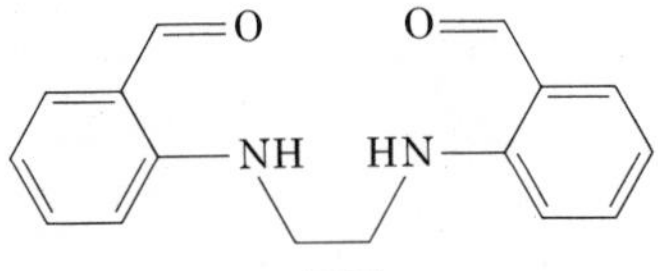

(XL)

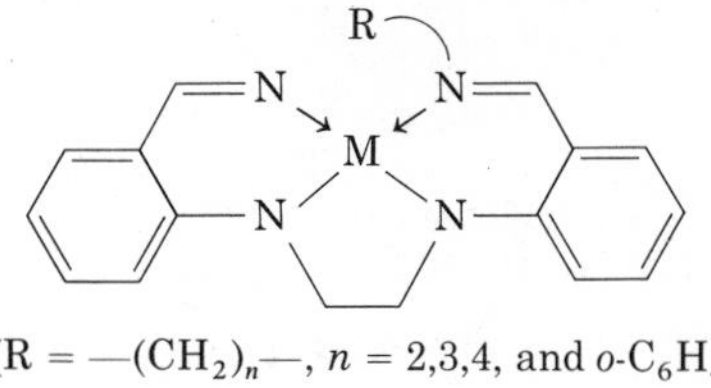

$(R = —(CH_2)_n—, n = 2,3,4,$ and $o\text{-}C_6H_4)$

(XLI)

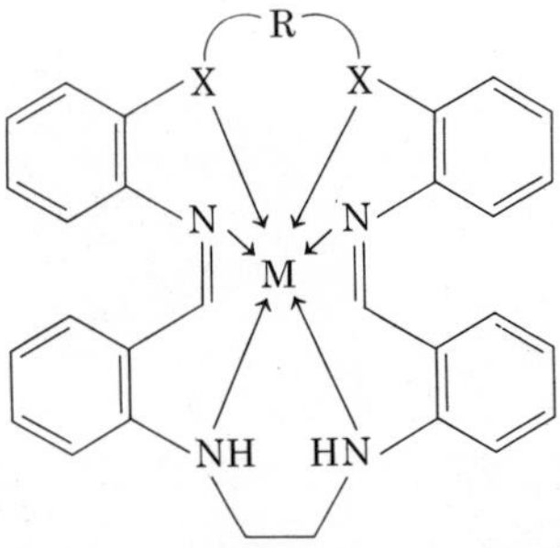

(XLII)

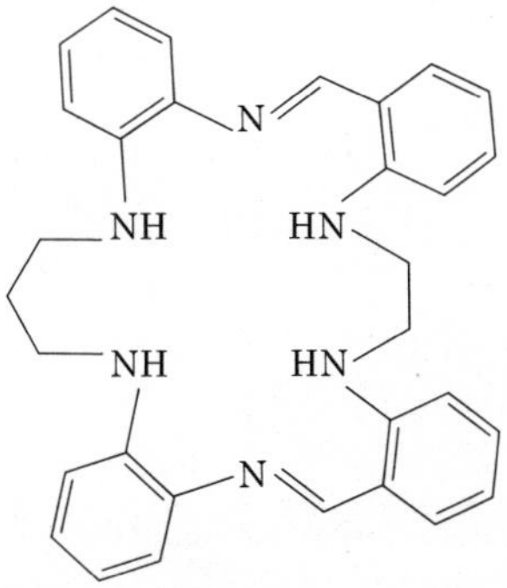

(XLIII)

presence of M'(II) (M' = Fe, Co, Ni, Zn) perchlorates (55, 133). In the absence of metal ions, the dialdehyde (**XL**) gives no detectable amounts of macrocyclic ligands when condensation reactions with diamines are attempted, and, hence, a kinetic template effect (Section II,A) is operating in these reactions. A detailed study of the reactions leading to the formation of complexes of the type **XLI** presented confirmatory evidence for template effects and also the occurrence of Lewis acid catalysis (62). The free ligands, e.g., **XLIII** and **XLIV**, may be liberated by reacting acetone solutions of the complexes with excess pyridine (54, 122).

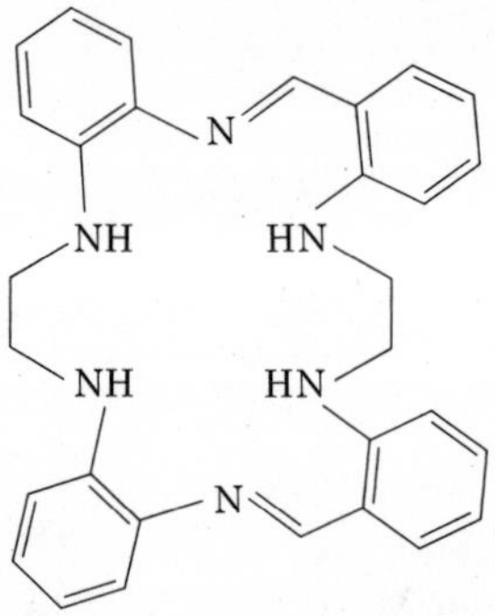

(XLIV) 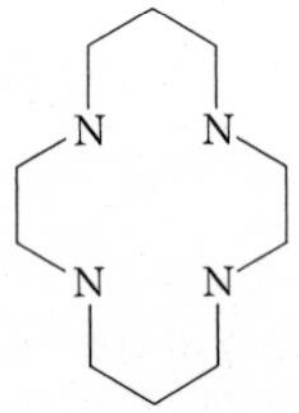   (XLV)

2. Nonbenzenoid Macrocycles

One of the simplest examples of a nitrogen-containing macrocycle is cyclam (**XLV**), which was synthesized with a 3% yield from the condensation of 1,3-bis(2'-aminoethylamino)propane and 1,3-dibromoethane in alcoholic KOH at high dilution (18). A more recent synthesis,

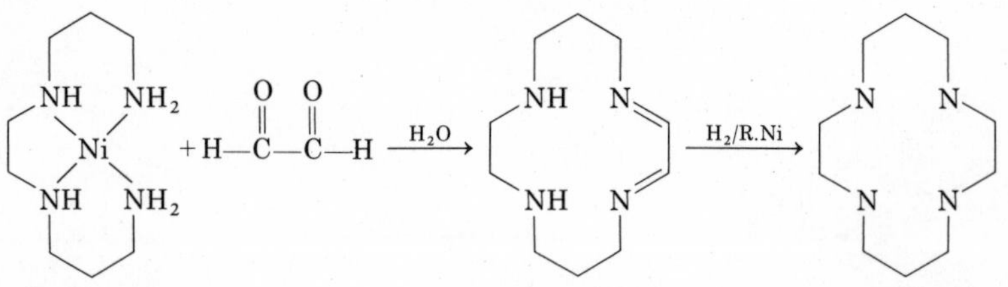

(9)

(XLV)

still only achieved a 20% yield (7). These moderate yields are typical of syntheses carried out in the absence of templating metals, although

some elegant syntheses have been designed to give macrocycles, e.g.,
XLVI–XLVIII, which have subsequently formed complexes with

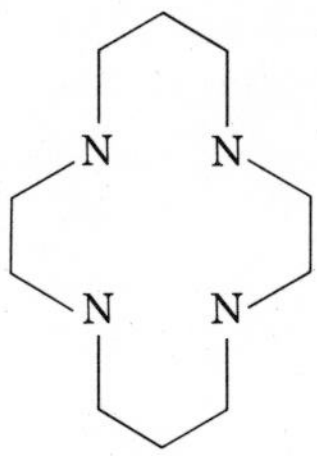

(XLV)

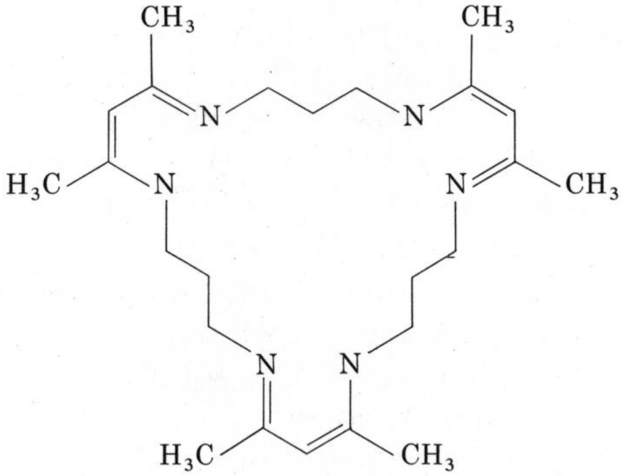

(XLVI)

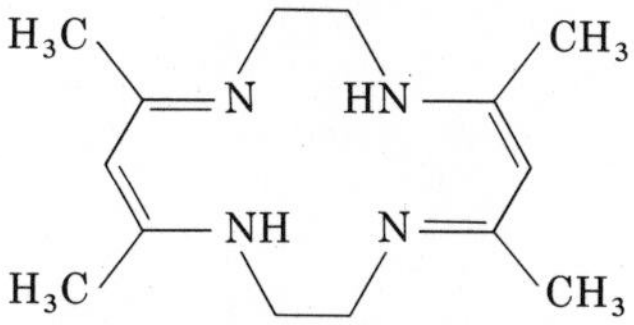

(XLVII)

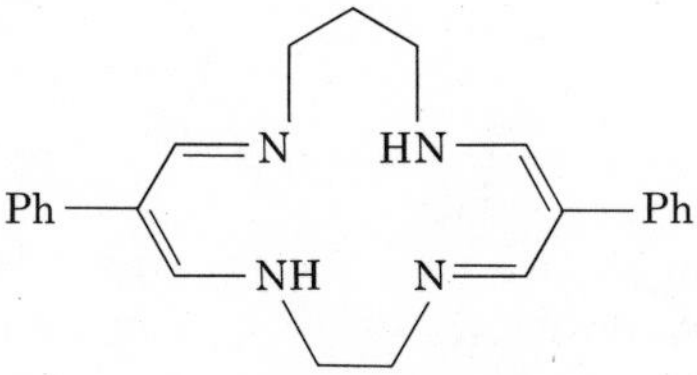

(XLVIII)

transition metals (*132*). Nitrogen analogs of crown ethers (Section
III,B) containing 9- to 21-membered rings with 3–7 heteroatoms have

also been synthesized by nontemplate means (*117*), e.g.,

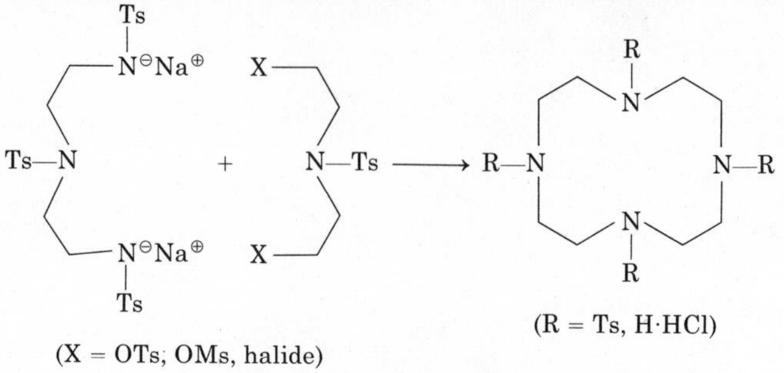

Ts
|
—N⊖Na⊕   X—

Ts—N   +   N—Ts   ⟶   R—N

—N⊖Na⊕   X—
|
Ts

(X = OTs, OMs, halide)

R—N ⟶ (R = Ts, H·HCl)

(10)

and in some cases, depending on the nature of X, very respectable yields (40–90%) have been obtained.

Most of the template syntheses of nonbenzenoid macrocycles originated with Curtis (*39*) and involve the condensation of metal–amine complexes with aliphatic carbonyl compounds, e.g., the reaction of acetone with tris(diaminoethane)nickel(II) perchlorate at ambient temperature leads to the isolation of three products, two of which may be represented as *cis*-**XLIX** and *trans*-**L** and the other is formed by a further interconversion of complex **L** in solution (*39, 143*). With Cu(II) diaminoperchlorates, a mixture of cis and trans complexes analogous to **XLIX** and **L** is formed, but with Co(II) only the trans analog of **L** has been isolated. When ketones containing bulky groups are used, the reaction is much slower, e.g., there is only a small yield of **LI** from

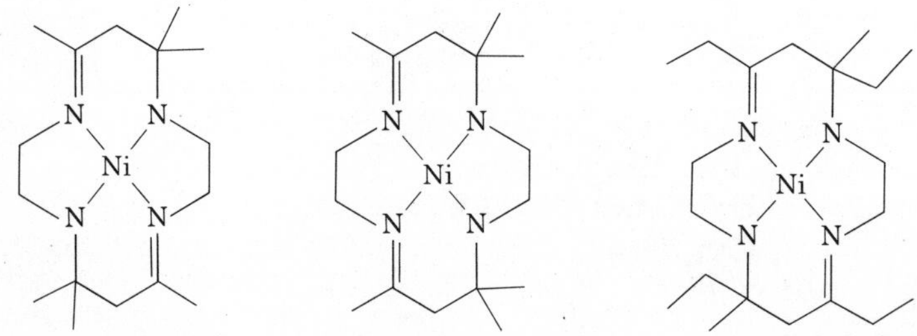

(L)   **(XLIX)**   **(LI)**

tris(diaminoethane)nickel(II) perchlorate and 2-butanone after heating at 100°C for many hours. A variety of carbonyl compounds may be used for condensation, e.g., isobutyraldehyde gives a 13-membered ring macrocycle (**LII**). A Curtis-type macrocycle (**LIII**) has been reported to form low-spin five-coordinate Co(II) and six-coordinate

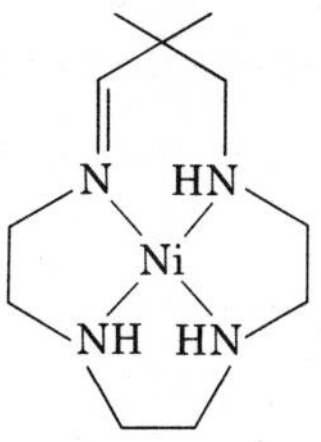

(LII) 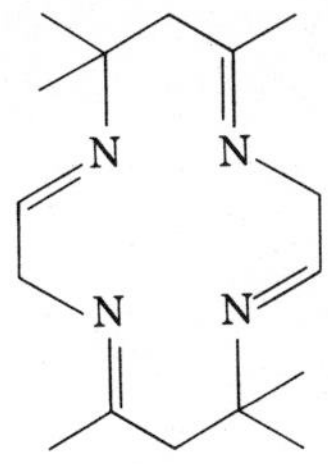   (LIII)

Co(III) complexes and also Fe(II) complexes of 13-, 14-, 15-, and 16-membered ring compounds related to macrocycle **LIII** have been prepared (131).

It has recently been reported (36, 37) that the condensation of triethylenetetramine with a β-diketone in the presence of Ni(II) ions gives rise to macrocyclic complexes (**LIV**) that undergo protonation at the methine carbon in acid solution to give a complex with a neutral macrocycle (**LV**), characterized as the $PF_6{}^-$ salt. Analogous Cu(II)

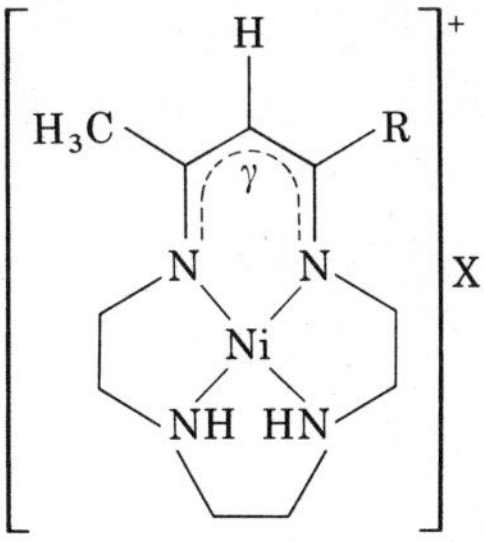

(LIV) 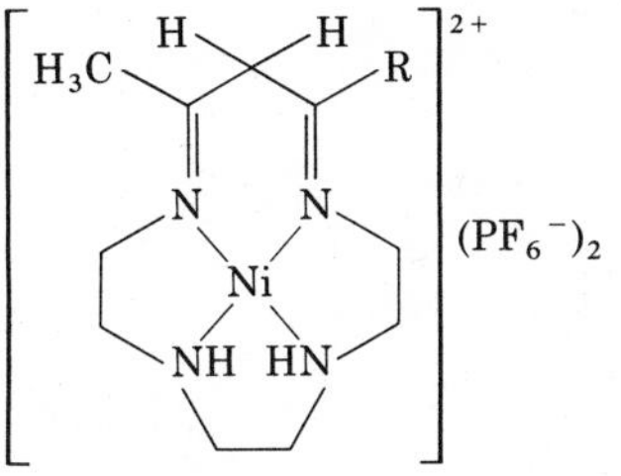   (LV)

complexes, [Cu(AT)]X and [Cu(ATH)]X$_2$ (R = CH$_3$), have been obtained (92) from the reaction of equimolar amounts of triethylenetetramine and acetylacetone in the presence of Cu(II) ions. Reduction of **LIV** and **LV** with Raney Ni/H$_2$ affords saturated macrocycle complexes of Ni(II) (71).

3. Unusual Macrocycles

Nitrogen-donor macrocycles have also been prepared with other heteroatoms in the ring but not coordinated to the metal. Schrauzer (124) observed that chelate complexes (**LVI**) could be obtained when certain metal salts were present in the reaction of borinic acid ester or anhydride with oximes. Umland and Thierig (*141*) obtained a macro-cycle (**LVII**) by mixing solutions of aqueous nickel(II) salts, amyl

(**LVI**)   (**LVII**)

(**LVIII**)

alcohol, methanolic dimethylglyoxime, and diphenylborinic acid aminoethyl ester and boiling for 1 to 2 min. The same macrocycle can be obtained as complexes with Fe(II), Cu(II), and Pd(II) ions, and these reactions can be described as kinetic template reactions. Ligand **LVIII** was synthesized (*104*) from tris(2-aldoximo-6-pyridyl)phosphine re-acting with $M(BF_4)_2$ [where M = Fe(II), Co(II), Ni(II), Zn(II)] and, on distilling with BF_3, a cation is encapsulated. This synthesis was modeled on that for the clathro-chelate (**LIX**), where Co(III) was the encapsulated ion (*19*).

Nitrogen-donor macrocycles can bind two metal ions, e.g., the re-action between acetone and 1,4-dihydrazinophthalazine is promoted by Ni(II) ions and can give two products: $[Ni_2(daph)]^{4+}$ (**LX**) and $[Ni_2(taph)]^{4+}$ (**LXI**), depending on the solvent and temperature used for the reaction (*119*).

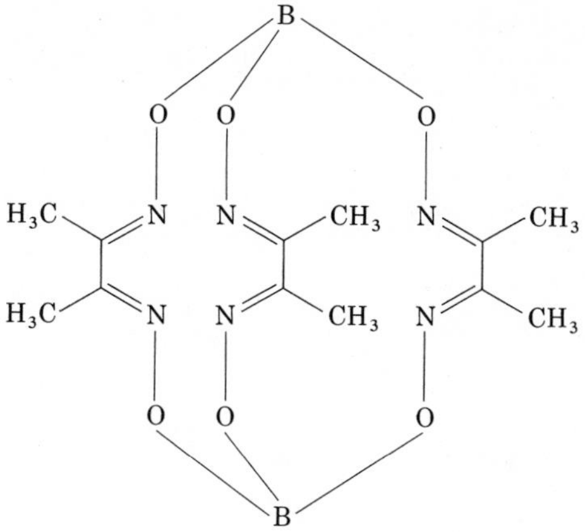

(LIX)

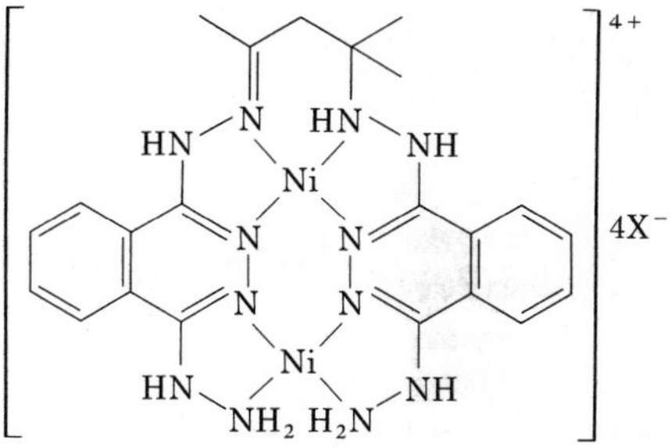

$(X^- = NCS^-, ClO_4^-)$

(LX)

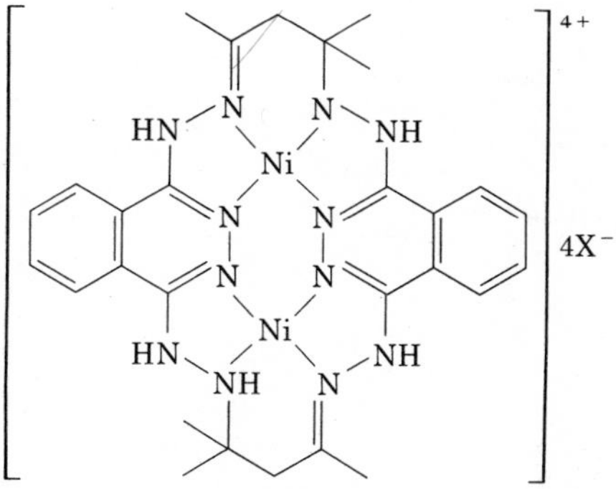

$(X^- = BF_4^-, NCS^-)$

(LXI)

B. OXYGEN-DONOR MACROCYCLES

A great variety of polyether macrocycles, containing from 3 to 20 oxygen atoms and with ring sizes from 9 to 60, have been prepared by Pedersen (105–108). The synthetic routes used in cyclic polyether formation all require NaOH, but a template role for the Na^+ ion has not been established:

$$\text{(OH)}\ + 2NaOH + Cl\text{---}R\text{---}Cl \longrightarrow \text{(O---R---O)} + 2NaCl + 2H_2O \quad (11)$$

e.g., 2,3-benzo-1,4,7,10,13-pentaoxacyclopentadeca-2-ene [R = —$(CH_2CH_2O)_3CH_2CH_2$]

$$+ 2NaOH + Cl\text{---}T\text{---}Cl \longrightarrow$$
$$+ 2NaCl + 2H_2O$$

e.g., 2,3,9,10-dibenzo-1,4,8,11-tetraoxacyclotetradeca-2,9-diene [S = T = —$(CH_2)_3$—]

$$(12)$$

$$2\ \text{(OH)} + 4NaOH + 2Cl\text{---}U\text{---}Cl \longrightarrow + 4NaCl + 4H_2O$$

e.g., 2,3,11,12-dibenzo-1,4,7,10,13,16-hexaoxacyclooctadeca-2,11-diene [U = —$(CH_2CH_2O)_3CH_2CH_2$]

$$(13)$$

$$2\ \text{(OH, V---Cl)} + 2NaOH \longrightarrow + 2NaCl + 2H_2O \quad (14)$$

e.g., 2,3,16-17-dibenzo-1,4,15,18-tetraoxacyclooctadeca-2,16-diene [V = —$(CH_2)_{10}$—]

An interesting type of macrocycle containing ether–ester ligands (**LXII**) has been synthesized by Bradshaw (20), and other nontemplate

(**LXII**)

routes to oxygen donor macrocycles, e.g., **LXIII** (*3*) and **LXIV** (*100*) are shown by the following reactions:

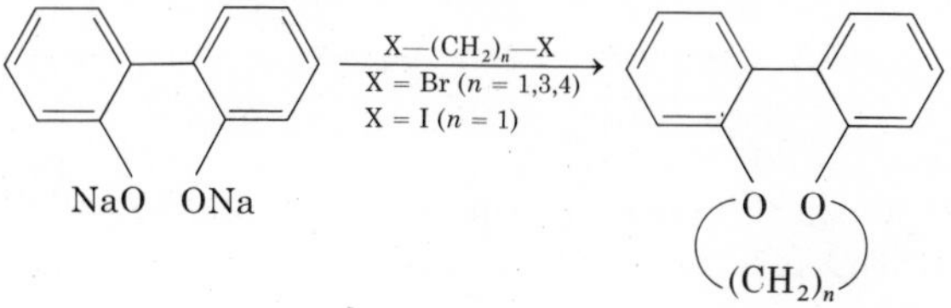

$$\text{(15)}$$

(LXIII)

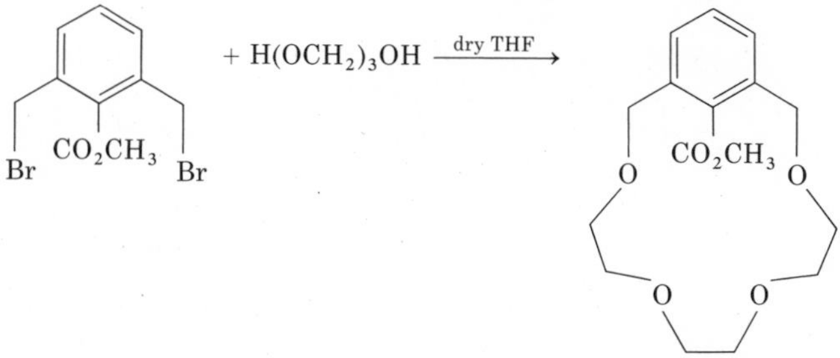

$$\text{(16)}$$

(LXIV)

In contrast to the nontemplate function of Na$^+$ ions, the K$^+$ ion acts as a template when an equimolar mixture of diol and a ditosylate (Ts) of the same or a different diol dissolved in benzene is added slowly to potassium-*t*-butoxide to give polyethers (*42, 43*) with $n = 5$ (20% yield), $n = 6$ (33% yield), $n = 7$ (26% yield), and $n = 8$ (15% yield), e.g.,

$$\text{HO(CH}_2\text{CH}_2\text{O)}_{n-m}\text{—H} + \text{TsO—(CH}_2\text{CH}_2\text{O)}_m\text{—Ts} \xrightarrow[\text{benzene}]{\text{KOBu}^t} \overline{\text{(CH}_2\text{CH}_2\text{O)}_n} + 2\text{KOTs}$$

$$(m > 1)$$

$$\text{(17)}$$

The K$^+$ ion has been shown (*63*) to act as a template in the synthesis of three crown ethers, e.g., 18-crown-6 (**V**) according to the reaction

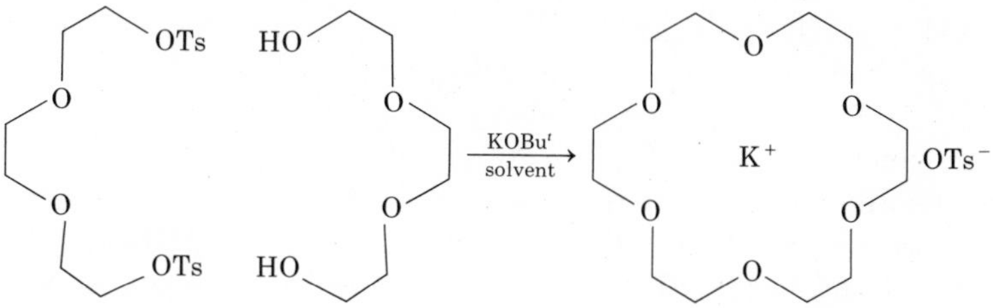

$$\text{(18)}$$

(V)

Because complex **V** binds strongly to K^+, the template synthesis may be considered to occur via **XLV** and **XLVI** that undergo ring-closure reactions. An alternative template synthesis of complex **V** has been described (*59*), which uses more readily available reagents and eliminates the need to synthesize tosylates as starting materials. Triethylene glycol and triethylene glycol dichloride are condensed together in a 10% aqueous solution of tetrahydrofuran in the presence of KOH, with the K^+ ion as the template.

(LXV)   (LXVI)

An improved synthesis of crown ethers that involves the cyclization of ethylene oxide using BF_3 in the presence of fluoroborate, fluorophosphate, and fluoroantimonate salts of the alkali, alkaline earth, and transition metals has been described (*41*). Cyclic tetramers, pentamers, and hexamers are formed in a template reaction where the ring sizes favored are those which complex most strongly with a particular cation, e.g., $Ca(BF_4)_2$ gives 50% tetramer, $Cu(BF_4)_2$ and $Zn(BF_4)_2$ gives 90% pentamer, and $Rb(BF_4)$ and $Cs(BF_4)$ give exclusively hexamer. Presumably the cation coordinates to the chain to facilitate cyclization and, because such a complex is already positively charged, complex formation prevents secondary reactions via oxonium salts that would degrade the growing chain to dioxane.

The reaction between acetone and furan in the presence of strong acid has been shown to exhibit a pronounced increase in the yield of complex **XII** with the addition of Group I and Group II metal salts (*27*). Yields of **XII** have been improved, e.g., 40–45% by the addition of $LiClO_4$ and 39–43% by the addition of perchlorates of calcium, magnesium, and zinc, from the 18–20% yield obtained (*1*) without added metal salts. No increase in yield was obtained when tetrabutylammonium perchlorate was added, thus proving the template action of the Group I and Group II metal ions (*27*). Hydrogenation of **XII** gives the saturated macrocycle **LXVII** that shows much stronger donor properties than the unsaturated macrocycle (**XII**). Recently, transition metal templates have been found to give enhanced yields of **XII** with $Ni(ClO_4)_2$ affording a yield comparable to $Mg(ClO_4)_2$ (*115*).

(LXVII)

The use of a Li^+ ion template (*27*) has been extended (*31*) to the synthesis of a crown compound,

$$(19)$$

and the same workers have also described the use of a Na^+ ion template,

$$(20)$$

A template synthesis using Group I metal ions has recently been developed (*114*) for producing crown ether compounds from substituted benzenes, furans, and thiophenes, e.g., the reactions of 1,2-bis(bromo-ethyl)benzene with disodium or dipotassium glycolates produced polyethers (**LXVIII** and **LXIX**) together with polymeric material,

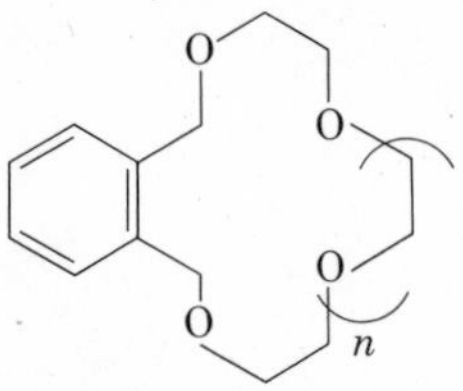

(n = 0–6)
yields 1–50%

(LXVIII)

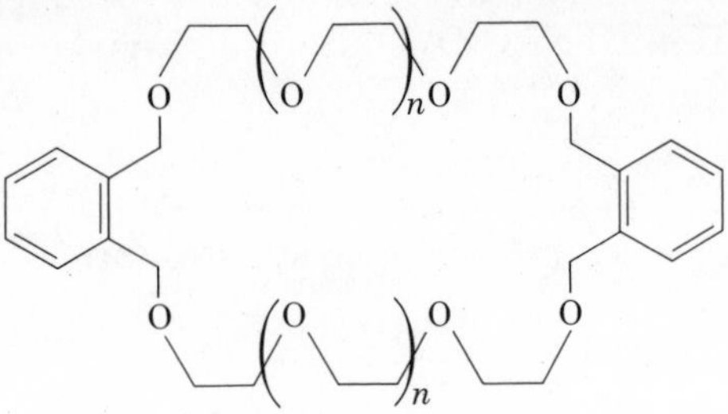

(n = 0–2)
maximum yield 24%

(LXIX)

whereas with dilithium glycolates only complex **LXVIII** was obtained and the reaction was much slower.

C. Sulfur-Donor Macrocycles

A wide range of macrocyclic polythioethers are known and have been obtained by elegant synthetic routes (*102*), yet relatively few of these macrocycles have been by template methods, unless the sodium ion serves to coordinate the sulfur atom in the reactions of disodium salts of thiols with dihalides. Some examples of macrocycles prepared by this route are **LXX–LXXII** (*3*), **LXXIII** and **LXXIV** (*120*), and **LXXV** (*15*).

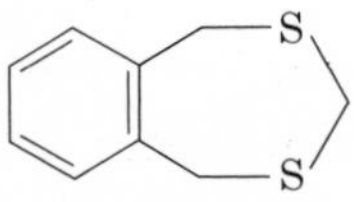

(LXX)

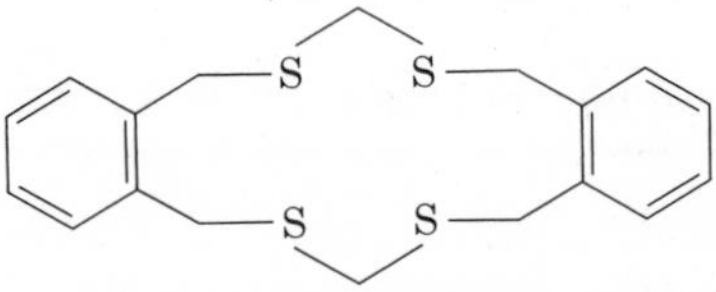

(LXXI)

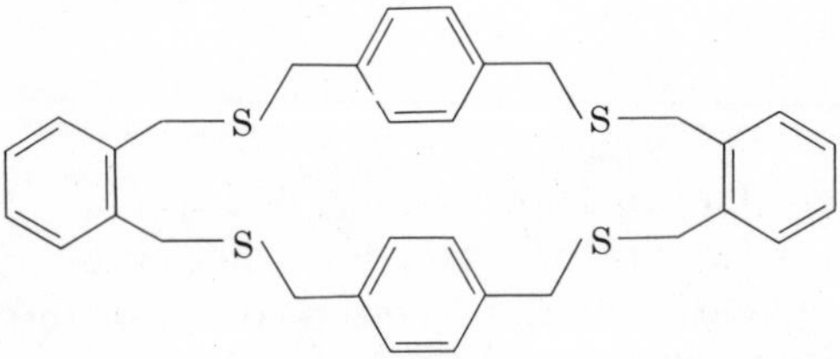

(LXXII)

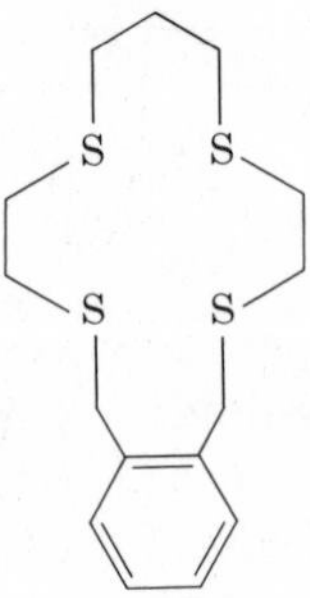

(LXXIII)

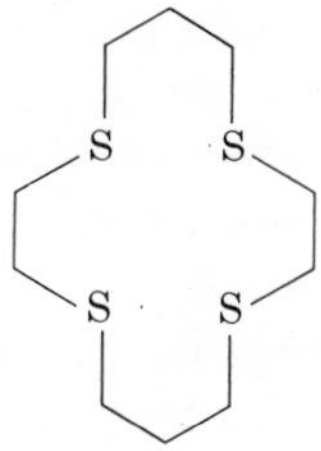

(LXXIV)

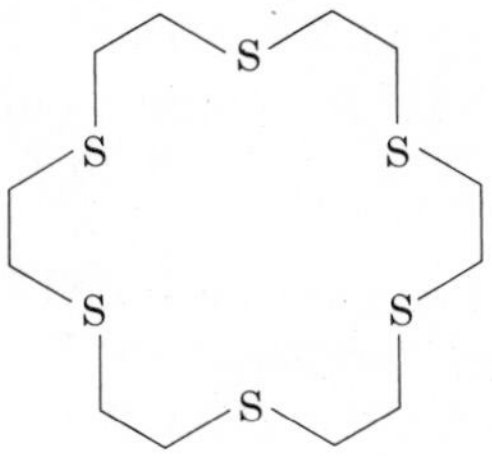

(LXXV)

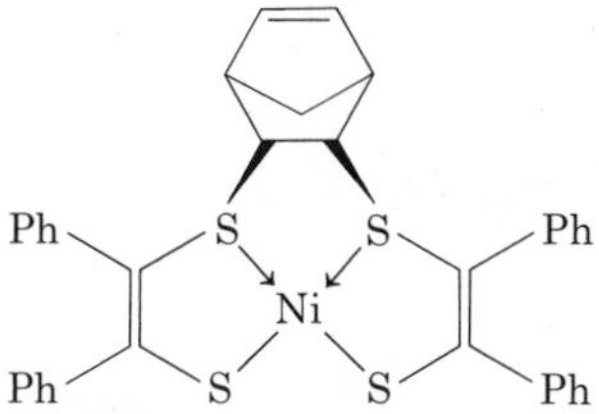

(LXXVI)

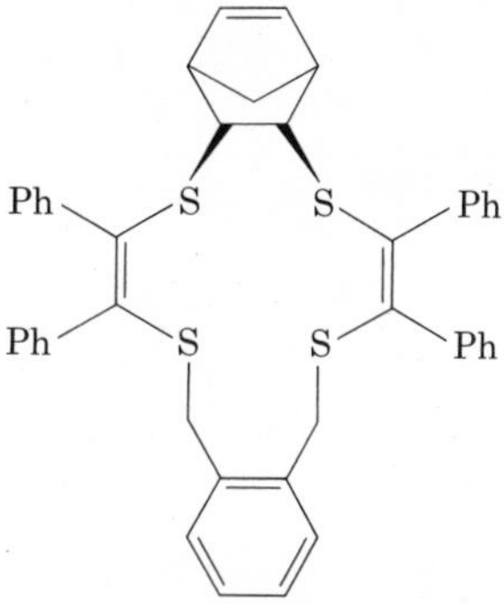

(LXXVII)

A definite template requirement does, however, exist in one case (*125*). The intermediate **LXXVI** had first to be obtained, and it was refluxed with a calculated amount of α,α′-dibromo-o-xylene to give a green crystalline solid. The free macrocycle (**LXXVII**) was liberated by dissolving the solid in methanol.

D. Mixed Donor Macrocycles

1. Nitrogen and Oxygen

The template synthesis of mixed nitrogen and oxygen donor macrocycles are mainly based on substituted salicylaldehydes or salicylaldehyde–metal complexes (*82*), e.g.,

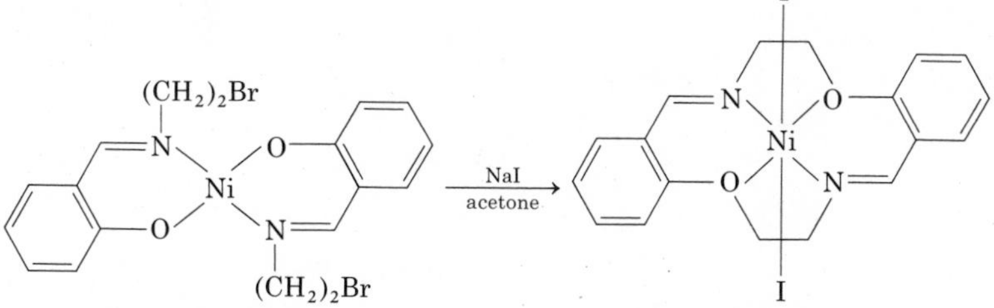

(21)

(LXXVIII)

(LXXIX)

The imine ether (**LXXIX**) can also be formed by heating the iodo analog of **XXVIII** in acetone, but the scope of the reaction is limited by proximity of the side-chain halide to the coordinated phenoxide and to ethyl and *n*-propyl side chains. The free macrocycle (**LXXX**) is obtained by hydrolyzing **LXXIX** with water—a reaction that occurs rapidly at room temperature. The macrocyclic ligand (**LXXXI**) was obtained (*55, 133*) using Zn(II) perchlorate as the template for reacting 4,7-diaza-2,3:8,9-dibenzodiene-1,10-dione with 1,2-di(*o*-aminophenoxy)-ethane. Reaction of the dialdehyde (**LXXXII**) with diamines in the presence of Ni(II) ions followed by hydrolysis of the intermediate complex, which is analogous to **LXXIX**, affords **LXXXIII** (*4, 5*). Ligand **LXXXIV** can be obtained according to Equation (22)

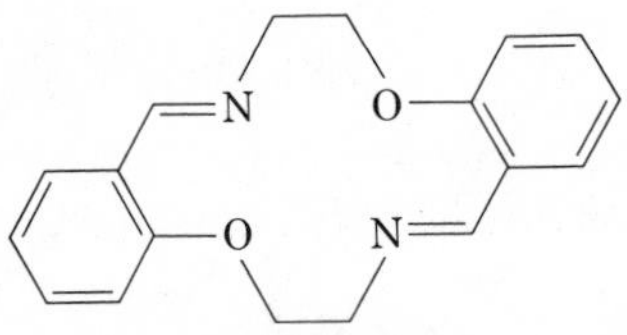

(**LXXX**)

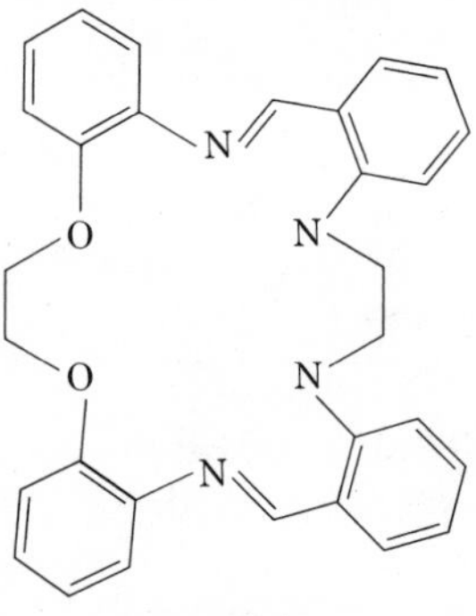

(**LXXXI**)

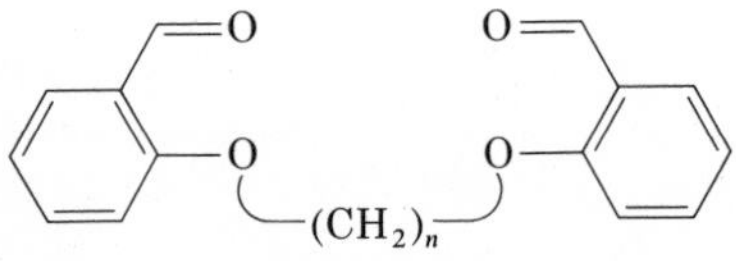

($n = 2,3$)

(**LXXXII**)

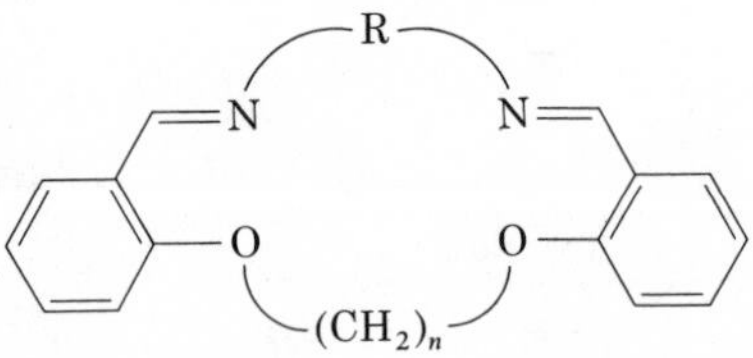

($n = 2,3$; R = —$(CH_2)_3$—, —$CH_2CH(CH_3)$—, C_6H_4—)

(**LXXXIII**)

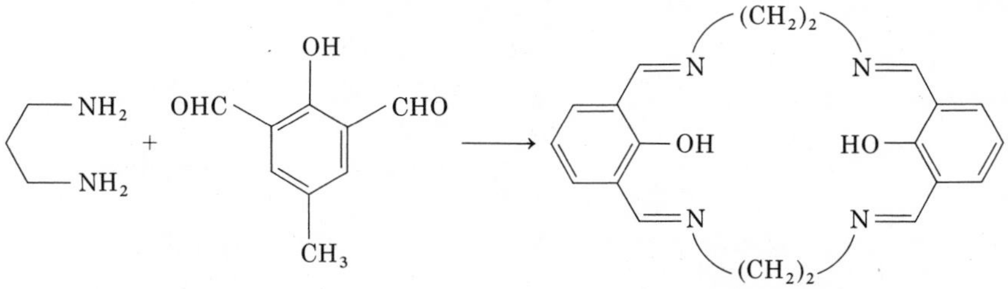

(LXXXIV)

$$(22)$$

only when a metal salt is present for complexation (*112*). In the form
shown, it forms mononuclear complexes with Ni(II), but when the
hydroxyl groups are deprotonated it forms binuclear complexes (*73*)
with the metal ions Mn(II), Fe(II), Ni(II), Cu(II), and Zn(II).

2. Nitrogen and Sulfur

The most common template syntheses have been for macrocycles containing the donor atoms in the ratio of 2:2, e.g., reaction of the Ni(II)
complexes (**LXXXV**), formed by condensation of α-diketones with
β-mercaptoamines in the presence of Ni(II) ions, with α,α′-dibromo-
o-xylene affords **LXXXVI** through a kinetic template effect (see Section II,A) (*136–138*). Reaction with acetone links the coordinated
amine groups of dithiodiamine to form the macrocycle complex
LXXXVII (*23*).

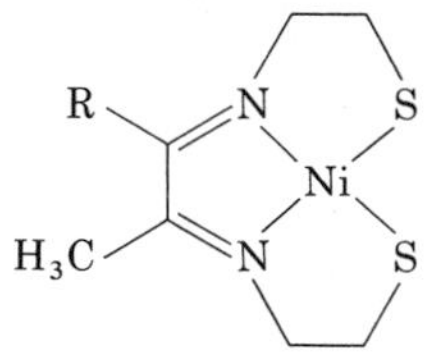

$$[R = CH_3, \quad Ni(BE)$$
$$= C_5H_{11}, \quad Ni(OE)$$
$$= C_2H_5, \quad Ni(PE)$$
$$= C_6H_5, \quad Ni(PPE)]$$

(LXXXV)

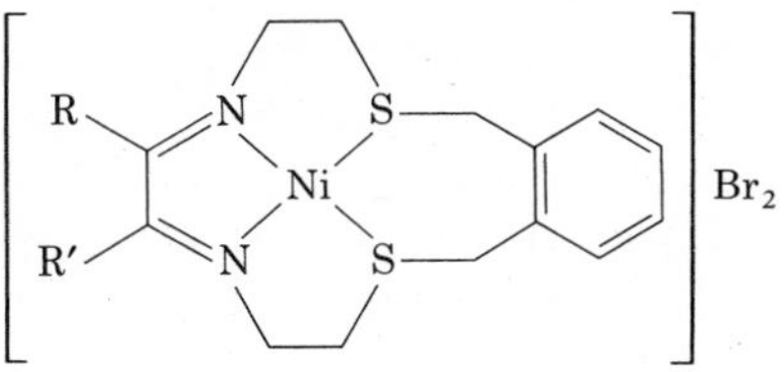

$$Br_2$$

(LXXXVI)

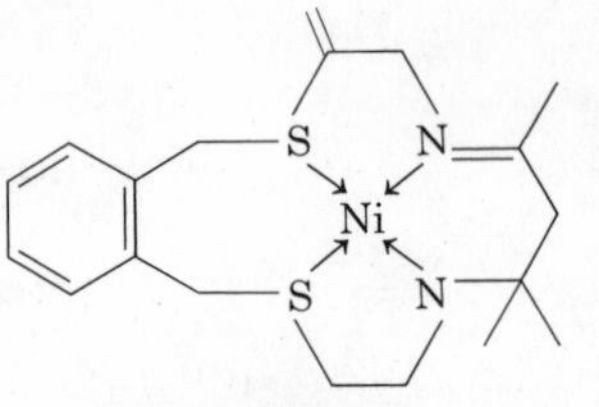

(LXXXVII)

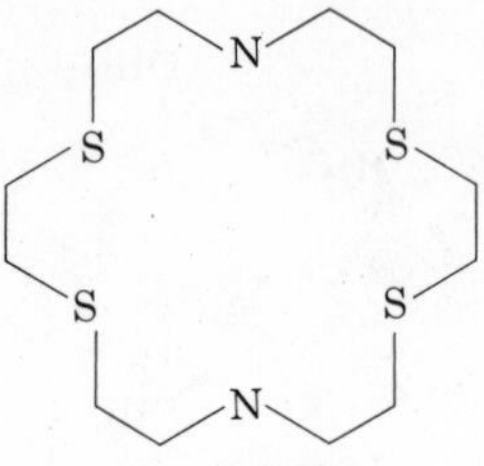

(LXXXVIII)

Macrocycle **LXXXVIII** was prepared (*13*) in an 8% yield from the reaction of the disodium salt of ethane-1,2-dithiol with di(2-bromo-ethyl)amine in ethanol at high dilution, and it was found to complex with Ni(II) and Co(II) ions when these were added as salts. A macrocycle containing the same donor atoms (**LXXXIX**) has been obtained in the form of complexes (*87*) by the template reactions of 1,2-bis(2-aminophenylthio)ethane and 1,4-bis(2-formylphenyl)-1,4-dithiabutane with Ni(II) and Co(II) perchlorates. Iron, cobalt, nickel, and zinc as their M(II) perchlorates have been used as templates in the formation of **XC** (*55, 133*).

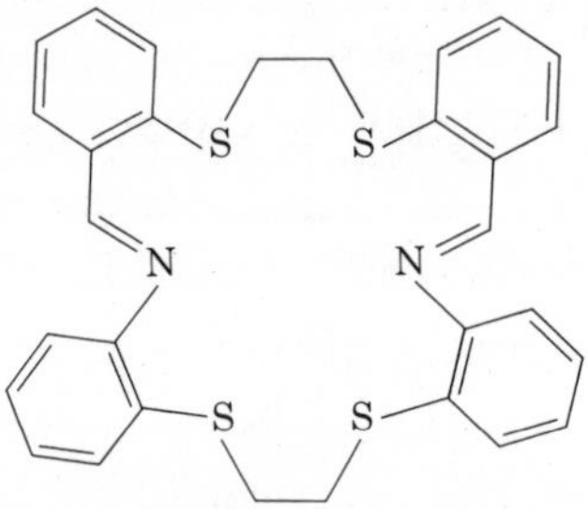

(LXXXIX)

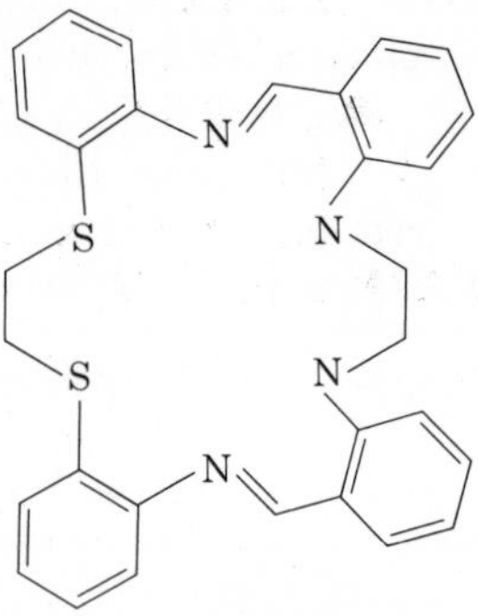

(XC)

Macrocycles with 2 nitrogen and 3 sulfur donors have been prepared (*14*) by a template synthesis in which the dialdehyde (**XCI**) is condensed with primary diamines, e.g., ethylenediamine gives **XCII** in boiling acetonitrile containing Fe(II) perchlorate. The reaction is typical of template condensation between carbonyl compounds and primary amines. An unusual monanionic macrocyclic ligand was produced (*2*) when formaldehyde was condensed with the hydrazine (**XCIII**) instead of a primary amine in the presence of Ni(II) salts, and complexes **XCIV** have been characterized.

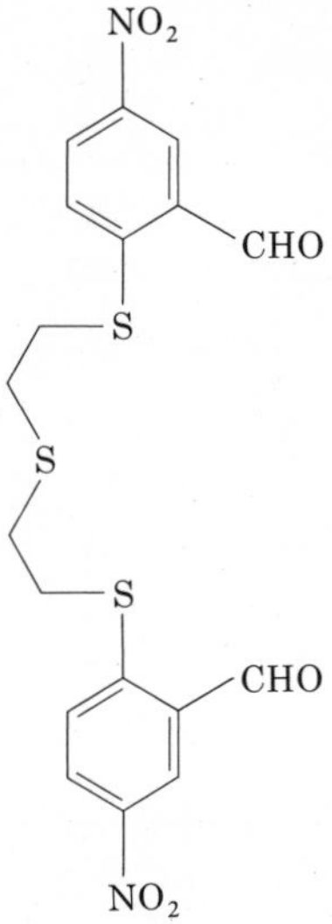

(XCI)

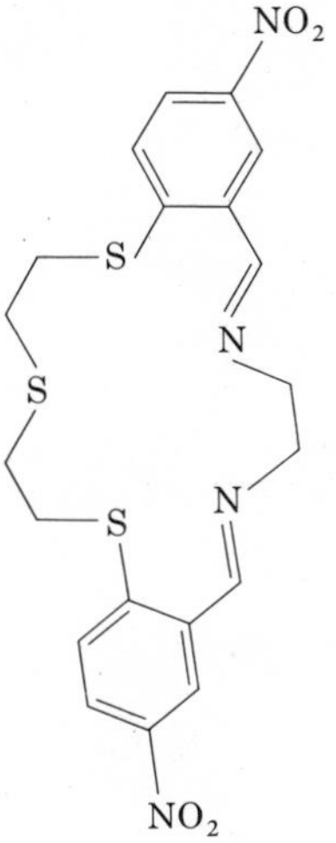

(XCII)

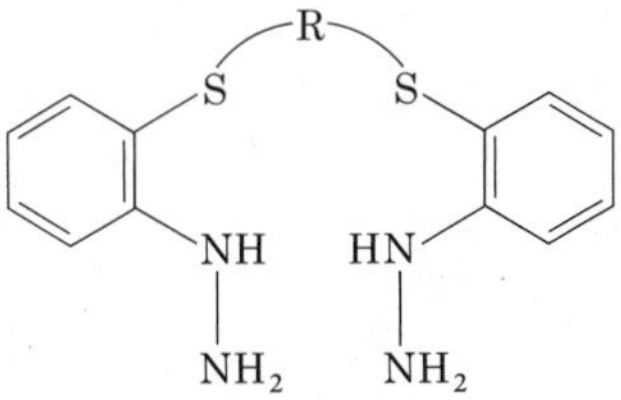

(XCIII)

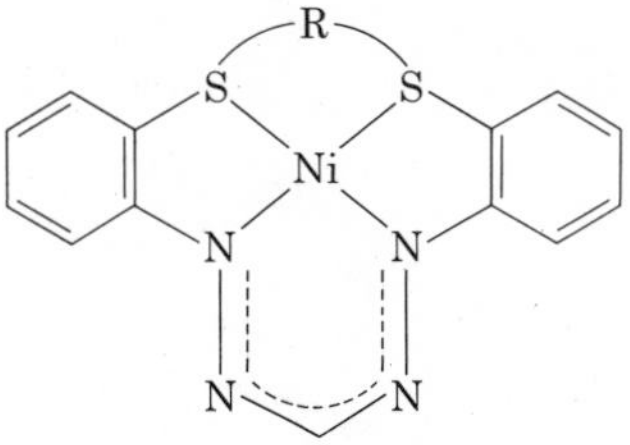

$(R = -(CH_2)_2-, -(CH_2)_3-)$

(XCIV)

3. Nitrogen and Phosphorus

The first macrocyclic ligand containing nitrogen and phosphorus donors has been prepared (118) by refluxing bis(3-aminopropyl)phenylphosphine with 2,6-diacetylpyridine in an ethanolic solution containing $NiCl_2 \cdot 6H_2O$. A four-coordinate complex $[Ni(pn_3)]$ (**XCV**) is obtained, but, if $NiBr_2 \cdot 6H_2O$ or $NiI_2 \cdot 6H_2O$ are used as templates followed by addition of NH_4PF_6, a five-coordinate complex $[Ni(pn_3)X]$ (X = Br, I) results.

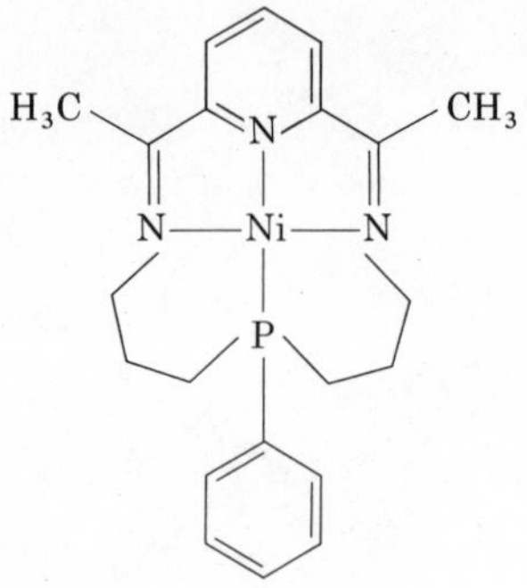

(XCV)

4. Oxygen and Sulfur

A number of mixed macrocycles have been prepared. A template requirement has not been established unless the sodium ion serves to coordinate to sulfur and oxygen in the reaction of cyclic vicinal mercaptophenols or dithiols with equivalent proportions of terminal-substituted ether dichlorides in the presence of sodium hydroxide (*109*) and in the reaction of 1,2-dibromoethane with the disodium salt of 3-oxapentane-1,5-dithiol at high dilution (*15*). Macrocycles **XCVI–XCVIII** were prepared by the former route, and **XCIX**, in low yield due to extensive polymerization, by the latter.

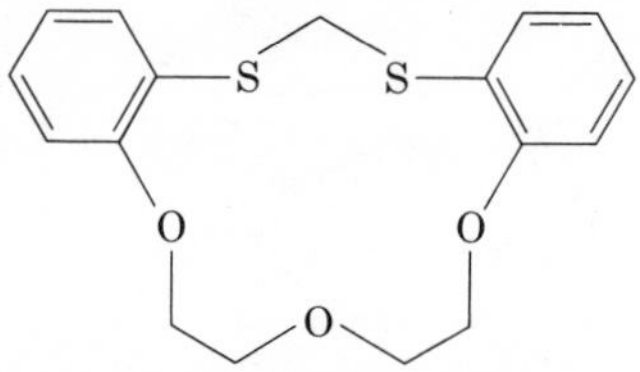

(XCVI)

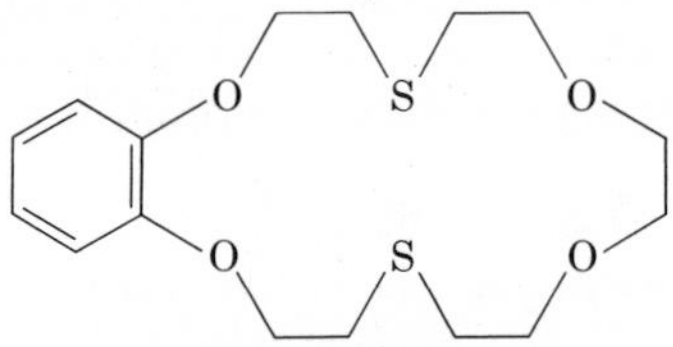

(XCVII)

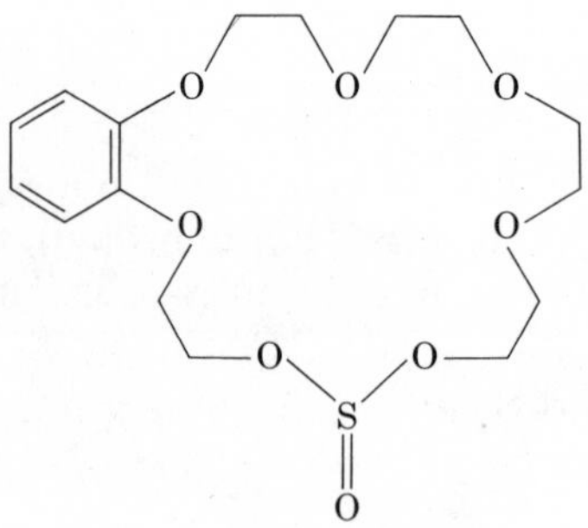

(XCVIII)

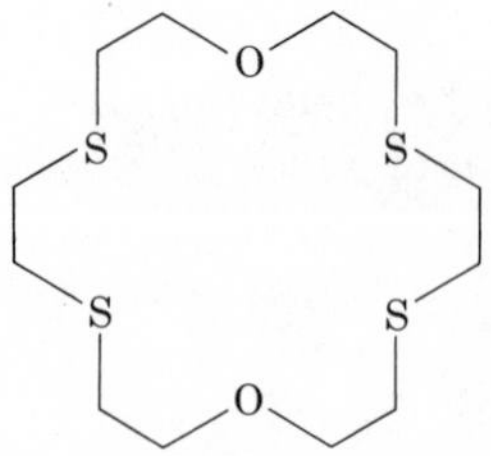

(XCIX)

5. Nitrogen, Oxygen, and Sulfur

A few compounds containing N, O, and S atoms have been synthesized (29, 47, 48), but no template requirement has been established. The following is a typical preparation (111):

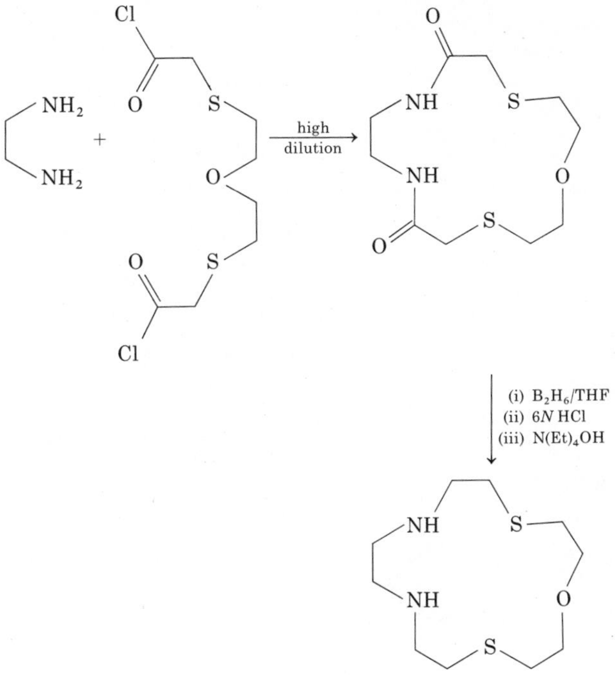

(23)

(i) B_2H_6/THF
(ii) $6N$ HCl
(iii) $N(Et)_4OH$

IV. Choice of Templates for Specific Syntheses

The metals that have been most widely used as templates are shown in Table I. Two correlations arise from the syntheses described in Section III.

1. An effective template metal ion binds strongly to the donor atoms of the macrocycle or its precursors, e.g., K^+ is the most common template for synthesis of crown ethers and forms definite complexes with a wide variety of crown ethers.

2. The preferential coordination geometry of the template metal ion determines the nature of the macrocycle that is formed, e.g., Ni^{2+} and Cu^{2+} are particularly effective template ions for the synthesis of N-donor macrocycles containing 4 nitrogen atoms that can be arranged in a plane and, hence, form a square planar macrocycle-metal complex.

The preference of a metal for a particular ligand is not always the overriding consideration as was demonstrated (115) for **XII** where

TABLE I

<small>SECTION OF THE PERIODIC TABLE SHOWING METALS KNOWN TO ACT AS TEMPLATES (ENCIRCLED)</small>

IA	IIA	IIIA	IVA	VA	VIA	VIIA	VIII			IB	IIB
H											
(Li)	Be										
(Na)	(Mg)										
(K)	(Ca)	Sc	Ti	V	Cr	(Mn)	(Fe)	(Co)	(Ni)	(Cu)	(Zn)
(Rb)	Sr	Y	Zr	Nb	Mo	Tc	Ru	Rh	Pd	Ag	Cd

Ni(II) ions, most commonly used in the synthesis of N-donor macrocycles, were found to be as effective as Mg(II) ions.

The significance of the counteranion in metal template reactions has received little attention. Where reactions are carried out in solvents with low polarities and low dielectric constants, namely, most organic solvents, the effectiveness of the metal template ion will be related to its availability in solution, i.e., dissociation of the salt. For the Li^+ ion in the template synthesis of **XII**, the order of yields, $LiClO_4 > LiCl > Li_2SO_4$ (27, 115), has been correlated with the dissociation energies of the salts (115).

V. Physical Studies of Template Reactions

Claims that metal ions are exercising a template effect are obviously justified in situations where enhanced yields of macrocycle are obtained or partially cyclized, macrocycle–metal complexes, which react further to give macrocycles, are isolated. In other situations, especially the formation of crown ethers, whether the metal ion is exercising a template effect is hard to determine, e.g., Eq. (18) (Section III,B). In this case the template action of the K^+ ion was established (63) by allowing 1 mole each of tri- and tetraethylene glycol to compete for 1 mole of the tosylate in the presence of either potassium tertiary butoxide or tetra-n-butylammonium hydroxide. In both reactions, the same ratio of the two possible macrocycles was obtained but the yield was greatly diminished in the reaction containing tetra-n-butylammonium hydroxide. The fact that the ratio of products was the same for the nontemplate reaction and the reaction where the K^+ ions acted as templates was interpreted (63) as showing that the oxygen cannot be bound to the potassium ion when the first tosyl group is lost because such a situation would not lead to the equal probability of macrocycles as observed. The metal acts by complexing to the oxygen before the second tosyl group is lost so that cyclizing groups are held in close proximity and ring closure is aided.

Template reactions have been monitored by NMR spectroscopy (126) for the reaction of Na^+ ions with dibenzo-18-crown-6 in N,N-dimethylformamide. One drawback of this technique is that the presence of paramagnetic ions, i.e., many transition metal ions, would cause appreciable broadening of the resonances.

The kinetic nature of the template effect for the reaction of 2,3-pentanedionebis(mercaptoethylimino)Ni(II) with α,α'-dibromo-o-xylene and benzyl bromide, i.e., the final step in Scheme 1 (see Section II,A) has been demonstrated using UV/visible spectroscopy (17). Spectrophotometric scans as a function of time while the reaction was taking place showed a single isosbestic point for the dibromoxylene with no evidence of any other absorption. For the benzyl bromide reaction, the scans showed the formation of an intermediate in a large enough amount to dominate the spectrum. The lack of intermediates in the dibromoxylene reaction was taken to suggest that a single rate-determining step dominates the reaction; it is, thus, an example of the kinetic template effect.

The driving force toward formation of a macrocycle has been related to stability constants by Cabbiness and Margerum (25). In aqueous solution it was found that compound **C** formed a more stable complex with Cu(II) than did compound **CI**, and it was concluded that differences in configuration and solvation properties of the free ligands

(C)   (CI)

must in some way contribute to the difference in stability of the two complexes. Similar studies were carried out for the complexation of cyclam (**XLV**) and **CI** with Ni(II) ions, and again an enhanced stability of the cyclic ligand (**XLV**) was observed (69, 70). Comparison between any conformations, bond strengths, ΔH and ΔS values for the two ligands were made. In terms of ΔS changes, complexation of the open-chain ligand (CI) is more favorable since a greater loss of entropy results for this ligand than for the cyclic ligand. Equilibrium and calorimetric studies show that a very negative ΔH for the cyclam complex is enough to overcome the ΔS handicap. The extra stability of the macrocyclic–metal complex, which has been described as the "macrocyclic effect" (25), has been explained in terms of solvation differences: For cyclam, steric hindrance limits hydrogen bonding of the solvent molecules to the N-donors, whereas in CI the open chain makes the N-atoms relatively more accessible. In consequence of the hydrogen-bonding differences, cyclam is less solvated and less energy is required to break the hydrogen bonds with the solvent than for the open-chain ligand. If the macrocyclic effect depends on the solvation properties, it should be independent of the metal ions, providing the metal ion coordination geometry is suitable, as was found for Cu(II) and Ni(II) ions which both have a tendency to form square planar complexes. Similarly, if the hydrogen-bonding potential of the solvent is reduced then the solvation of the macrocycle should diminish giving only a small macrocyclic effect for this situation. Experiments with the sulfur analog of XLV and CI complexing to Ni(II) ions in nitromethane, which is a relatively poor hydrogen-bonding solvent, indeed showed a decreased macrocyclic effect (128).

VI. Applications of Template Reactions

The use of metal ions as templates for macrocycle synthesis has an obvious relevance to the understanding of how biological molecules are formed *in vivo*. The early synthesis of phthalocyanins from phthalonitrile in the presence of metal salts (89) has been followed by the use of Cu(II) salts as templates in the synthesis of copper complexes of etioporphyrin-I (32), tetraethoxycarbonylporphyrin (26), etioporphyrin-II (78), and coproporphyrin-II (81). Metal ions have also been used as templates in the synthesis of corrins, e.g., nickel and cobalt ions in the synthesis of tetradehydrocorrin complexes (64) and nickel ions to hold the two halves of a corrin ring system while cyclization was effected (51), and other biological molecules (67, 76, 77).

The ability of crown ethers to bind selectively to particular Group IA and Group IIA metal ions, because of the relationship between hole size and metal ion radius, has led to considerable interest in them in relation to membranes (models for selective ion transport), antibiotics (similar polyether structure), organic synthesis [solubilization of inorganic reagents leading to milder routes for oxidation (122), nucleophilic substitution (123), fluoridation (90)] and extraction of alkali cations (*140*). Crown ethers have also been used catalytically by solubilizing the active agent (*93, 103*). A crown ether, dibinaphthyl-22-crown-6 (**CII**), has recently been synthesized by a template synthesis

(**CII**)

that is able to extract α-phenylethylammoniumhexafluorophosphate from chloroform and has the ability to distinguish between *R* and *S* enantiomers of phenylethylamine. The fact that such separations can be achieved in a separating funnel emphasizes the considerable potential of the chiral crown ethers (*33, 60, 61, 65, 66, 83, 129, 139*).

The variety of macrocycles that have already been prepared by template reactions (see Sections III and VI) suggest that, in designing a macrocycle synthesis to meet a future need, the use of a metal ion template should not be ignored.

REFERENCES

1. Ackman, R. G., Brown, W. H., and Wright, G. F., *J. Org. Chem.* **20**, 1147 (1955).
2. Alcock, N. W., and Tasker, P. A., *J. Chem. Soc., Chem. Commun.* p. 1239 (1972).
3. Allen, D. W., Brauton, P. N., and Millar, I. T., *J. Chem. Soc. C* p. 3454 (1971).
4. Armstrong, L. G., and Lindoy, L. F., *Inorg. Chem.* **7**, 1322 (1973).
5. Armstrong, L. G., and Lindoy, L. F., *Inorg. Nucl. Chem. Lett.* **10**, 349 (1974).
6. Bamberger, E., *Chem. Ber. B* **60**, 314 (1927).
7. Barefield, E. K., *Inorg. Chem.* **11**, 2273 (1972).
8. Beals, R. E., and Brown, W. H., *J. Org. Chem.* **21**, 447 (1956).
9. Black, D. St. C., and Hartshorn, A. J., *Coord. Chem. Rev.* **9**, 219 (1972).
10. Black, D. St. C., and Kortt, P. W., *Aust. J. Chem.* **25**, 281 (1972).
11. Black, D. St. C., and Lane, M. J., *Aust. J. Chem.* **23**, 2027, 2039, 2055 (1970).
12. Black, D. St. C., and Markham, E., *Rev. Pure Appl. Chem.* **15**, 109 (1965).
13. Black, D. St. C., and McLean, I. A., *Chem. Commun.* p. 1004 (1968).
14. Black, C. St. C., and McLean, I. A., *Inorg. Nucl. Chem. Lett.* **6**, 675 (1970).
15. Black, D. St. C., and McLean, I. A., *Tetrahedron Lett.* p. 3961 (1969).
16. Black, D. St. C., and Srivastava, R. C., *Aust. J. Chem.* **24**, 287 (1971).
17. Blinn, E. L., and Busch, D. H., *Inorg. Chem.* **7**, 820 (1968).
18. Bosnich, B., Poon, C. K., and Tobe, M. L., *Inorg. Chem.* **4**, 1102 (1965).
19. Boston, D. R., and Rose, N. J., *J. Am. Chem. Soc.* **90**, 6859 (1968).
20. Bradshaw, J. S., Hansen, L. D., Nielsen, S. F., Thompson, M. D., Reeder, R. A., Izatt, R. M., and Christensen, J. J., *J. Chem. Soc., Chem. Commun.* p. 874 (1975).
21. Brown, W. H., and French, W. N., *Can. J. Chem.* **36**, 371 (1958).
22. Brown, W. H., and French, W. N., *Can. J. Chem.* **36**, 537 (1958).

23. Busch, D. H., *Helv. Chim. Acta (Fasciculus Extraordinaries Alfred Werner)*, p. 174 (1967).

24. Busch, D. H., *Rec. Chem. Progr.* **25**, 107 (1964).

25. Cabbiness, D. K., and Margerum, D. W., *J. Am. Chem. Soc.* **91**, 6540 (1969).

26. Caughey, W. S., Corwin, A. H., and Singh, R., *J. Org. Chem.* **25**, 290 (1960).

27. Chastrette, M., and Chastrette, F. *J. Chem. Soc., Chem. Commun.* p. 534 (1973).

28. Chave, P., and Honeybourne, C. L., *Chem. Commun.* p. 279 (1969).

29. Cheney, J., and Lehn, J. M., *J. Chem. Soc., Chem. Commun.* p. 487 (1972).

30. Christensen, J. J., Eatough, D. J., and Izatt, R. M., *Chem. Rev.* **74**, 351 (1974).

31. Cook, F. L., Thomas, C. C., Byrne, M. P., Bowers, C. W., Speck, D. H., and Liotta, C. L., *Tetrahedron Lett.* p. 4024 (1974).

32. Corwin, A. H., and Sydow, V. L., *J. Am. Chem. Soc.* **75**, 4484 (1953).

33. Cram, D. J., and Cram, H. J., *Science* **183**, 803 (1974).

34. Cummings, S. C., and Busch, D. H., *Inorg. Chem.* **10**, 1220 (1971).

35. Cummings, S. C., and Busch, D. H., *J. Am. Chem. Soc.* **92**, 1924 (1970).

36. Cummings, S. C., and Sievers, R. E., *Inorg. Chem.* **9**, 1131 (1970).

37. Cummings, C. S., and Sievers, R. E., *J. Am. Chem. Soc.* **92**, 215 (1970).

38. Curry, J. D., and Busch, D. H., *J. Am. Chem. Soc.* **86**, 592 (1964).

39. Curtis, N. F., *Coord. Chem. Rev.* **3**, 3 (1968).

40. Curtis, N. F., and Hay, R. W., *Chem. Commun.* p. 524 (1966).

41. Dale, J., and Daasvatn, K., *J. Chem. Soc., Chem. Commun.* p. 295 (1976).

42. Dale, J., and Kristiansen, P. O., *Acta Chem. Scand.* **26**, 1471 (1972).

43. Dale, J., and Kristiansen, P. O., *Chem. Commun.* p. 670 (1971).

44. Dennstedt, M., *Chem. Ber.* **23**, 1370 (1890).

45. Dennstedt, M., and Zimmermann, J., *Chem. Ber.* **20**, 850, 2449 (1887).

46. Dennstedt, M., and Zimmermann, J., *Chem. Ber.* **21**, 1478 (1888).

47. Dietrich, B., Lehn, J. M., and Sauvage, J. P., *Chem. Commun.* p. 1055 (1970).

48. Dietrich, B., Lehn, J. M., and Sauvage, J. P., *Tetrahedron Lett.* p. 2885, 2889 (1969).

49. Drew, M. G. B., and Othman, A. H. B., *J. Chem. Soc., Chem. Commun.* p. 818 (1975).

50. Eichhorn, G. L., and Latif, R. A., *J. Am. Chem. Soc.* **76**, 5180 (1954).

51. Eschenmoser, A., Scheffold, R., Bertele, E., Pesaro, M., and Gschwand, H., *Proc. Roy. Soc., Ser. A* **288**, 306 (1965).

52. Farmery, K., and Busch, D. H., *Chem. Commun.* p. 1091 (1970).

53. Fleischer, E. B., and Hawkinson, S., *J. Am. Chem. Soc.* **89**, 720 (1967).

54. Fleischer, E. B., and Klem, E., *Inorg. Chem.* **4**, 637 (1965).

55. Fleischer, E. B., and Tasker, P. A., *Inorg. Nucl. Chem. Lett.* **6**, 349 (1970).

56. Goedken, V. L., Molin-Case, J., and Whang, Y. A., *J. Chem. Soc., Chem. Commun.* p. 337 (1973).

57. Goedken, V. L., and Park, Y., *J. Chem. Soc., Chem. Commun.* p. 214 (1975).

58. Goedken, V. L., Park, Y., Peng, S., and Norris, J. M., *J. Am. Chem. Soc.* **96**, 7693 (1974).

59. Gokel, G. W., Cram, D. J., Liotta, C. L., Harris, H. P., and Look, F. L., *J. Org. Chem.* **39**, 2445 (1974).

60. Gokel, G. W., and Durst, H. D., *Synthesis* p. 168 (1976).

61. Gokel, G. W., Timko, J. M., and Cram, D. J., *J. Chem. Soc., Chem. Commun.* pp. 394, 444 (1975).

62. Green, M., Smith, J., and Tasker, P. A., *Inorg. Chim. Acta* **5**, 17 (1971).

63. Greene, R. N., *Tetrahedron Lett.* p. 1793 (1972).

64. Harris, R. L. N., Johnson, A. W., and Kay, I. T., *Q. Rev. Chem. Soc.* **20**, 211 (1966).

65. Helgeson, R. C., Koga, K., Timko, J. M., and Cram, D. J., *J. Am. Chem. Soc.* **95**, 3021 (1973).

66. Helgeson, R. C., Timko, J. M., Moreau, P., Peacock, S. C., Mayer, J. M., and Cram, D. J., *J. Am. Chem. Soc.* **96**, 6762 (1974).

67. Hill, H. A. O., Pratt, J. M., and Williams, R. J. P., *Chem. Ber.* **5**, 156 (1969).

68. Hiller, H., Dimroth, P., and Pfitzner, H., *Ann.* **717**, 137 (1968).

69. Hinz, F. P., and Margerum, D. W., *Inorg. Chem.* **13**, 2941 (1974).

70. Hinz, F. P., and Margerum, D. W., *J. Am. Chem. Soc.* **96**, 4993 (1974).

71. Holtman, M. S., and Cummings, S. C., *Inorg. Chem.* **15**, 660 (1976).

72. Honeybourne, C. L., *Inorg. Nucl. Chem. Lett.* **11**, 191 (1975).

73. Hoskins, B. F., McLeod, N. J., and Schaap, H. A. S., *Aust. J. Chem.* **29**, 515 (1976).

74. Hurley, T. J., Robinson, M. A., and Trotz, S. I., *Inorg. Chem.* **6**, 389 (1967).

75. (a) Jager, E. G., *Z. Anorg. Allg. Chem.* **364**, 177 (1969); (b) *Z. Chem.* **8**, 30, 392, 470 (1968); (c) Jager E. G., and Uhlig, E., *Ibid.* **4**, 437 (1964).

76. Johnson, A. W., *Chem. Ber.* **3**, 253 (1967).

77. Johnson, A. W., *Chem. Soc. Rev.* **4**, 1 (1975).

78. Johnson, A. W., and Kay, I. T., *J. Chem. Soc.* **2**, 2418 (1961).

79. Karn, J. L., and Busch, D. H., *Inorg. Chem.* **8**, 1149 (1969).

80. Katovic, K., Taylor, L. T., and Busch, D. H., *J. Am. Chem. Soc.* **91**, 2122 (1969).

81. Kay, I. T., *Proc. Acad. Nat. Sci. U.S.* **48**, 901 (1962).

82. Kluiber, R. W., and Sasso, G., *Inorg. Chim. Acta* **4**, 226 (1970).

83. (a) Kyba, E. P., Siegel, M. G., Sousa, L. R., Sogah, G. D. Y., and Cram, D. J., *J. Am. Chem. Soc.* **95**, 2691 (1973).

    (b) Kyba, E. P., Koga, K., Sousa, L. R., Siegel, M. G., and Cram, D. J., *J. Am. Chem. Soc.* **95**, 2692 (1973).

84. Lewis, J., and Wainwright, K. P., *J. Chem. Soc., Chem. Commun.* p. 169 (1974).

85. Lindoy, L. F., *Chem. Soc. Rev.* **4**, 421 (1975).

86. Lindoy, L. F., *Q. Rev. Chem. Soc.* **25**, 379 (1971).

87. Lindoy, L. F., and Busch, D. H., *J. Am. Chem. Soc.* **91**, 4690 (1969).

88. Lindoy, L. F., and Busch, D. H., *in* "Preparative Inorganic Reactions" (W. L. Jolley, ed.), Vol. VI, p. 1. Wiley (Interscience), New York, 1971.

89. Lindstead, R. P., and Lowe, A. R., *J. Chem. Soc.* p. 1022 (1934).

90. Liotta, C. L., and Harris, H. P., *J. Am. Chem. Soc.* **96**, 2250 (1974).

91. Long, K. M., and Busch, D. H., *Inorg. Chem.* **9**, 505 (1970).

92. Martin, J. G., Wei, R. M. C., and Cummings, S. C., *Inorg. Chem.* **11**, 475 (1972).

93. Maskornick, M. J., *Tetrahedron Lett.* p. 1797 (1972).

94. McAuliffe, C. A., *Adv. Inorg. Radiochem.* **17**, 165 (1975).

95. McGeachin, S. G., *Can. J. Chem.* **44**, 2323 (1966).

96. Melson, G. A., and Busch, D. H., *J. Am. Chem. Soc.* **86**, 4834 (1964).

97. Melson, G. A., and Busch, D. H., *J. Am. Chem. Soc.* **87**, 1706 (1965).

98. Nelson, S. M., Bryan, P., and Busch, D. H., *Chem. Commun.* p. 641 (1966).

99. Neves, D. R., and Dabrowiak, J. C., *Inorg. Chem.* **15**, 129 (1976).

100. Newcomb, M., and Cram, D. J., *J. Am. Chem. Soc.* **97**, 1257 (1975).

101. Ochai, E., and Busch, D. H., *Inorg. Chem.* **8**, 1974 (1969).

102. Ochyrmowycz, L. A., Mak, C. P., and Michna, J. D., *J. Org. Chem.* **39**, 2079 (1974).

103. Orvik, J. A., *J. Am. Chem. Soc.* **98**, 3322 (1976).

104. Parks, J. E., Wagner, B. E., and Holm, R. H., *J. Am. Chem. Soc.* **92**, 3500 (1970).

105. Pedersen, C. J., *J. Am. Chem. Soc.* **89**, 2495 (1967).

106. Pedersen, C. J., *J. Am. Chem. Soc.* **89**, 7017 (1967).

107. Pedersen, C. J., *J. Am. Chem. Soc.* **92**, 386 (1970).

108. Pedersen, C. J., *J. Am. Chem. Soc.* **92**, 391 (1970).

109. Pedersen, C. J., *J. Org. Chem.* **36**, 254 (1971).

110. Pedersen, C. J., and Frensdorff, H. K., *Angew. Chem., Int. Ed. Engl.* **11**, 16 (1972).

111. Pelissard, D., and Louis, R., *Tetrahedron Lett.* p. 4589 (1972).
112. Pilkington, N. H., and Robson, R., *Aust. J. Chem.* **23**, 2225 (1970).
113. Prince, R. H., Stotter, D. A., and Wolley, P. R., *Inorg. Chim. Acta* **9**, 51 (1974).
114. Reinhoudt, D. R., Gray, R. T., Smit, C. J., and Veenstra, I., *Tetrahedron* **32**, 1161 (1976).
115. Rest, A. J., Smith, S. A., and Tyler, I. D., *Inorg. Chim. Acta* **16**, L1 (1976).
116. Rich, R. L., and Stucky, G. L., *Inorg. Nucl. Chem. Lett.* **1**, 61 (1965).
117. Richman, J. E., and Atkins, T. J., *J. Am. Chem. Soc.* **96**, 2268 (1974).
118. Rikker-Nappier, J., and Meek, D. W., *J. Chem. Soc., Chem. Commun.* p. 442 (1974).
119. Rosen, W., *Inorg. Chem.* **10**, 1832 (1971).
120. Rosen, W., and Busch, D. H., *J. Am. Chem. Soc.* **91**, 4694 (1969).
121. Rothemund, P., and Gage, C. L., *J. Am. Chem. Soc.* **77**, 3340 (1955).
122. Sam D. J., and Simmons, H. E., *J. Am. Chem. Soc.* **94**, 4024 (1972).
123. Sam, D. J., and Simmons, H. E., *J. Am. Chem. Soc.* **96**, 2252 (1974).
124. Schrauzer, G. N., *Chem. Ber.* **95**, 1438 (1962).
125. Schrauzer, G. N., Ho, R. K. Y., and Murillo, R. P., *J. Am. Chem. Soc.* **92**, 3508 (1970).
126. Shchori, G., Jagur-Grodzinski, J., Luz, Z., and Shporer, M., *J. Am. Chem. Soc.* **93** 7133 (1971).
127. Siedel, F., and Dick, W., *Chem. Ber. B* **60**, 2018 (1927).
128. Smith, G. F., and Margerum, D. W., *J. Chem. Soc., Chem. Commun.* p. 807 (1975).
129. Sousa, L. R., Hoffman, D. H., Kaplan, L., and Cram, D. J., *J. Am. Chem. Soc.* **96**, 7100 (1974).
130. Stotz, R. W., and Stoufer, R. C., *Chem. Commun.* p. 1682 (1970).
131. Tait, A. M., and Busch, D. H., *Inorg. Chem.* **15**, 197 (1976).
132. Tang, S., Koch, S., Weinstein, G. N., Lane, R. W., and Holm, R. H., *Inorg. Chem.* **12**, 2589 (1973).
133. Tasker, P. A., and Fleischer, E. B., *J. Am. Chem. Soc.* **92**, 7072 (1970).
134. Taylor, L. T., Urbach, F. L., and Busch, D. H., *J. Am. Chem. Soc.* **91**, 1072 (1969).
135. Taylor, L. T., Vergez, S. C., and Busch, D. H., *J. Am. Chem. Soc.* **88**, 3170 (1966).
136. Thompson, M. C., and Busch, D. H., *J. Am. Chem. Soc.* **84**, 1762 (1962).
137. Thompson, M. C., and Busch, D. H., *J. Am. Chem. Soc.* **86**, 213 (1964).
138. Thompson, M. C., and Busch, D. H., *J. Am. Chem. Soc.* **86**, 3651 (1964).
139. Timko, J. M., Helgeson, R. C., Newcomb, M., Gokel, G. W., and Cram, D. J., *J. Am. Chem. Soc.* **96**, 7097 (1974).
140. Tusêk, L. J., Danesi, R. P., and Chiarizia, R., *J. Inorg. Nucl. Chem.* **37**, 1538 (1975).
141. Umland, F., and Thierig, D., *Angew. Chem., Int. Ed. Engl.* **1**, 333 (1962).
142. Von Baeyer, A., *Chem. Ber.* **19**, 2184 (1886).
143. Warner, L. G., Rose, N. J., and Busch, D. H., *J. Am. Chem. Soc.* **90**, 6938 (1968).

CYCLOPHOSPHAZENES

S. S. KRISHNAMURTHY and A. C. SAU

Department of Inorganic and Physical Chemistry,
Indian Institute of Science, Bangalore, India

and

M. WOODS

Department of Chemistry, Birkbeck College, University of London, London, England

I. Introduction .	41
II. Synthetic Routes to Cyclophosphazenes .	43
A. Reaction of Ammonium Halides with Halophosphoranes	43
B. Other Ring-Forming Reactions .	44
III. Halogen Replacement Reactions of Cyclophosphazenes	46
A. Aminolysis Reactions	47
B. Hydrolysis	57
C. Reactions with Alkoxides, Aryloxides, and Thiolates	59
D. Metathetical Exchange Reactions .	61
E. Reactions with Organometallic Reagents	63
F. Friedel–Crafts Reactions .	65
IV. Other Reactions of Cyclophosphazenes	66
A. Complexes, Salts, and Adducts .	66
B. Reactions at Side Chains .	70
C. Thermal Polymerization of Cyclophosphazenes	71
D. Ring Degradation Reactions .	72
E. Miscellaneous Reactions	74
V. Physical Methods .	75
A. Nuclear Magnetic Resonance Spectroscopy	75
B. Vibrational Spectroscopy .	82
C. X-Ray Diffraction Studies .	88
D. Basicity Measurements .	91
E. Nuclear Quadrupole Resonance Spectroscopy .	93
F. Other Techniques .	93
VI. Bonding and Electronic Structure .	94
VII. Potential Applications	96
References	97
Appendix .	108
References	111

I. Introduction

Cyclophosphazenes occupy a prominent place among inorganic heterocyclic compounds. They contain an $-\!\!+\!N\!\!=\!\!PX_2\!+\!\!-$ repeating unit

41

in a valence-unsaturated skeleton. Two typical members are the cyclic trimer (1) and the cyclic tetramer (2). The chlorocyclophosphazene, $N_3P_3Cl_6$ (1), was first isolated by Liebig (285) in 1834. The cyclic struc-

(1)   (2)

ture for this compound was originally suggested by Stokes (426) who also identified the higher homologs, $(NPCl_2)_{4-7}$ (429) in the closing years of the nineteenth century. Research activity in the chemistry of cyclophosphazenes remained quiescent until the mid-1950s, but recently various aspects of this topic have received considerable attention. The major developments are largely due to the advent of new and improved synthetic techniques and powerful instrumental methods of structure determination, the emergence and evolution of quantum theory of chemical binding, and the growing interest in the potential of inorganic polymers.

Early comprehensive reviews of phosphazene chemistry by Audrieth, Steinman, and Toy (43), Gribova and Ban-Yuan' (214), Paddock and Searle (337), Shaw, Fitzsimmons, and Smith (401), and Schmulbach (388) were followed by reviews on specific aspects, such as preparative methods (402), structure and bonding (336, 407), and high polymers (254). Some excellent books dealing with the chemistry of cyclophosphazenes (13, 21, 216) have also appeared. The recent reviews of Allcock (22), Sowerby (418), and Keat and Shaw (249) describe the developments up to 1970–1971. Several short articles on this topic have also appeared from time to time (264, 403, 404, 408). Phosphazene chemistry is now reviewed annually (251).

During the past 5 years considerable progress has occurred in the structural chemistry of phosphazenes, substitution reactions, reaction mechanisms, synthetic procedures, and phosphazene high polymers. In this review, a broad outline of cyclophosphazene chemistry will be presented with an emphasis on the most recent work. The chemistry of phosphazene high polymers has been reviewed comprehensively in recent years (21, 22, 24, 412), and this topic will be mentioned only briefly. Cyclophosphazanes (21, 216) and phosphorines containing skeletal heteroatoms other than nitrogen and phosphorus (21, 249) are outside the scope of this review.

There has not been any general agreement on the nomenclature of phosphorus–nitrogen ring systems (21), but there is a discernible trend in recent literature toward adopting the more systematic "phosphazene" notation [originally proposed by Shaw, Fitzsimmons, and Smith (401)] in place of the older "phosphonitrilic" terminology. The phosphazene notation will be employed in this chapter. The full systematic name for cyclophosphazene derivatives is often very lengthy and suitable abbreviations will be used unless an exact description is essential. Compounds 1 and 2 will be referred to as the trimeric chloride and the tetrameric chloride, respectively, and their substituted derivatives will be described as cyclotriphosphazenes and cyclotetraphosphazenes, respectively.

II. Synthetic Routes to Cyclophosphazenes

There are a number of methods for the direct synthesis of chloro- and bromocyclophosphazenes and their alkyl and aryl derivatives. Other derivatives have invariably been prepared by replacement reactions of the halogenocyclophosphazene precursors. In this section, ring-forming reactions will be discussed and the following section will deal with the substitution reactions.

A. REACTION OF AMMONIUM HALIDES WITH HALOPHOSPHORANES

Cyclophosphazenes are conveniently prepared by the reaction of halophosphoranes with ammonium halides,

$$nR'RPX_3 + nNH_4X \rightarrow (NPRR')_n + 4nHX \tag{1}$$

$$(R,R' = Cl, Br, \text{ or organic group}; X = Cl, Br)$$

usually in boiling organic solvents such as sym-tetrachloroethane or chlorobenzene (21, 71, 126, 331, 345, 350, 352). In general, the cyclic trimer and tetramer are formed in greatest yields, although small amounts of higher oligomeric species can be isolated from the reactions of PCl_5 and PBr_5 with their corresponding ammonium halides. Mixed chlorobromocyclophosphazene derivatives are obtained by the use of an appropriate mixture of phosphorus halide and ammonium halide (129, 361, 370):

$$3PCl_3 + 3Br_2 + 3NH_4Br \xrightarrow[110°]{sym\text{-}C_2H_2Cl_4} N_3P_3Cl_2Br_4 + 12HX \tag{2}$$

The reaction of phosphorus pentachloride with ammonium chloride is tantalizingly complex and has been the subject of numerous investigations (21). Cyclic oligomers and linear products are formed, and these can be separated by exploiting the poor solubility of the latter in petroleum. The separation of individual cyclic compounds, $(NPCl_2)_{3-8}$, can be accomplished by a combination of selective extraction with sulfuric acid, fractional crystallization, and fractional distillation techniques (286). The yields of cyclic products can be maximized by carrying out the reaction under high dilution and by the use of finely divided ammonium chloride (52, 53, 165, 257) and surfactants (288). Various metal halides are found to catalyze the reaction (208, 232, 339, 340, 351, 457). Addition of phosphoryl chloride not only reduces the reaction time but also affords improved yields of cyclic compounds (162, 165). The reaction can also be carried out in the solid state (423, 456, 457). In the presence of 4 moles of pyridine, the reaction is reported to terminate within minutes and produce cyclic products in 65% yield (458).

The mechanism of ammonolysis of phosphorus pentachloride is not completely understood. It has been shown (164) that it proceeds in two steps. Initially, the intermediate $[Cl_3PNPCl_3]^+[PCl_6]^-$ is formed, probably via PCl_4NH_2 and PCl_3NH (48, 380). In the second step, this intermediate reacts with NH_4Cl to produce cationic phosphazene chains that undergo cyclization by the elimination of $[PCl_4]^+$ (164, 256).

For the preparation of bromocyclophosphazenes the use of a mixture of PBr_3 and bromine in place of PBr_5 and addition of bromine to the reaction mixture at frequent intervals leads to improved yields of $(NPBr_2)_{3,4}$ (128, 130, 236, 451). 1,2-Dibromoethane is preferred as the reaction solvent because tiny quantities of chlorobromocyclophosphazenes are formed in sym-tetrachloroethane (130). The preparation of bromocyclophosphazenes is difficult and tedious, and as yet no effective catalyst has been found to reduce the lengthy reaction periods (15–20 days). Fluorocyclophosphazenes cannot be prepared by this route but are obtained by metathetical reactions (Section III,D). Attempts to prepare iodophosphazenes have been unsuccessful (249).

B. OTHER RING-FORMING REACTIONS

1. Thermal Decomposition of Azidophosphines

The reaction of organohalophosphine with sodium or lithium azide

$$R_2PCl + MN_3 \rightarrow [R_2PN_3] \rightarrow (R_2PN)_3 + N_2 \tag{3}$$
$$(M = Na, Li, Me_3Si)$$

or Me_3SiN_3 gives organoazidophosphine which on heating eliminates nitrogen to form cyclo- or poly-phosphazenes (21). The azidophosphine intermediates are often highly explosive and careful handling is required (*431*). In many cases it is not necessary to isolate the intermediate (*226*). An equimolar mixture of Ph_2PCl and $PhPCl_2$ yields a nongeminal hexaphenylcyclotetraphosphazenederivative,$N_4P_4Ph_6Cl_2$ (*226*). This synthetic route has not found widespread use because the reactions often lead to high yields of polymeric materials.

2. Cyclization of Linear Phosphazenes

The air-stable linear phosphazene (**3**) prepared from the reaction of Ph_2PCl_3 with ammonia in chloroform (*57, 225*) has proved a useful intermediate for cyclization reactions (Fig. 1) (*379*). A noteworthy feature of this approach is the synthesis of cyclophosphazenes in which hydrogen is attached to phosphorus (*56, 378, 382*). The only other route (*383*) reported for the preparation of a hydridocyclophosphazene derivative is

$$3(Me_2N)_2PCl + 3NH_3 \xrightarrow[-72°]{Et_2O} N_3P_3H_3(NMe_2)_3 + 3Me_2NH_2Cl \qquad (4)$$

An optically active cyclotriphosphazene has been synthesized by the treatment of the optically active phosphazene salt $[(Ph)(C_6H_4Me)(NH_2)P\text{····}N\text{····}P(NH_2)(C_6H_4Me)(Ph)]^+Cl^-$ with PCl_5 (*391*).

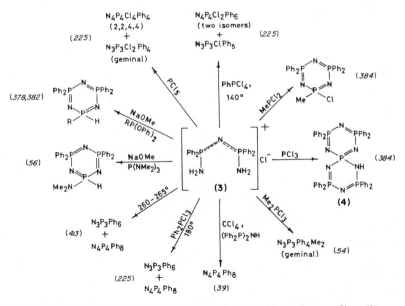

FIG. 1. Formation of cyclophosphazenes from the linear intermediate (**3**).

3. Miscellaneous Syntheses

Aminodichlorophosphoranes undergo dehydrohalogenation in the presence of a tertiary amine (21, 435) to yield cyclophosphazenes:

$$R_2PCl_2NH_2 \xrightarrow[Et_3N]{-2HCl} (R_2PN)_n \tag{5}$$

Diarylchlorophosphines can be transformed to cyclophosphazenes by treatment with chloramine (194) or with ammonia and chloramine (413). Fully substituted cyclotriphosphazenes result from the cyclo-condensation of bis(diphenylphosphino)amine and aminoiminodi-organylphosphoranes or their hydrohalides with CCl_4 in the presence of triethylamine (39). Dehydrohalogenation of fluorocyclodiphos-phazanes in the presence of CsF leads to the formation of cyclotriphos-phazene derivatives (393). An anomalous Kirsanov reaction [Eq. (6)] giving rise to $(NPCl_2)_{3,4}$ has been reported (198):

$$[PhCH_2NH_3]^+Cl^- + PCl_5 \rightarrow \tfrac{1}{n}(NPCl_2)_n + PhCH_2Cl + 3HCl \tag{6}$$
$$(n = 3, 4)$$

Fluorocyclophosphazenes can be directly prepared by the reaction of NF_3 or of CF_3SF_5 with N_5P_3 (293) or by the reaction of NF_3 with P_4S_3 and P_4S_{10} (430) at elevated temperatures, but these routes are unsuitable for routine synthesis.

III. Halogen Replacement Reactions
of Cyclophosphazenes

Halogenocyclophosphazenes undergo nucleophilic and electrophilic reactions that involve the replacement of halogen atoms by a variety of groups. In these reactions the ring is retained in most but not in all cases. Aminolysis reactions, particularly those of hexachlorocyclo-triphosphazene, $N_3P_3Cl_6$ (1), constitute a major area of investigation in cyclophosphazene chemistry. Alcoholysis, phenolysis, thioalcohol-ysis, metathetical replacement, and many other reactions have been studied to a lesser extent. The halogen replacement reactions of cyclophosphazenes normally give rise to complex mixture of products. The detection, isolation, and purification of derivatives with different degrees of halogen replacement and of different isomers have been greatly facilitated by the advent and adaptation of column, thin-layer and gas-liquid chromatographic techniques. Goldschmidt (203) has critically reviewed the various separation techniques commonly employed in cyclophosphazene chemistry.

A. Aminolysis Reactions

1. Reaction Pattern and Mechanism

Halogenocyclophosphazenes react with ammonia and primary and secondary amines to form an ammono- or amino-substituted cyclophosphazene:

$$N_3P_3X_6 + 2nNHR_2 \rightarrow N_3P_3X_{6-n}(NR_2)_n + nR_2NH.HX \tag{7}$$
$$(X = F, Cl, Br; R = H, alkyl, aryl)$$

Two equivalents of amine or ammonia are needed to replace one halogen atom: the amine also functions as the hydrogen halide acceptor. In some cases the aminophosphazene formed in the reaction has a base strength comparable to that of the parent amine, and it is subsequently isolated as a hydrohalide adduct (see Section IV, A,2). A tertiary base, such as triethylamine or pyridine, can also be employed as a hydrogen halide acceptor.

The halogen replacement may proceed either by a geminal or a nongeminal pathway. Figure 2 illustrates these two patterns with

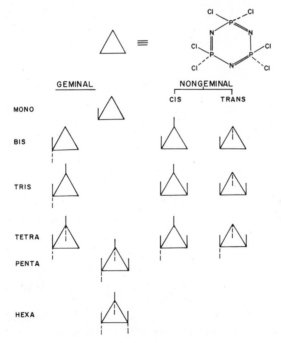

Fig. 2. Geminal and nongeminal modes of replacement of chlorine atoms from $N_3P_3Cl_6$ (1). (The corners of the triangle represent phosphorus atoms; the full and broken lines denote the orientation of the substituents above and below the N_3P_3 ring plane, respectively; chlorine and ring nitrogen atoms are not shown.)

respect to the trimeric (six-membered ring) system. In the nongeminal mode of replacement, the possibility of cis–trans isomerism at the bis, tris, and tetrakis stages of replacement must be considered. This type of isomerism depends on the orientation of the substituents with respect to the P—N ring plane. Similar replacement patterns can be envisaged for the cyclotetraphosphazene system and the possible products containing chlorine and another substituent are shown in Fig. 3.

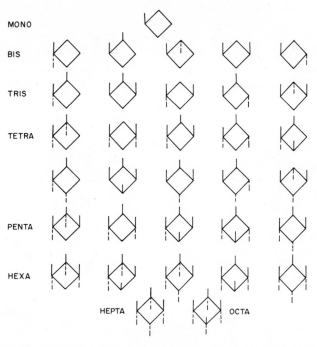

FIG. 3. Possible substitution products from $N_4P_4Cl_8$ (2). (The corners of the square represent phosphorus atoms; the full and broken lines denote the orientation of substituents above and below the N_4P_4 ring plane, respectively; chlorine and ring nitrogen atoms are not shown.)

The wealth of data available on the aminolysis reactions of $N_3P_3Cl_6$ (1) presents a complex picture (Table I). A comprehensive mechanistic theory explaining all the observations has not yet emerged. Some of the variations in the aminolysis patterns are undoubtedly due to a lack of awareness on the part of earlier investigators of the importance of reaction variables, such as temperature, solvent, and stoichiometry of reagents in determining the relative proportions of isomers formed.

TABLE I

PRODUCTS OF THE REACTIONS OF $N_3P_3Cl_6$ WITH VARIOUS AMINES[a,b]

Amine	$N_3P_3Cl_4R_2$			$N_3P_3Cl_3R_3$			$N_3P_3Cl_2R_4$			Ref.
	gem	cis	trans	gem	cis	trans	gem	cis	trans	
Ammonia[c]	+	−	−	−	−	−	−	−	−	181, 278, 426
Methylamine	+	+	+	−	−	−	−	−	−	201, 279, 312, 317
Ethylamine	−	(+)[g]	+	−	⁓⁓(+)⁓⁓		+	−	−	139
Isopropylamine	+	+	+	−	−	−	+	−	−	139, 142, 285a
t-Butylamine	+	−	−	−	−	−	+	−	−	140
Aniline	+	⁓⁓(+)⁓⁓		(+)	−	−	+	−	−	74, 153, 277
Benzylamine	+	−	+	−	−	−	+	−	−	221
Dimethylamine[d]	(+)	+	+	+	(+)	+	−	+	(+)	47, 121, 201, 211, 214, 240, 358
Diethylamine	−	+	+	+	+	+	−	(+)	+	283, 453
N-Methylaniline[e]	−	+	+	+	+	−	−	−	−	266
Aziridine[d]	+	−	−	+	−	−	+	−	−	259, 268, 334, 357
Piperidine[f]	−	+	+	+	(+)[g]	+	−	+	−	244, 267, 272
Pyrrolidine[f]	−	+	+	+	+	+	−	+	+	270, 271
Morpholine[f]	−	+	+	−	⁓⁓ + ⁓⁓		−	⁓⁓ + ⁓⁓		269, 319

[a] Isolated in good/modest yield, + ; isolated in trace quantities, (+); the structure of the geometrical isomer (cis or trans) is not known with certainty, ⁓⁓ + ⁓⁓ and ⁓⁓(+)⁓⁓; not isolated, −.

[b] In all cases mono-($N_3P_3Cl_5R$) and hexakis derivative ($N_3P_3R_6$) were isolated.

[c] The mono compound ($N_3P_3Cl_5NH_2$) is isolated by the reaction of $N_3P_3Cl_4(NH_2)_2$ with hydrogen chloride.

[d] Pentakis derivative ($N_3P_3ClR_5$) isolated.

[e] A mixed derivative, $N_3P_3Cl_2(NHPh)(NMePh)_3$, is also isolated.

[f] Pentakis derivative ($N_3P_3ClR_5$) reported but not authenticated.

[g] Obtained by the isomerization of the geometrical isomer.

Most secondary amines react predominantly by the nongeminal mode of replacement. Assumption of an $S_N2(P)$ mechanism (228) would be consistent with this observation. The phosphorus atom carrying the amino substituent acquires a partial negative charge because of

(5)

the flow of lone pair of electrons on the adjacent exocyclic nitrogen atom (5) and, thus, nucleophilic attack at the same phosphorus atom is retarded. In the reactions of dimethylamine, diethylamine and piperidine, the trans-bis isomer is always formed in greater yield than the cis isomer. A "cis effect" has been postulated to explain this observation (244). An alternative hypothesis considers a substituent solvating effect and has been suggested on the basis of kinetic data (204, 205). A notable feature of the N-methylaniline system is the formation of both cis- and trans-bis-N-methylanilino isomers, $N_3P_3Cl_4(NMePh)_2$, in comparable amounts (266). In the diethylamine system, the proportion of the cis-bis isomer relative to the trans compound is higher than in the dimethylamine system even though diethylamine is bulkier than dimethylamine (453). It appears that at the bis stage of chlorine replacement, the steric effect of the nucleophile exerts only a minor role in determining the structure of the products. The nucleophilic displacement reactions of tetracoordinate phosphorus(V) compounds by an S_N2-type process are accompanied by both retention and inversion of configuration (228). With the weaker nucleophiles, such as N-methylaniline and possibly dibenzylamine (221), the cis effect may be of little significance and there is no preferential formation of the trans isomer.

At the tris stage of chlorine replacement, the trans-nongeminal isomer predominates, although the relative proportion of the geminal compound can be significantly enhanced in aromatic reaction media (408). The use of methyl cyanide as solvent for the secondary amine reactions results in almost exclusive formation of nongeminal tris compounds: the trans isomer usually predominates over the cis isomer. However, in the N-methylaniline–$N_3P_3Cl_6$ system, the geminal tris compound is the major product even in methyl cyanide. This reaction is considerably retarded after the bis stage of chlorine replacement and unreacted amine is invariably present in the reaction medium. The unreacted N-methylaniline could fulfill the role of an aromatic reaction medium, and thereby favor the formation of the geminal tris compound even in methyl cyanide (266). It is believed that aromatic solvents preferentially solvate a $\equiv PCl_2$ group rather than a $\equiv PClR$ group and thus facilitate attack at the latter (408).

At the tetra stage of chlorine replacement, the structure of the nongeminal derivatives appears to be dominated by steric considerations. In the dimethylamine and diethylamine systems, the major tetrakis isomer has a cis and a trans structure, respectively, and only minor quantities of the other stereoisomer are obtained in each case (211, 240, 283, 453). The exceptional behavior of aziridine in yielding exclusively geminal compounds (259, 268, 334, 357) is difficult to explain, but its small size and weak nucleophilic and basic character may be contributing factors (21, 408).

The reactions of $N_3P_3Cl_6$ with bulky secondary amines [dicyclo-hexylamine (358) and dibenzylamine (221)] as well as with the nitro-genous base 2,2,2-triphenylmonophosphazene, $HN{=}PPh_3$ (66), terminate at the bis stage of chlorine replacement. The increasing difficulty of substitution has been attributed mainly to steric factors (358), although strong electron supply from the $NPPh_3$ group could deactivate the ring (66).

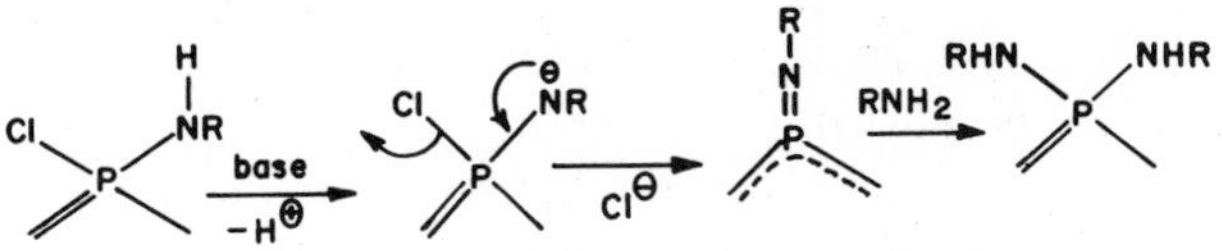

(6)

Fig. 4. Proton abstraction mechanism.

The reactions of $N_3P_3Cl_6$ (1) with ammonia and primary amines reveal additional features of interest. Reactive primary amines [methylamine (279) and ethylamine (139)] afford mainly nongeminal products at the bis stage of chlorine replacement and presumably an $S_N2(P)$ mechanism is involved. The geminal mode is predominant at all stages of the reactions with primary amines that react slowly whether for steric [e.g., t-butylamine (140)] or for polar reasons [e.g., aniline (153)]. Ammonia gives exclusively the geminal bis derivative (181, 278). The geminal replacement occurs also at the later stages of the reactions with the more reactive ethylamine (139) and isopropyl-amine (142). The geminal tris* and nongeminal tetrakis derivatives have not been isolated, although substantial amounts of the former must be present as reactive intermediates in the formation of the geminal tetrakisalkylamino derivatives. A proton abstraction mech-anism (Fig. 4) has been invoked to explain the geminal pattern observed for ammonia and t-butylamine (140). This mechanism appears plausible after the recent isolation of a three-coordinate phosphorus(V) compound, $(Me_3Si)_2NP({=}NSiMe_3)_2$ (329, 348, 376), which is clearly of the same type as the "metaphosphorimidate" intermediate (6). Shaw and co-workers (139) have attempted to rationalize the experimental observations in the reaction of $N_3P_3Cl_6$ with primary amines on the basis of chelated hydrogen bonding. These reactions could involve six-membered, cyclic, hydrogen-bonded complexes (7–9). The inter-mediate (7), in which a ring nitrogen (most basic site) and a nongeminal

* A notable exception is the isolation of the geminal trisanilino derivative, $N_3P_3Cl_3$-$(NHPh)_3$ (Table I).

(7)   (8)   (9)

chlorine atom (most ionic in character) are the sites for chelated hydrogen bonds, is most likely (*408*).

Authentic pentakisaminomonochlorocyclotriphosphazene derivatives, $N_3P_3(NRR')_5Cl$, are rare and derivatives containing only primary amino groups have not been isolated. The strong electron supply from amino groups probably causes a changeover to a relatively facile ionization of the P—Cl bond in an S_N1-type process, with the result that the remaining chlorine atom is replaced rapidly. However, Sowerby and co-workers (*121, 211*) have isolated the pentakisdimethylaminomonochloro derivative, $N_3P_3(NMe_2)_5Cl$. It is formed in low yield and is readily hydrolyzed by atmospheric moisture. The pentakisaziridino derivative, $N_3P_3(NC_2H_4)_5Cl$ (*259, 334*), and the mixed amino derivatives, $N_3P_3Cl(NHR)(NMe_2)_4$ (*410*), have also been satisfactorily characterized. A stable monochloropentakisamino compound, $N_3P_3Cl(NMe_2)_4[N(CH_2Ph)_2]$, has been reported recently, and probably steric shielding by the bulky dibenzylamino group contributes to its hydrolytic stability (*221*).

The aminolysis reactions of fluoro- and bromocyclotriphosphazenes have not been investigated systematically. Compound $N_3P_3F_6$ reacts with ammonia (*362*) and with primary (*196, 327*) and secondary amines (*114, 196, 327, 328*) to give the monoaminopentafluoro derivatives, $N_3P_3RF_5$. Some examples of nongeminal bis- and trisamino derivatives, $N_3P_3R_2F_4$ and $N_3P_3R_3F_3$, are also known (*114, 197, 328*). Aminofluoro derivatives are usually prepared by indirect methods (Section III,D). The reaction of $N_3P_3Br_6$ with ammonia has been reported and only the geminal bisammono derivative, $N_3P_3Br_4(NH_2)_2$, has been obtained (*156*). The reaction of $N_3P_3Br_6$ with dimethylamine in diethyl ether (*419, 421*) gives the derivatives $N_3P_3Br_{6-n}(NMe_2)_n$ [$n = 1, 2$ (three isomers), 3 (gem and trans)]. It has not been possible to isolate a tetrakisdimethylamino derivative $N_3P_3Br_2(NMe_2)_4$, although thin-layer chromatography indicates its presence. The cis-tris derivative, $N_3P_3Br_3(NMe_2)_3$, has been obtained only from a deaminolysis reaction of $N_3P_3(NMe_2)_6$ with hydrogen bromide in boiling xylene (*322*). The halogen replacement pattern observed in the reaction of $N_3P_3Br_6$

with ethylamine is essentially similar to that found in the analogous reaction of $N_3P_3Cl_6$ (355).

The aminolysis reactions of halogenocyclotetraphosphazenes, $N_4P_4X_8$, and the higher homologs, $(NPX_2)_n$, have received little attention and only the reactions of the octachloride, $N_4P_4Cl_8$ (2), have been investigated in some detail. The paucity of information can be attributed to the practical problems associated with the separation of complex mixture of products and the subsequent difficulties in making unambiguous structural assignments to the pure isomers. The number of isomers that can arise in the tetrameric system is much larger than in the corresponding trimeric system (Figs. 2 and 3). Millington and Sowerby (304) have investigated the reaction of $N_4P_4Cl_8$ with dimethylamine in diethyl ether at $-78°$ and isolated the derivatives, $N_4P_4Cl_{8-n}(NMe_2)_n$ [$n = 2$, 3 (three isomers), 4 (four isomers), 5 (two isomers), 6 and 8]. The reaction proceeds via the nongeminal path; geminal products are formed in poor yields. Evidence is presented for the presence of $N_4P_4Cl(NMe_2)_7$ but the pure compound could not be obtained. $N_4P_4Cl_8$ (2) reacts with N-methylaniline to yield the partially substituted derivatives, $N_4P_4Cl_{8-n}(NMePh)_n$ [$n = 1$, 2 (two isomers), 3, 4 (five isomers) and 6]. A notable feature is the isolation of five tetrakis, $N_4P_4Cl_4(NMePh)_4$, and two bis, $N_4P_4Cl_6(NMePh)_2$, derivatives (355). The complexity of the system was not apparent to earlier workers (233, 235).

The reactions of $N_4P_4Cl_8$ (2) with ethylamine (265) and t-butylamine (373) have many features in common. In the former system, the mono, two bis, a tris, two tetrakis, and the octakis derivatives have been isolated. The t-butylamine system gives the mono, two bis, a tris, and the octakis derivatives. In addition, the hydrochloride adduct, $N_4P_4(NHBu^t)_8 \cdot HCl$ has been obtained. The isolation of two distinct bis-t-butylaminohexachloro derivatives, $N_4P_4Cl_6(NHBu^t)_2$, m.p. = 171° and 128° [which have been assigned a 2,6- and 2,4-structure, respectively (373)] reconciles some of the earlier observations (359, 233). All the chloroamino derivatives have nongeminal structures. This observation may be contrasted with the exclusive formation of geminal products in the reaction of $N_3P_3Cl_6$ with t-butylamine (140) and the isolation of a geminal tetrakis derivative, $N_3P_3Cl_2(NHEt)_4$, with ethylamine (139). The difference in the behavior between the trimeric and tetrameric systems is probably due to the greater reactivity of the tetrameric chloride, $N_4P_4Cl_8$ (2) toward nucleophilic reagents (Section III,A,4). Reactions involving higher stoichiometries (particularly 1:10 and 1:12, $N_4P_4Cl_8$/amine) give only copious quantities of sticky resinous materials, and chloroaminocyclotetraphosphazenes could not be detected (265). These resins appear to contain cross-linked tetrameric units. Resin formation is also a prominent feature of the reactions of primary amines with $N_3P_3Cl_6$ (1) (139).

The reactions of halogenocyclophosphazenes with diamines is less well-documented. Hydrazine (335) and phenylhydrazine (375) react with $N_3P_3Cl_6$ (1) to yield the hexakis compounds, $N_3P_3(NHNHR)_6$ (R = H,Ph). o-Phenylenediamine (11) and N,N'-dimethylethylenediamine (117) form spirocyclic derivatives. The reactions of $N_3P_3Cl_6$ (1) with many aliphatic diamines (49, 102, 103) and amino alcohols (102, 450) have been reported, but the structures of the products have not been established with certainty. With meta- and para-aromatic diamines (and probably with aliphatic diamines also), ring-coupling reactions occur (21). o-Aminophenol brings about an unusually rapid degradation of the ring in halogenocyclophosphazenes and certain organocyclophosphazenes (see Section IV,D).

2. Mixed Amino Derivatives

A number of mixed aminochloro derivatives of cyclotri- (221, 249, 408) and cyclotetraphosphazene (371) have been prepared in order to assess the roles of substituent and nucleophile in determining the structure of products. The reactions of $N_3P_3Cl_5(NHEt)$ and $N_3P_3Cl_5(NHBu^t)$ with two equivalents of ethylamine yield nongeminal derivatives in both cases, whereas geminal products are obtained in the analogous reactions with two equivalents of t-butylamine (Scheme 1) (247). The results indicate that the attacking nucleophile determines the course of these aminolysis reactions and not the substituent already present.

There are many examples of this kind of behavior in the trimeric system (152, 153, 249). It appears that the reactions of the octachloride, $N_4P_4Cl_8$ (2), with primary amines are similarly influenced by the nucleophile (Scheme 2) (373). The substituent already present in the cyclophosphazene ring can sometimes counteract the influence of the entering nucleophile. At present, the only example of this type is provided by the triphenylphosphazenyl ($-N=PPh_3$) substituent in the trimeric system (Scheme 3) (66, 323).

$$N_3P_3Cl_5(NHEt) \underset{2Bu^tNH_2}{\overset{2EtNH_2}{\diagup\diagdown}} \begin{array}{l} nongem\text{-}N_3P_3Cl_4(NHEt)_2 \\ \\ gem\text{-}N_3P_3Cl_4(NHEt)(NHBu^t) \end{array}$$

$$N_3P_3Cl_5(NHBu^t) \underset{2Bu^tNH_2}{\overset{2EtNH_2}{\diagup\diagdown}} \begin{array}{l} nongem\text{-}N_3P_3Cl_4(NHEt)(NHBu^t) \\ \\ gem\text{-}N_3P_3Cl_4(NHBu^t)_2 \end{array}$$

SCHEME 1

$N_4P_4Cl_7(NHEt)$ $\xrightarrow{\text{2EtNH}_2}$ $2,6\text{-}N_4P_4Cl_6(NHEt)_2$, m.p. 116°

$\xrightarrow{\text{2Bu}^t\text{NH}_2}$
$\begin{cases} 2,6\text{-}N_4P_4Cl_6(NHEt)(NHBu^t), \text{ m.p. } 145°\text{--}46° \\ + \\ 2,4\text{-}N_4P_4Cl_6(NHEt)(NHBu^t)(\text{liquid}) \end{cases}$

$N_4P_4Cl_7(NHBu^t)$ $\xrightarrow{\text{2EtNH}_2}$ $2,6\text{-}N_4P_4Cl_6(NHEt)(NHBu^t)$, m.p. 145°--46°

$\xrightarrow{\text{2Bu}^t\text{NH}_2}$
$\begin{cases} 2,6\text{-}N_4P_4Cl_6(NHBu^t)_2, \text{ m.p. } 171° \\ + \\ 2,4\text{-}N_4P_4Cl_6(NHBu^t)_2, \text{ m.p. } 128° \end{cases}$

The numbers alongside each compound refer to the
disposition of amino substituents.

SCHEME 2

$N_3P_3Cl_5R$ $\xrightarrow{\text{2RH}}$ $\text{nongem-}N_3P_3Cl_4R_2$

$\xrightarrow{\text{2Ph}_3\text{PNH}}$ $\text{nongem-}N_3P_3Cl_4R(NPPh_3)$

$N_3P_3Cl_5(NPPh_3)$ $\xrightarrow{\text{2RH}}$ $\textit{gem-}N_3P_3Cl_4R(NPPh_3)$

$\xrightarrow{\text{2Ph}_3\text{PNH}}$ $\text{nongem-}N_3P_3Cl_4(NPPh_3)_2$

$[R = NMe_2, NC_5H_{10} \text{ (piperidino)}]$

SCHEME 3

3. Cis-Trans Isomerizations

Many nongeminal chlorocyclophosphazenes undergo cis–trans
isomerization under a variety of conditions. Amine hydrochlorides
(239, 242), aluminum chloride (241), hydrogen halides (320, 322), and
bases [408] have been used as inverting agents. Isomerization can also
be brought about by purely thermal methods (289, 408). Generally,
isomerization takes place in boiling chloroform, methyl cyanide, or
pyridine. Isomerization apparently does not occur in petroleum ether,
benzene, or diethyl ether: the poor solubility of amine hydrochlorides
and other inverting agents in these solvents may be a contributing
factor. Several mechanisms have been advanced to explain the isomerization (21, 289). It must be emphasized that not all nongeminal compounds undergo isomerization (241, 242, 403); the reasons for this
remain obscure.

Amine hydrochloride is one of the products of aminolysis reactions
of chlorocyclophosphazenes and, hence, cis—trans isomerization may
occur during aminolysis. Consequently, generalizations on the distribution of nongeminal isomeric products in an aminolysis reaction

(particularly carried out at elevated temperature) and assignment of structures to aminochlorocyclophosphazenes by subsequent substitution steps must be made with caution.

There does not appear to be any reliable information on the isomerization reactions of chloroaminocyclotetraphosphazenes. Lehr (281) claims to have observed the isomerization of 2-trans-6- and 2-trans-4-$N_4P_4(NMe_2)_2Cl_6$ derivatives to the corresponding cis compounds in the presence of pyridinium hydrochloride in chloroform. Details of the experimental procedures and the evidence for the structural assignments have not been stated.

4. Kinetic Studies

Studies on the kinetics of aminolysis reactions of halogenocyclophosphazenes are limited (45, 101, 202, 204, 205, 314). The rate of aminolysis by n-propylamine (314) follows the order $N_3P_3F_6 <$ $N_3P_3Cl_6 \approx N_4P_4F_8 < N_3P_3Br_6 \approx N_4P_4Cl_8 < N_4P_4Br_8$. Capon et al. (101) have found that during replacement of the first chlorine atom by diethylamine, $N_4P_4Cl_8$ reacts faster than $N_3P_3Cl_6$ by a factor of 10^2 to 10^3. Several explanations have been suggested for the observed trends in reactivity (21, 101, 314).

Recently, Goldschmidt and Licht have reported detailed investigations of the kinetics of the reactions of cyclotriphosphazenes with dimethylamine (202) and methylamine (204) in tetrahydrofuran (THF). A mechanism involving the participation of a solvent (THF) molecule in the transition state has been suggested (202). The same authors have also studied the kinetics of the reactions of $N_3P_3Cl_5(NHMe)$ with dimethylamine and that of $N_3P_3Cl_5(NMe_2)$ with methylamine forming the same nongeminal product, $N_3P_3Cl_4(NHMe)(NMe_2)$ (205). The rate of the former reaction is almost equal to that of the reaction of $N_3P_3Cl_5(NMe_2)$ with dimethylamine and the rate of the latter reaction roughly equals that of the reaction of $N_3P_3Cl_5(NHMe)$ with methylamine. These results reveal the dominant role of the nucleophile in determining the nature of products of the aminolysis reactions (Section III,A,2).

More extensive kinetic studies on different systems are needed for a better understanding of the mechanism of aminolysis reactions.

5. Solvent Effects

The precise role of the solvent in the reactions of halogenocyclophosphazenes with nitrogenous bases is not clearly understood. However, it has been recognized that the yields of specific isomers can be significantly altered in certain reaction media. Some examples in the aminolysis reactions of $N_3P_3Cl_6$ have been mentioned in Section

III,A,1. In the reaction of the octachloride, $N_4P_4Cl_8$, with t-butylamine (1:4 stoichiometry), 2,6- and 2,4-$N_4P_4(NHBu^t)_2Cl_6$ are obtained in comparable amounts in benzene as solvent. The 2,4-isomer is formed almost exclusively in chloroform, whereas in methyl cyanide the 2,6-isomer is the major product (*373*). Similar effects are observed in the reactions of $N_4P_4Cl_8$ with N-methylaniline (*355*) and benzylamine (*353*). In the tetrameric system, the choice of solvent not only influences the relative yields of isomeric products but sometimes gives rise to an entirely different class of compounds (Fig. 5). Thus, the 2-*trans*-6-bisethylaminohexachloro derivative, $N_4P_4(NHEt)_2Cl_6$, reacts with an excess of dimethylamine or ethylamine in diethyl ether to give the fully substituted cyclotetraphosphazene derivatives, $N_4P_4(NHEt)_2(NRR')_6$ (**10** and **11**) in very high yield (80%). If the reactions are carried out in chloroform, bicyclic compounds, $N_4P_4(NRR')_5(NHEt)(NEt)$ (**12** and **13**) are formed as the major products in addition to the expected compounds (**10** and **11**) (*100*, *372*). A mechanism involving an intramolecular transannular nucleophilic substitution reaction has been suggested for the formation of bicyclic compounds (*371*).

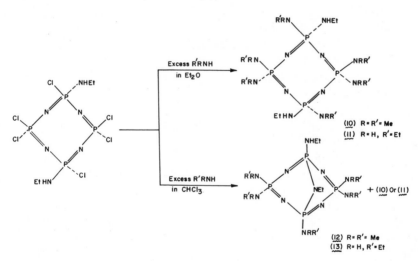

FIG. 5. Formation of bicyclic phosphazenes.

B. HYDROLYSIS

Hydrolysis of halogenocyclophosphazenes occurs rapidly in acidic, basic, and neutral solution. The ease of hydrolysis increases in the order F < Cl < Br. The initial step is the formation of a hydroxyphosphazene that undergoes tautomeric shift to give a hydroxyoxophosphazane (**14**). In an acidic medium, the formation of the hydroxyoxophosphazane (**14**) is quickly followed by ring cleavage and skeletal

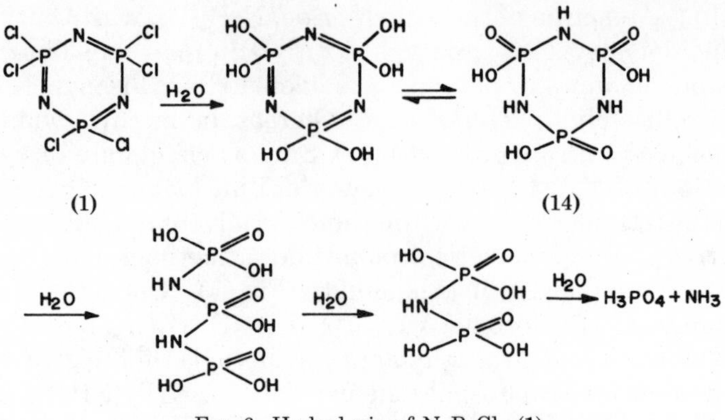

FIG. 6. Hydrolysis of $N_3P_3Cl_6$ (1).

degradation (427, 452) (Fig. 6). In alkaline media, salts of the hydroxyoxophosphazane (14) can be isolated (427). The octachloride is hydrolyzed (428) much more rapidly than $N_3P_3Cl_6$, but the hydroxyoxophosphazane derivative, $(NH)_4P_4P_4(OH)_4$, is much more stable to hydrolysis than 14. The greater stability of the tetrameric ring is also evident in the different hydrolytic behavior of $N_3P_3(NH_2)_6$ (159, 258) and $N_4P_4(NH_2)_8$ (159). The tetrameric fluoride, $N_4P_4F_8$, reacts with methanolic alkali at room temperature, whereas the trimeric fluoride, $N_3P_3F_6$, requires heating in a sealed tube at 100° for comparable hydrolysis (398).

Chlorocyclophosphazenes are rapidly hydrolyzed by aqueous pyridine to give labile pyridine salts (374, 424), and this reaction forms the basis of an analytical method for the determination of chlorine in chlorocyclophosphazenes (424).

Hydrolytic degradation of chlorocyclophosphazenes containing phenyl substituents provides valuable structural information (1, 62, 73, 226). Pentaphenylchlorocyclotriphosphazene, $N_3P_3ClPh_5$, is hydrolyzed by aqueous pyridine to the hydroxyphosphazene (15) which, on treatment with $N_3P_3ClPh_5$ in the presence of pyridine, gives the oxygen-bridged compound $Ph_5N_3P_3—O—P_3N_3Ph_5$ (389). By contrast, the hydrolysis of pentaphenoxy derivative, $N_3P_3Cl(OPh)_5$ (190),

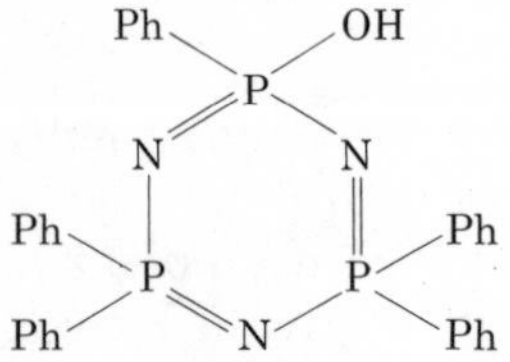

(15) 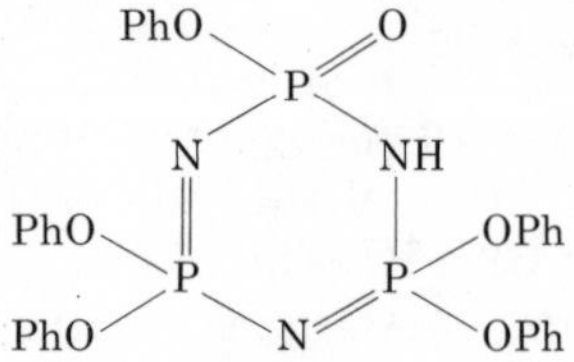   (16)

yields a phospha*diene* (16). Recently, a cyclophosphazadiene, $HN_3P_3OCl_2(NEt_2)_3$, has been isolated from the reaction of $N_3P_3Cl_6$ with diethylamine in aqueous benzene (91). The compound exists as a hydrogen-bonded dimer (17) in the solid state.

(17)

The alkaline hydrolysis of fluoroalkoxy- (19), aryloxy-, and spiroaryl-enedioxycyclophosphazenes (20) proceeds by the initial cleavage of the P—O bond. The removal of trifluoroethoxy groups from $[NP(OCH_2CF_3)_2]_3$ occurs by a nongeminal pathway (19). Cyclic tetramers undergo hydrolysis 2–4 times faster than the corresponding trimers. An S_N2-type mechanism involving a pentacoordinate trigonal bipyramidal transition state has been postulated. The hydrolysis rate is markedly accelerated by the presence of a five-membered exocyclic ring at phosphorus, probably as a result of release of steric strain (20).

C. REACTIONS WITH ALKOXIDES, ARYLOXIDES, AND THIOLATES

Alkoxy(aryloxy)cyclophosphazene derivatives are usually prepared by the reaction of halogenocyclophosphazenes with alcohols (phenols) in organic solvents in the presence of a hydrogen halide acceptor (pyridine or triethylamine) or by reaction with a metal alkoxide(aryl-loxide) (189). A large number of fully substituted alkoxy(aryloxy)cyclo-phosphazenes (21, 249) have been prepared by these methods. Only a few mixed halogenoalkoxy(aryloxy) derivatives have been charac-terized because of difficulties encountered in separation and purification procedures. The *n*-butoxy derivatives, $N_3P_3Cl_{6-n}(OBu^n)_n$ ($n = 1, 2, 3, 5$) (416, 454, 455), nongem-$N_3P_3Cl_3(OMe)_3$ (77, 185), $N_3P_3(OMe)_5Cl$ (77), and the fluoro derivatives, $N_3P_3F_5(OR)$ (R = Me, Et, Ph) (326) have been reported. Recently, Schmutz and Allcock (392) have prepared and separated nine trifluoroethoxy derivatives, $N_3P_3Cl_{6-n}(OCH_2CF_3)_n$ ($n = 1$–6) by gas-liquid chromatography. It has been shown that the replacement of chlorine atoms by fluoroalkoxy (392, 399), phenoxy (148), and *p*-bromophenoxy groups (147) proceeds by the nongeminal pathway. This substitution pattern is consistent with an $S_N2(P)$-type mechanism.

The reaction of $N_3P_3Cl_6$ with catechol (8), 2,3-dihydroxynaphthalene (11), 1,8-dihydroxynaphthalene (17), 2,2-dihydroxybiphenyl (11), and several aliphatic diols (21, 399) gives spirocyclic derivatives. A typical example is

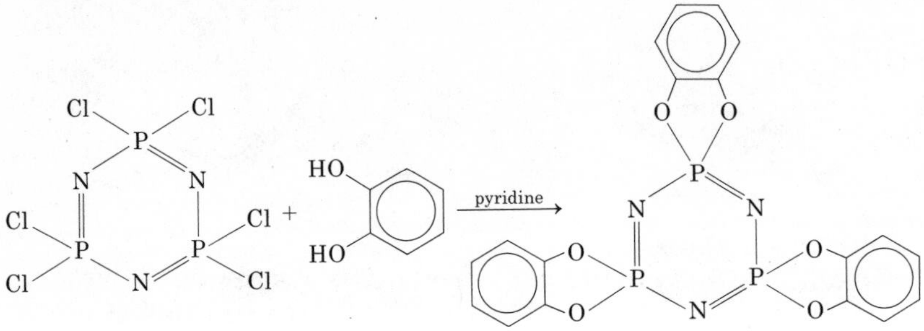

$$(8)$$

(1)   (18)

The reaction of catechol with $N_4P_4Cl_8$ does not give the tetrameric analog of 18 but leads to the degradation of the phosphazene skeleton (17). With diols, such as hydroquinone or resorcinol, cyclolinear and cyclomatrix polymers are formed (21).

A notable feature of the chemistry of alkoxycyclophosphazenes is their ability to rearrange to oxocyclophosphazanes (20) when heated alone or in the presence of an alkyl halide (187). The rearrangement

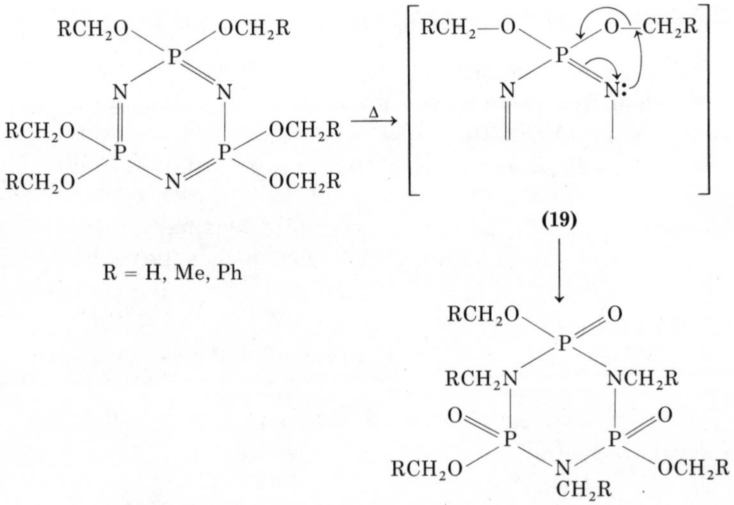

(19)

R = H, Me, Ph

(20)

reaction is believed to proceed by an inter- or intramolecular attack of a ring nitrogen atom on the α-carbon atom of the alkoxy group (19). This mechanism is supported by the observation that phenoxy and trifluoroethoxy derivatives do not undergo this rearrangement.

Chlorothioalkoxy(aryloxy)cyclophosphazenes are prepared from the interaction of a chlorocyclophosphazene with an excess of sodium thiolate (104, 105) in an anhydrous organic solvent; the reactions are apparently heterogeneous. The degree of chlorine atom replacement obtained is governed by the steric requirements of the thiolate and by the choice of the reaction solvent. The alkanethio derivatives, $N_3P_3Cl_4(SR)_2$ (R = Et, Pr^n, Bu^n), are prepared in diethyl ether at room temperature, whereas boiling benzene is needed to obtain the isopropylthio derivative, $N_3P_3Cl_4(SPr^i)_2$. The hexa-alkanethio compounds, $N_3P_3(SR)_6$ (R = Et, Pr^n, Bu^n), are obtained in boiling tetrahydrofuran, but the isopropanethiolate reaction terminates at the tetra stage of chlorine replacement (104). Recently, the preparation of methanethiolate derivatives, $N_3P_3Cl_{6-n}(SMe)_n$ (n = 1, 2, 3, 4, 6) has been reported (438). The replacement of chlorine (104, 438) or fluorine atoms (324) by thiolate groups gives rise to products with geminal structures, and compounds containing two, four, or six substituents predominate.

The reactions of the octachloride, $N_4P_4Cl_8$, with sodium thiolates in diethyl ether give only the geminal 2,2,6,6-derivative, $N_4P_4Cl_4(SR)_4$, and organic disulfides (105). Attempts to prepare the fully substituted compounds, $N_4P_4(SR)_8$, by utilizing higher reaction temperature were unsuccessful: cleavage of the P—N ring occurred to give phosphorotrithioites, $P(SR)_3$. The exclusive geminal replacement pattern observed for thioalcoholysis reactions of the trimeric and tetrameric chlorides may be a consequence of the high polarizability of a $\equiv$PCl.SR group compared to that of a $\equiv$PCl$_2$ group (104, 105).

D. METATHETICAL EXCHANGE REACTIONS

The most extensively studied metathetical exchange reaction in cyclophosphazene chemistry is the replacement of chlorine by fluorine. A variety of fluorinating agents—sodium fluoride, antimony trifluoride, potassium fluorosulfite, lead fluoride, and silver monofluoride—have been used (21, 22). Fluorocyclophosphazenes, $(NPF_2)_{3-9}$, are conveniently prepared by the reaction of the corresponding chlorocyclophosphazenes with NaF suspended in acetonitrile (394) or KSO_2F in nitrobenzene (107, 398). The mixed chloride fluorides, $N_3P_3F_nCl_{6-n}$ (n = 1–3) are formed when $N_3P_3Cl_6$ (1) is fluorinated with NaF in nitromethane; higher fluorinated derivatives are obtained in nitrobenzene (*163*). Tetrameric fluoride chlorides are prepared by the reaction of $N_4P_4Cl_8$ with KSO_2F in the absence of a solvent (*163*). Fluorination of $(NPCl_2)_{3,4}$ by KSO_2F follows a geminal pathway. In the tetrameric system fluorination can take place by two geminal routes, but proceeds almost exclusively by attack on the phosphorus atom closest to that already fluorinated (*163*):

$$2,2\text{-}N_4P_4F_2Cl_6 \longrightarrow \begin{cases} 2,2,6,6\text{-}N_4P_4F_4Cl_4 \\ (1\%) \\ 2,2,4,4\text{-}N_4P_4F_4Cl_4 \\ (99\%) \end{cases} \longrightarrow 2,2\text{-}N_4P_4Cl_2F_6 \qquad (9)$$

The reaction of the pentameric chloride, $N_5P_5Cl_{10}$, with KSO_2F has been studied (*343*). The chloride fluorides, $N_5P_5Cl_{10-n}F_n$ ($n = 1-9$) have been identified mainly by gas-liquid chromatography. Nuclear magnetic resonance spectroscopic data indicate that compounds containing more than one $\equiv$PClF group are not formed. The preferential fluorination of a $\equiv$PFCl group rather than a $\equiv$PCl$_2$ group indicates that the electron withdrawal by fluorine considerably enhances the electrophilicity of the phosphorus atom to which it is bonded (*163*). The ratio of the rate of replacement of the first fluorine (k_1) to the rate of replacement of a second fluorine (k_2) is ca. 8 for $N_3P_3Cl_6$, ca. 100 for $N_4P_4Cl_8$, and ca. 7 for $N_5P_5Cl_{10}$. This result indicates that the π-inductive effect may also be important in determining the replacement pattern (*163*, *343*).

Compounds $N_3P_3Cl_6$ and $N_3P_3Cl_5(NMe_2)$ do not undergo fluorination with antimony trifluoride in boiling 1,1,2,2-tetrachloroethane (*210*), but the latter can be fluorinated at 200° if no solvent is used and traces of $SbCl_5$ are present (*213*). Paddock and Patmore (*344*) have shown that fluorination of the cyclic chlorophosphazenes, $(NPCl_2)_{3-6}$, with a mixture of SbF_3 and $SbCl_5$ gives reasonable yields of the mono-fluoro compounds. Mixtures of nongeminal chloride fluorides can also be obtained by this route. A small amount of geminal substitution occurs with $N_6P_6Cl_{12}$.

Fluorination of dimethylaminochlorocyclophosphazenes has been studied by Sowerby and co-workers. Reaction of *cis*- and *trans*-$N_3P_3Cl_4(NMe_2)_2$ with KSO_2F takes place at the $\equiv$PCl$_2$ site rather that at the $\equiv$PCl(NMe$_2$) site (*209*). In contrast, the tetrameric derivatives, $N_4P_4Cl_6(NMe_2)_2$ and $N_4P_4Cl_5(NMe_2)_3$, undergo complete fluorination with this reagent to yield pairs of isomeric hexa- and pentafluoro compounds (*309*). Fluorination with SbF_3 proceeds preferentially (although not exclusively) at a $\equiv$PCl(NMe$_2$) group (*209*, *210*, *307*).

Antimony trifluoride not only replaces chlorine by fluorine in dimethylaminochlorocyclophosphazenes but also brings about deamination (119, 213, 307, 308). The reaction of SbF_3 with $N_4P_4(NMe_2)_8$ yields fourteen compounds, $N_4P_4F_n(NMe_2)_{8-n}$ [n = 1, 2 (four isomers), 3 (three isomers), 4 (three isomers), and 5 (three isomers)]. The reaction follows a nongeminal path (308). The analogous reaction with $N_3P_3(NMe_2)_6$ takes place much less rapidly (119) and provides a route to the pentaaminofluoro derivative, $N_3P_3(NMe_2)_5F$ (121). The reactions of SbF_3 are believed to proceed via the initial formation of an adduct by coordination of SbF_3 to the most basic, ring nitrogen atom (209, 308). Replacement of dimethylamino groups by chlorine and bromine can be effected by anhydrous hydrogen chloride and bromide (114, 120, 181, 197, 322). Deamination is not observed with hydrogen iodide but adducts, $N_3P_3(NMe_2)_6 \cdot HI$ and $N_3P_3(NMe_2)_6 \cdot HI_3$, are obtained (322).

Chlorocyclophosphazenes, $(NPCl_2)_{3-6}$, undergo exchange reactions with chloride ion. The reaction has been followed by using tetraethylammonium chloride containing radioactive chlorine (417). The rate of exchange varies in the order $N_4P_4Cl_8 > N_5P_5Cl_{10} > N_6P_6Cl_{12} > N_3P_3Cl_6$. The overall second-order kinetics suggest a bimolecular mechanism. The slower exchange rate of the trimer is probably a consequence of the considerable restraint on its skeletal flexibility which hinders the formation of the pentacoordinate transition state.

Other inorganic metathetical exchange reactions studied include the replacement of bromine by fluorine and chlorine in $N_3P_3Br_6$ and the preparation of azido-, cyano-, and isothiocyanatocyclophosphazenes (21, 22, 249). In the reactions of $N_3P_3Cl_6$ with KF and KSCN, addition of [18-Crown-6]ether considerably enhances the yields of substituted cyclophosphazenes, $N_3P_3R_6$ (R = F, NCS) (449).

Exchange reactions in which an organic group at phosphorus is replaced by a related anion are also known. For example, the fluoroalkoxycyclophosphazenes, $[NP(OCH_2CF_3)_2]_{3,4}$, undergo ligand exchange with phenoxide or ethoxide ions. In general, a particular group is displaced by a less electron-withdrawing group (15).

E. Reactions with Organometallic Reagents

The alkylation and arylation reactions of halogenocyclophosphazenes are often very complex, and ring cleavage and/or polymerization occurs in many cases. The cleavage of the phosphazene ring appears to increase in the order F < Cl < Br. Cyclic products have not been isolated from the reactions of $N_3P_3Br_6$ with Grignard reagents (21).

Alkylation is found to proceed smoothly when an amino group is present at the phosphorus atom undergoing the reaction (402, 432).

The reaction of phenylmagnesium bromide with $N_3P_3Cl_6$ in diethyl ether (61, 73) gives a small quantity of the cyclic hexaphenyl compound, $N_3P_3Ph_6$. The main reaction products are acyclic phenylated magnesium complexes (61) from which the hydrogen halide derivatives, $Ph_3P=NPPh_2=NH\cdot HX$ (X = Cl, Br), have recently been characterized (68). With $N_4P_4Cl_8$, the reaction is somewhat slower (62), and two isomeric products, $N_4P_4Cl_4Ph_4$, which have structures (21) and (22), have been obtained together with small quantities of $N_4P_4Ph_8$. The reaction mixture also contains acyclic phosphazenyl magnesium complexes.

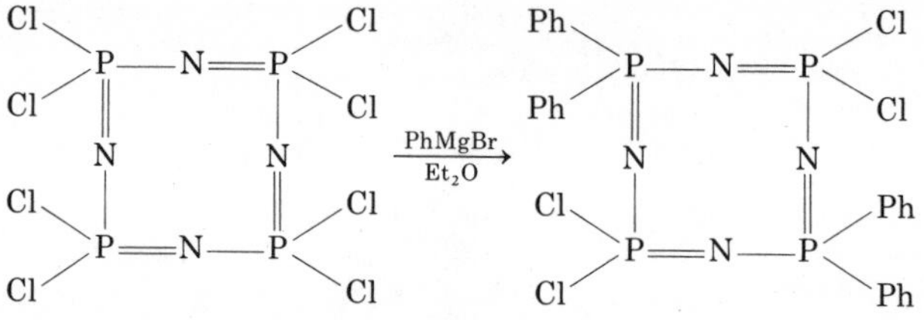

(2)   (21)

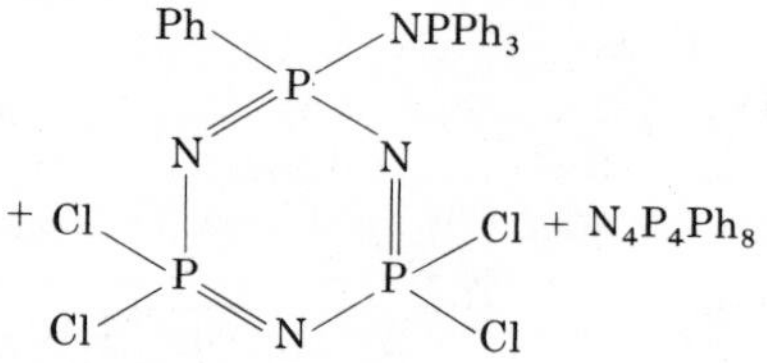

(22)

Cleavage of the tetrameric ring followed by cyclization of the phenylated acyclic species, $Ph_3P=N-PPhCl=(NPCl_2)_2=NMgBr$, would give the ring-contracted derivative (22).

The reaction of $N_3P_3Cl_6$ with diphenylmagnesium in 1,4-dioxane (60, 63, 69) follows an unusual course. Cyclic compounds ($N_3P_3Ph_6$,

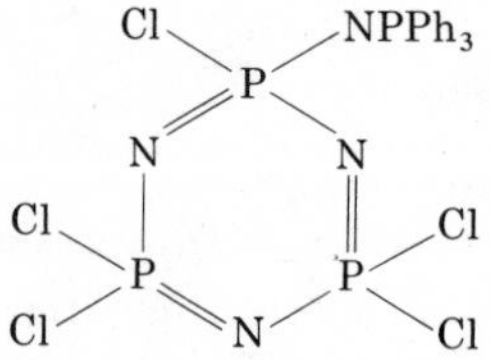

(23) 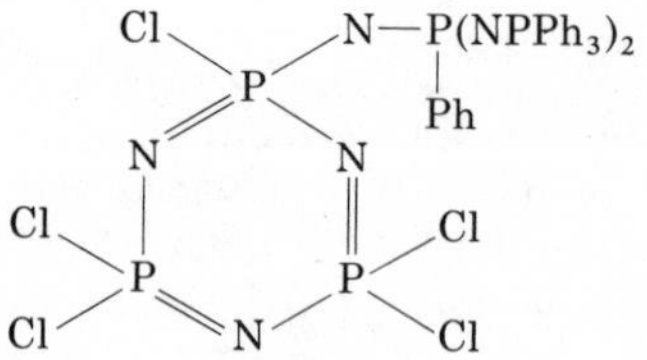   (24)

(25)

23, **24**, and **25**) as well as linear phosphazenylmagnesium products are formed. Compound (**24**) is the major product. Because the products obtained from this reaction differ from those formed in the corresponding reaction with phenylmagnesium bromide in diethyl ether, some of the phosphazenylmagnesium intermediates involved must also be different. Formation of products (**23**) and (**24**) suggests that the species $Ph_3P{=}N{-}Mg$ and $(Ph_3P{=}N)_2P(Ph){=}N{-}Mg$, play an important role. Phenyllithium (*64*), *n*-butyllithium (*402*) and R_3SnLi (*349*) behave similarly to phenylmagnesium bromide and bring about extensive ring cleavage of the trimer (**1**). The nongeminal trisdimethylamino derivative, $N_3P_3Cl_3(NMe_2)_3$, reacts with methylmagnesium iodide to give $N_3P_3Me_3(NMe_2)_3$ in 81% yield (*432*).

The reaction of $N_3P_3F_6$ with phenylmagnesium bromide (*34*) gives the mono and the geminal diphenyl derivatives, $N_3P_3F_5Ph$ and $N_3P_3F_4Ph_2$, in modest yields. The analogous reaction with phenyllithium (*32, 311*) affords the mono compound ($N_3P_3F_5Ph$) and the three bis isomers ($N_3P_3F_4Ph_2$); nongeminal derivatives predominate. The reaction of *gem*-$N_3P_3F_4Ph_2$ with phenyllithium gives the geminal tris derivative, $N_3P_3F_3Ph_3$ (*33*). Methyllithium reacts with $N_3P_3F_6$ (*341*) to give the geminal derivative, $N_3P_3F_4Me_2$. The hexamethyl compound has been isolated as the quaternary salt, $N_3P_3Me_7I$ (*341*). The corresponding reaction with $N_4P_4F_8$ gives the derivatives, $N_4P_4F_{8-n}Me_n$ ($n = 1, 2, 3, 4, 8$). Fluorine atoms are replaced in a predominantly geminal sequence and the tetra derivative, $N_4P_4F_4Me_4$, has a 2,2,6,6-structure. This replacement pattern has been interpreted in terms of a π-inductive effect of the substituents (*354*). The preparation of $N_5P_5Me_{10}$ by this route has also been reported (*341*).

F. FRIEDEL–CRAFTS REACTIONS

Although the reaction of $N_3P_3Cl_6$ with boiling benzene in the presence of aluminum chloride is slow, the geminal derivatives, $N_3P_3Cl_4Ph_2$ and $N_3P_3Cl_2Ph_4$, have been isolated (*1, 73*). The yield of the diphenyl derivative, $N_3P_3Cl_4Ph_2$, can be improved by the addition of triethylamine (*301*). The addition of the base also reduces the lengthy reaction times (*332*). Only very small quantities of the hexaphenyl compound, $N_3P_3Ph_6$, have been isolated from this reaction and drastic conditions (150°, stainless steel autoclave) are necessary to obtain modest yields (*1*). Chlorobenzene reacts with $N_3P_3Cl_6$ more readily than either benzene or toluene to give the geminal derivatives, $N_3P_3(p\text{-}C_6H_4Cl)_nCl_{6-n}$ ($n = 2, 4, 6$), and also the mono compound, $N_3P_3Cl_5(p\text{-}C_6H_4Cl)$ (*1*). Phenylation of 2-*trans*-4,6-$N_3P_3Cl_3Ph_3$ in the presence of aluminum chloride yields the cis and trans isomers, 2,2,4,6-$N_3P_3Ph_4Cl_2$, and the pentaphenyl derivative, $N_3P_3Ph_5Cl$ (*215*). Analogous reactions of $N_3P_3F_5Ph$ and $N_3P_3F_4Ph_2$ (mixture of three isomers) provide a route for the preparation of the geminal derivatives, $N_3P_3F_4Ph_2$ and $N_3P_3F_2Ph_4$, respectively (*33*). The octachloride, $N_4P_4Cl_8$, reacts with boiling benzene in the presence of aluminum chloride (*151*) to give the ring-contracted derivative, 2,2,4,4:6:6-$N_3P_3Cl_4Ph(NPPh_3)$ (**22**) in 0.6% yield. The formation of a pentaphenyl-cyclotetraphosphazene compound in this reaction has been indicated by the isolation of its dimethylamino derivative, $N_4P_4Ph_5Cl_2(NMe_2)$ (0.6% yield).

The Friedel–Crafts reaction probably proceeds by the initial formation of an ionic complex (e.g., $[N_3P_3Cl_5]^{\oplus}[AlCl_4]^{\ominus}$) followed by electrophilic attack on the aromatic molecule by the phosphazenium cation. The observed geminal replacement pattern can be rationalized because the heterolysis of a P—Cl bond is more likely to occur at the $\equiv$PClPh group than at the $\equiv$PCl$_2$ group. The postulated mechanism is also in agreement with the observation that phenylation of the amino derivatives, $N_3P_3Cl_5R$ (R = NMe_2, piperidino, $NPPh_3$), proceeds preferentially at a $\equiv$PClR site rather than at a $\equiv$PCl$_2$ group (*58, 144, 145, 191, 245*).

IV. Other Reactions of Cyclophosphazenes

A. COMPLEXES, SALTS, AND ADDUCTS

The skeletal nitrogen atoms in cyclophosphazenes possess a lone pair of electrons and, hence, they have long been viewed as potential donor sites to bind a proton or to form complexes with electron-acceptor molecules. The possibility of formation of anion–cation complexes by release of a halogen ion to a Lewis acid and charge-transfer complexes has also been studied. In addition, some cyclophosphazene derivatives form crystalline inclusion clathrates with a variety of guest molecules. Allcock (*21, 22*) has reviewed these aspects in detail.

1. Metal Complexes

Halogenocyclophosphazenes appear to form two types of complexes with metal halides. The first category is typified by the complex, $N_3P_3Cl_6 \cdot AlCl_3$, which is believed to have the cation–anion structure, $[N_3P_3Cl_5]^{\oplus}[AlCl_4]^{\ominus}$ and is presumably an intermediate in Friedel–Crafts arylation of cyclophosphazenes (73). Similar anion–cation complexes with other metal chlorides have been reported (22). A non-ionic fluorine-bridged structure has been proposed for complexes of the type $(NPF_2)_n \cdot 2SbF_5 (n = 3$ to 6) (112). The second category of complexes comprises those which are formed by coordination of the metal to a ring nitrogen atom as in $N_3P_3Br_6 \cdot nAlBr_3$ ($n = 1, 2$) and $N_3P_3Cl_6 \cdot AlBr_3$ (133).

The availability of the lone pair of electrons on skeletal nitrogen atoms of cyclophosphazenes can be greatly enhanced by replacing halogen atoms by less electronegative groups (e.g., Me) or by substituents capable of conjugative electron release (e.g., NRR′). The transition metal complexes of methyl- and dimethylaminocyclophosphazenes have been described (98, 274, 342, 397), and in a few cases the structures have been determined by X-ray crystallography (94–96, 127, 217, 296, 300, 439, 440). Cupric chloride reacts with $N_4P_4Me_8$ to give the complex, $[N_4P_4Me_8H]^{\oplus}[CuCl_3]^{\ominus}$, in which a proton and a $[CuCl_3]^{\ominus}$ group are *bonded* to opposite nitrogen atoms of the eight-membered phosphazene ring. The phosphazene ring has a "tub" conformation. The geometry around copper is distorted square planar (439). The cobalt chloride complex of the same ligand has the composition $[N_4P_4Me_8H]_2^{\oplus}[CoCl_4]^{2-}$ and contains two protonated phosphazene rings with tub and "saddle" conformations and a $[CoCl_4]^{2-}$ ion. The association is ionic and there is no evidence of cobalt–nitrogen bonding (440). Complex $[N_5P_5Me_{10}H_2]^{2+}[CuCl_4]^{2-} \cdot H_2O$ consists of a distorted tetrahedral $[CuCl_4]^{2-}$ ion and a ten-membered P—N ring with 2 protonated nitrogen atoms (96). The interaction of $N_6P_6(NMe_2)_{12}$ (L) with $CuCl_2$, CuCl, or an equimolar mixture of the two halides yields the same complex, $[LCu(II)Cl]^{\oplus}[Cu(I)Cl_2]^{\ominus}$ (300). The cobalt chloride complex formed with the same ligand has been isolated as the chloroform solvate, $[(LCoCl)^{\oplus}]_2[Co_2Cl_6]^{2-} \cdot 2CHCl_3$ (217). In both the complexes, the metal ion is coordinated to 4 of the 6 ring nitrogen atoms and a chlorine atom in a distorted trigonal bipyramidal geometry (Fig. 7).

Complexes of $N_6P_6(NMe_2)_{12}$ with other metal chlorides and nitrates have essentially similar structures as indicated by infrared and electronic spectroscopic data (98). Compounds $N_4P_4Me_8$ and $N_5P_5Me_{10}(L')$ react with molybdenum and tungsten hexacarbonyls

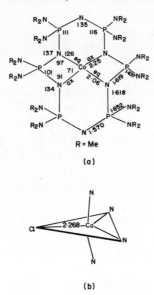

R = Me

(a)

(b)

FIG. 7. Metal atom coordination in $[N_6P_6(NMe_2)_{12}CoCl]^{\oplus}$. [Reproduced from Harrison and Trotter (*217*) by permission of The Chemical Society, London.]

to give $L'M(CO)_3$ (M = Mo and W) (*342*); $N_4P_4(NMe_2)_8$ forms a tetra-carbonyl, $N_4P_4(NMe_2)_8 \cdot W(CO)_4$, in which the phosphazene ligand is coordinated to tungsten through a ring nitrogen atom and an exo-cyclic nitrogen atom (Fig. 8). The geometry around the metal is distorted octahedral with the nitrogen atoms occupying cis positions (*94*). The bond length and bond angle variations in the phosphazene rings caused by coordination can be explained in terms of π-bonding theory (Section VI). The quaternary iodide, $[N_4P_4Me_9]^{\oplus}I^{\ominus}$, reacts with $M(CO)_6$ (M = Cr, Mo) to give the crystalline compounds, $[N_4P_4Me_9]^{\oplus}[M(CO)_5I]^{\ominus}$ (*342*). The crystal structure of the chromium complex has been reported (*95*). The hexachloride reacts with

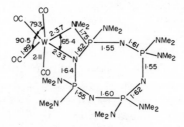

FIG. 8. The structure of $[N_4P_4(NMe_2)_8 \cdot W(CO)_4]$. [Adapted from Calhoun *et al.* (*94*) by permission of The Chemical Society, London.]

$Cr(CO)_3(CH_3CN)_3$ to give $[N_3P_3Cl_6][Cr(CO)_3]$, which is believed to be a π-complex (227).

There are also other reports of metal complexes of cyclophosphazene derivatives, but in most cases the structures of the complexes have not been established with certainty (21, 150, 229, 231, 357, 386).

2. Salts and Adducts

Halogenocyclophosphazenes form salts with perchloric acid (74) and hydrofluoric acid (387). Alkyl- and alkylaminocyclophosphazenes are much more basic than halogenocyclophosphazenes, and their hydrogen chloride adducts have been well-characterized. The adducts are prepared by the direct reaction of the phosphazene derivatives with the acid in an organic solvent (149, 220, 322, 397, 410). Many aminocyclophosphazenes have base strengths comparable to or greater than that of the parent amine from which they are derived and, hence, a hydrohalide adduct of the aminophosphazene may be formed in the aminolysis reactions of halogenocyclophosphazenes (140, 142, 149, 192, 313, 355, 371). It appears that in the aminolysis reactions, the (primaryamino)cyclophosphazene derivatives have a much more pronounced tendency to form hydrochloride adducts than secondary amino derivatives (140, 142, 355, 371). Hydrogen halide adducts of aminocyclotetraphosphazenes have been isolated only recently (e.g., $N_4P_4(NMe_2)_6(NHEt)_2 \cdot 2HCl$; $N_4P_4(NHBu^t)_8 \cdot HCl$; $N_4P_4(NMe_2)_6^-$ $(NHBu^t)_2 \cdot HCl$ (two isomers)) (371). The free phosphazene base can be generated from the adducts by treating the latter with a strong tertiary base (e.g., triethylamine, pyridine) in an organic solvent.

Basicity measurements and infrared and NMR spectroscopic data (Section V) show that the site of protonation for most cyclophosphazenes is a ring nitrogen atom rather than an exocyclic nitrogen atom (for aminocyclophosphazenes). X-Ray crystal structure analyses of $N_3P_3Cl_2(NHPr^i)_4 \cdot HCl$ (292), $N_4P_4Et_2Me_6 \cdot 2HCl(2\text{-}trans\text{-}6)$ (99), $[HN_3P_3(NMe_2)_6]^{\oplus}[Mo_6O_{19}]^{2-}$ (23), $[HN_3P_3(NMe_2)_6]^{\oplus}[CoCl_4]^{\ominus}$ (287), and some metal derivatives of cyclophosphazenes containing protonated species (Section IV,A,1) confirm that protonation occurs at the skeletal nitrogen atom(s). Recent basicity measurements suggest that for cyclophosphazene derivatives bearing a triphenylphosphazenyl ($-N{=}PPh_3$) substituent, protonation may occur either at a ring nitrogen atom or at the exocyclic phosphazenyl nitrogen atom (323). The bicyclic phosphazene (13) forms a hydrochloride adduct, and spectroscopic data suggest that the proton is probably attached to the bridgehead nitrogen atom (372). It is known that the bridgehead nitrogen atom of the base (12) has considerable sp^3 character (100).

Methylcyclophosphazenes react with alkyl iodides to form N-alkylphosphazenium iodides (397). The iodide can be exchanged for other anions, such as Cl^- or HgI_3^-. Dimethylaminocyclotriphosphazene derivatives react with trimethyloxonium tetrafluoroborate ($Me_3O^+BF_4^-$) to give onium ions in which methylation takes place at the exocyclic nitrogen atom(s). However, ring alkylation occurs under similar conditions with gem-$N_3P_3Cl_2(NHPr^i)_4$ and $N_3P_3Ph_6$ (356).

The reaction of $N_3P_3Me_6$ with iodine results in the formation of a 1:1 adduct in which charge transfer interaction takes place by the donation of nitrogen lone pair of electrons into an antibonding orbital of the iodine molecule (295). Recently, the formation of 1:3 adduct of $N_3P_3F_6$ and PF_3 has been reported from the reaction of $[Cl_3P{=}N{-}PCl_3][BCl_4]$ with AsF_3 (72). Molecular addition compounds between cyclophosphazene derivatives [e.g., $N_3P_3Cl_4(NHPr^i)_2$: $N_3P_3Cl_2(NHPr^i)_4$ and $N_4P_4Cl_4(NHEt)_4$: $N_4P_4(NHEt)_8$] (141, 143, 265) are also known.

3. Inclusion Clathrates

When some cyclotriphosphazenes are recrystallized from organic solvents, they often tend to hold the solvent molecules tenaciously in their crystal lattice to form molecular inclusion clathrates. The clathrates of tris(o-phenylenedioxy)- (18) (9), tris(2,3-dioxynaphthyl)- (11), and tris(1,8-naphthalenedioxy)cyclotriphosphazenes (17) have been investigated in some detail. X-Ray crystallographic analysis (9, 25, 411) indicates that the guest molecules are trapped in channels that exist between the bulky substituent side groups in the host lattice. This behavior may be contrasted with that of tris(2,2'-dioxydiphenyl)-cyclotriphosphazene which does not form an inclusion clathrate, presumably because the greater bulkiness of the side groups prevents the formation of channels (16). The clathration process can be used to separate the components of a mixture of organic solvents (9). Tris-(o-phenylenediamino)cyclotriphosphazene specifically retains esters and ketones, and it is suggested that hydrogen bonding is responsible for the specificity (11). Other cyclophosphazenes also form inclusion adducts (21, 56) and particular mention may be made of the triphenylphosphazenylcyclotriphosphazene derivatives, $N_3P_3Cl_5$-$[N{=}P(Ph)(N{=}PPh_3)_2]$ (24) (63) and nongeminal $N_3P_3Cl_4(NPPh_3)_2$ (66).

B. REACTIONS AT SIDE CHAINS

There are many reactions of cyclophosphazene derivatives involving substitution or transformations at side chains that do not affect the ring skeleton. The Kirsanov reaction of ammonocyclophosphazenes,

$$RNH_2 + X_3'PX_2 \rightarrow RN{=}PX_3' + 2HX$$

$$(R = \text{cyclophosphazenyl unit}; X' = Cl, Br, Ph; X = Cl, Br)$$

(10)

belongs to this category and has been investigated in some detail (55, 181, 245, 278, 282, 363, 415, 442). The side chain in $N_3P_3F_5(NPX_3)$ (X = F, Cl) can be lengthened by treating these compounds alternately with hexamethyldisilazane and PCl_5 (364).

The ammono derivatives react with $COCl_2$ (434), $SOCl_2$ (325, 368), and oxalyl dichloride [436], and add to isocyanates and isothiocyanates (366, 433). Compound $N_4P_4F_6(NSO)_2$ eliminates SO_2 in the presence of pyridine to give a bicyclic phosphazene containing a sulfur diimido bridging group (368). Under similar conditions, $N_3P_3F_5(NSO)$ gives the derivative, $N_3P_3F_5{-}N{=}S{=}N{-}N_3P_3F_5$ (325).

Treatment of $N_4P_4Me_8$ with methyllithium gives the anion $N_4P_4Me_4(CH_2{}^-)_4$, which reacts with MeI and Me_3XCl (X = Si, Ge, Sn) to give $N_4P_4Me_4(CH_2R)_4$ (R = Me, XMe_3) [96a].

Ring closure occurs in the reaction of gem-$N_3P_3Cl_4(NPCl_3)_2$ with heptamethyldisilazane (280) and of gem-$N_3P_3(NH_2)_2(OCH_2CF_3)_4$ with $MeSiHCl_2$ (237). Trans-esterification reactions take place between alkoxycyclophosphazenes and R_3SiCl (21). The reaction of 2,6-bis-(azido)hexaphenylcyclotetraphosphazene, $N_4P_4(N_3)_2Ph_6$, with Ph_3P yields a bis(triphenylphosphazenyl)hexaphenyl derivative; with cis-1,4-bis(diphosphino)butane, cage-type compounds can be obtained (400). Cleavage of the aziridine ring occurs when hexakisaziridino-cyclotriphosphazene reacts with protic species (357). Photochemical chlorination, addition of bromine to double bonds, polymerization of olefinic units, and reduction of nitro groups have also been observed with organic side chains in cyclophosphazene derivatives (21).

Brief reports of many other reactions involving side chains are scattered in the literature (115, 182, 327, 364, 367, 414).

C. THERMAL POLYMERIZATION OF CYCLOPHOSPHAZENES

The thermal polymerization of chlorocyclophosphazenes to give an "inorganic rubber" of high molecular weight was discovered by Stokes (429) in 1897. Allcock and co-workers (10, 12) have shown that poly(dichlorophosphazene) free from cross-links can be obtained by careful control of the reaction conditions employed in the polymeri-zation of $N_3P_3Cl_6$. Poly(dichlorophosphazene) is a key intermediate in the preparation of high molecular weight poly(organophosphazenes) (18, 21, 22, 24, 37, 412). The mechanism of the polymerization of halogenocyclophosphazenes is very complex and has not yet been completely unraveled. Recently, Allcock and co-workers (29) have established that traces of water function as a powerful catalyst, whereas PCl_5 is a powerful inhibitor for the polymerization of $N_3P_3Cl_6$. The polymerization and copolymerization of phenylhalogenocyclotriphosphazenes have been studied (28). Attempts to polymerize fully substituted organocyclotri- and tetraphosphazenes have been unsuccessful. Poly(bisorganophosphazenes), prepared indirectly by the nucleophilic displacement of chlorine atoms from poly(dichlorophosphazene), depolymerize on heating to organocyclophosphazene oligomers. This behavior has been rationalized in terms of the thermodynamic stability of cyclic oligomers and the kinetic stability of poly(bisorganophosphazenes) at ambient temperatures (21, 22, 24, 26, 27).

The ammonocyclophosphazenes, $[NP(NH_2)_2]_{3,4}$ and $N_4P_4Cl_4(NH_2)_4$ pyrolyze to phosphams, $(PN_2H)_n$, by elimination of NH_3 or HCl, respectively (146, 303). Anilinocyclophosphazenes, $[NP(NHPh)_2]_{3,4}$, can also be converted to phosphams on heating by the elimination of aniline (75). It is believed that the phosphams contain six or eight-membered P—N rings linked by N—H or N—Ph bridges (303). The formation of resins in the reaction of halogenocyclophosphazenes with primary amines is well authenticated, although their exact nature is not clear (139, 221, 265, 355, 371). It is possible that these resins may have a structure similar to that of phosphams (265). The thermal stability of several spirocyclophosphazenes toward ring-opening polymerization seems to depend on the steric strain of the exocyclic rings (21).

D. RING DEGRADATION REACTIONS

Several reactions of cyclophosphazenes that involve cleavage of the phosphorus–nitrogen skeleton have been described. The formation of ring-degraded products in the hydrolysis (Section III,B) and Friedel–Crafts reactions (Section III,F) and also in the reactions with organometallic reagents (Section III,E) has already been mentioned. Compound 1 reacts with catechol in the presence of tertiary amines to give the spirophosphorane (26) in addition to the spirocyclophosphazene (18) (7, 8). A similar ring-degradation occurs in the reactions of o-aminophenol with fluoro-, chloro-, and bromocyclophosphazenes to yield the phosphorane (27), but in contrast to the reactions of catechol, the spirocyclophosphazene intermediate has not been isolated (14). o-Aminophenol also reacts with spirocyclophosphazenes that contain a five-membered arylenedioxy, arylenedithio, or arylenediamino group at phosphorus to give the same phosphorane (27).

(26)   (27)

Recently, Allcock and co-workers have discussed the mechanism of these reactions (*30*).

Chlorocyclophosphazenes react with organic acids or their salts (*21, 346, 448*), acid amides (*206*), and acid halides (*207*) to yield the nitrile of the acid; the phosphazene ring is degraded. The reaction of benzoyl chloride with $[NP(OEt)_2]_{3,4}$ gives ethyl chloride, ethyl phosphenates, and triphenyl-*s*-triazine (*188*). The trifluoroethoxy derivatives, $[NP(OCH_2CF_3)_2]_{3,4}$, react with diphenyl ketone to afford the acid, $(CF_3CH_2O)_2P(O)(OH)$, and organic imines (*405, 406*).

A novel reaction of quaternary *N*-methylphosphazenium iodides, $[N_nP_nMe_{2n+1}]^{\oplus}I^{\ominus}(n = 3, 4)$, with sodium bis(trimethylsilyl)amide has been reported. Phosphorines with exocyclic methylamino groups are formed as a result of phosphazene ring cleavage. Similar cleavage occurs in the reaction of quaternary phosphazenium iodides with potassium-*t*-butoxide (*97*).

The reaction of the tricyclic compound $N_7P_6Cl_9$ (**28**) with dimethylamine involves the rupture of one of the central bonds leading to the formation of a bridged-ring imide (**29**) (*330*).

(28)   (29)

Ring degradation is also reported to occur in the reactions of $N_3P_3Cl_6$ with formamide and thioformamide (*230*), potassium bromide in the presence of [18-Crown-6]ether (*449*), dimethylsulfoxide (*447*), and metal hydrides (*262*).

E. MISCELLANEOUS REACTIONS

Trimethylamine reacts with $(NPCl_2)_{3,4}$ at room temperature to give chlorodimethylaminocyclophosphazenes. Similar dealkylation does not occur with triethylamine (92).

Dealkylation of aromatic secondary and tertiary amines has also been reported (106, 266). The reaction of fluorocyclophosphazenes with (dimethylamino)trimethylsilane affords dimethylamino-substituted derivatives (113). A similar approach has been followed to prepare spirocyclic aminohydrazino- and dihydrazinocyclotriphosphazenes by the reactions of $N_3P_3Cl_6$ with cyclic silicon–nitrogen compounds (200). Hexamethyldisilylazane undergoes a very slow reaction with $N_3P_3Cl_6$ to give the geminal product, $N_3P_3Cl_4(NHSiMe_3)_2$ (183). Epoxides undergo ring opening on treatment with $(NPCl_2)_{3,4}$ to yield chloroalkoxy derivatives (275, 276). The reaction of $N_3P_3Cl_6$ with aldoximes is a useful route to nitriles (369). A novel cyclophosphazene-substituted Sn-N ring compound has been isolated from the reaction of $N_3P_3F_5[N(SnMe_3)_2]$ and $(CF_3CO)_2O$ (365).

Reaction of P-hydridocyclophosphazenes (e.g., **30**) with CCl_4 results in the replacement of hydrogen by chlorine (383–385). Metalation of the hydridocyclophosphazene (**30**) and subsequent treatment with the chlorocyclophosphazene (**31**) leads to the formation of the two-ring assembly phosphazene (**32**), which can also be prepared by the direct reduction of **31** with potassium or by thermal disproportionation of the phosphinocyclophosphazene (**33**) (385).

(31)   (30)   (32)   (33)

Chlorocyclophosphazenes function as peptide coupling agents, aromatic coupling reagents, and substrates for photochemical and electron-induced alkylation and arylation (21). The reaction of $N_3P_3Cl_6$ with sodium salt of O,O-diethyl-N-methylamido-phosphate (193) gives the derivatives $N_3P_3Cl_{6-n}[NMeP(O)(OEt)_2]_n$ ($n = 1, 2, 3$). The bis compound is nongeminally substituted, whereas the tris derivative possesses a geminal structure (437). The reaction of chlorocyclophosphazenes with dimethylformamide (425) gives "onium"-type salts.

V. Physical Methods

A. Nuclear Magnetic Resonance Spectroscopy

Nuclear magnetic resonance spectroscopy has emerged as the most powerful tool for elucidating the molecular structures of cyclophosphazene derivatives in solution. Proton NMR spectroscopy has been widely used because of its easy accessibility. The recent development of sophisticated instrumental facilities and the application of broadband proton decoupling have greatly improved the quality and usefulness of the ^{31}P spectra (252) of cyclophosphazenes, and it is likely that this technique will become increasingly popular in the future. Fluorine NMR studies are useful for deducing the structures of fluorocyclophosphazenes, and the potential of this technique has been demonstrated in recent years (209, 210, 213, 307, 308, 343).

1. Proton Magnetic Resonance

Proton NMR spectroscopy is particularly useful for determining the disposition of substituents in aminohalogenocyclophosphazenes. Four criteria are utilized for this purpose: (a) the number of proton environments, (b) the value of $^3J^*(P—H)$,* (c) the relative chemical shifts (geminal versus nongeminal and cis versus trans), and (d) the presence or absence of "virtual coupling" (see detailed discussion in the following).

The three isomeric trisdimethylaminotrischlorocyclotriphosphazenes, cis-, trans-, and gem-$N_3P_3Cl_3(NMe_2)_3$, have been identified by the observation of one, two, and three dimethylamino doublets (coupling to ^{31}P), respectively, in their proton NMR spectra. In addition, the magnitude of $^3J^*(P—H)$ in $\equiv PCl(NMe_2)$ and $\equiv P(NMe_2)_2$

---

* The value of P—H coupling constant determined from the spectrum is slightly different from the true value of $^3J(P—H)$ because of second-order effects (see later) and, hence, is designated as the apparent P—H coupling constant, $^3J^*(P—H)$.

groupings lies in the range 16–17 and 11–13 Hz, respectively (240, 243). Similar trends in $^3J^*$(P—H) help distinguish a geminally substituted group, $\equiv PR_2$, from a nongeminal one, $\equiv PXR$ [X = Cl and R = —NHMe (192, 279), —NHEt (139, 265), —NMePh (266, 355), —NEt$_2$ (283, 453); X = Br and R = —NMe$_2$ (166, 246), —NHEt (355)]. In general, the geminal coupling constant is 3–4 Hz lower than the nongeminal one.

The chemical shifts of N—H protons can often provide a diagnostic test for geminal and nongeminal structures in cyclophosphazene derivatives with primary amino substituents (139, 221, 265, 355, 371). The N—H resonances for the nongeminal compounds occur at 6.1–6.4τ, whereas the geminal compounds display resonances at 7.1–7.8τ. The proton NMR spectra of cis- and trans-bisaminotetrahalogeno derivatives, $N_3P_3R_2X_4$, (X = Cl, Br) are essentially similar, but a distinction is possible because of the greater shielding of protons (three bonds away from phosphorus) in the cis isomer [R = —NMe$_2$ (166, 240, 246), —NEt$_2$ (283, 453), —NMePh (266), —NHMe (279), and NHEt (139, 355)].

Unambiguous structural assignments for cyclotetraphosphazene derivatives by proton NMR spectroscopy are difficult because it is often possible to interpret the data on the basis of more than one isomeric configuration. In a number of cases, the spectra are poorly resolved but the problem can be overcome to some extent by recording the spectra at higher field strength (220 MHz) (265, 304, 355). Figure 9 shows the spectrum of 2-trans-6-$N_4P_4Cl_2$(NMePh)$_6$ recorded at 100 and 220 MHz and illustrates the effect of field strength in simplifying the spectrum. Recent crystallographic studies of chloro- and fluoro-dimethylaminocyclotetraphosphazene derivatives (51, 86, 87, 89, 306) confirm the structures proposed on the basis of the proton NMR data except in the case of $N_4P_4Cl_3$(NMe$_2$)$_5$, m.p. = 154° (46).

The proton NMR spectra of many cyclophosphazenes are complicated by second-order effects characteristic of multispin systems. These second-order effects give rise to additional lines or a broad hump among the signals expected on the basis of first-order considerations and are referred to as "virtual coupling." One of the essential conditions for the occurrence of virtual coupling is that the chemical shift between the ^{31}P nuclei involved in these effects is small or zero (184, 185, 252). Unfortunately, reliable guidelines for predicting the trends in the chemical shifts of ^{31}P nuclei of cyclophosphazene derivatives containing different substituents have not yet been established (see Section V,A,2). However, the strength of virtual coupling effects or their absence can sometimes yield structural information on isomeric compounds (65, 88, 142, 355, 371). For example, the geminal 2,2,6,6-isomer of $N_4P_4Ph_4$(NMe$_2$)$_4$ is distinguished from the 2,2,4,4-isomer by

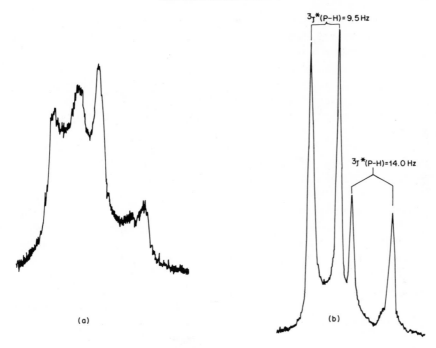

FIG. 9. The ^{1}H NMR spectrum of 2-*trans*-6-$N_4P_4Cl_2(NMePh)_6$ at (*a*) 100 MHz and (*b*) 220 MHz in $CDCl_3$ (methyl signals only).

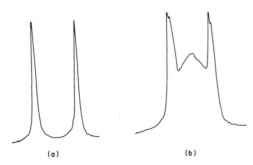

FIG. 10. The ^{1}H NMR spectrum (100 MHz) of (*a*) 2,2,6,6-$N_4P_4Ph_4(NMe_2)_4$ and (*b*) 2,2,4,4-$N_4P_4Ph_4(NMe_2)_4$ in $CDCl_3$ (methyl signals only). [Reproduced from Biddlestone *et al.* (*65*) by permission of Gordon and Breach Science Publishers Ltd., London.]

the absence of the virtual coupling effect in the former derivative (Fig. 10) (*65*).

The ^{1}H NMR study of dimethylamino and/or methoxy derivatives of partially substituted cyclophosphazenes has been utilized to obtain structural information on the halogeno precursors (*140, 142, 148, 153, 334, 355, 371*). This indirect approach must be used with caution in view of the tendency of aminohalogenocyclophosphazenes to undergo cis–trans isomerization in the presence of amine hydrochlorides (see Section III,A,3) and of methoxy compounds to rearrange to oxocyclophosphazanes on heating (see Section III,C).

Proton resonance data for the hydrogen halide adducts of aminocyclophosphazenes provide evidence for ring protonation (*220, 313, 355, 373*). Flow of electron density from the exocyclic nitrogen atoms results in the deshielding of the exocyclic N—H or N—R protons in the adducts compared to the corresponding free bases. In most cases, the detection of the signal arising from the acidic proton(s) attached to the ring nitrogen is difficult (*313, 355, 371*). A study of the proton NMR spectrum of the hydrochloride adduct, $N_4P_4(NHBu^t)_8 \cdot HCl$, at different temperatures shows that the exchange of the proton among the four, equivalent, ring nitrogen atoms is slow on the NMR time scale (*373*). The proton NMR spectrum of the dihydrochloride adduct, 2-*trans*-6-$N_4P_4(NHEt)_2(NMe_2)_6 \cdot 2HCl$ (*372*), can be interpreted on the basis of protonation occurring at two far off ring nitrogen atoms [cf. crystal structure of 2-*trans*-6-$N_4P_4Et_2Me_6 \cdot 2HCl$ (*99*)].

2. ^{31}P NMR

Many unsymmetrically substituted cyclophosphazene derivatives constitute a complex multispin system and the ^{31}P spectra of these compounds can rarely be analyzed on a first-order basis. However, in many cases ^{31}P NMR data have served to confirm the structural assignments made on the basis of ^1H NMR spectra. In several instances ^{31}P NMR spectroscopy can provide independent structural evidence where proton NMR spectra are inadequate for this purpose [e.g., the assignment of geminal structures to $N_3P_3Cl_2(NHBu^t)_4$ (*140*), $N_3P_3Cl_4(NHPh)_2$ (*153*), $N_3P_3Cl_4(NCS)_2$ (*155*), and $N_3P_3Br_4(NH_2)_2$ (*156*)]. In principle, ^{31}P NMR spectroscopy can distinguish the isomeric structures (**34–36**). The spectra of these cyclotetraphosphazene isomers

(34)   (35)   (36)

R′   R′
\P/
‖
R′   N   ⊕NH···Cl⊖
\P/   \P—R′
R′   N   N   R′
\P/
R′   R′

(R′ = NHBuᵗ)

(37)

would be of the types, AB_2C, $AA'BB'$, and A_2B_2, respectively, and these can be recognized readily.* It is not possible to differentiate cis and trans isomers in this way because the spin system will remain unchanged, assuming that possible differences in ring conformations are small and do not affect the ^{31}P spectra. The two bis-t-butylamino isomers, $N_4P_4Cl_6(NHBu^t)_2$, m.p. = 171° and 128°, give rise to symmetrical A_2B_2 and $AA'BB'$ ^{31}P spectra, respectively: a 2,6- and a 2,4-disposition of t-butylamino groups is established (Fig. 11) (373). Other examples of A_2B_2 [2-$trans$-6-$N_4P_4(NHEt)_2R_6$ (R = Cl, NMe_2)

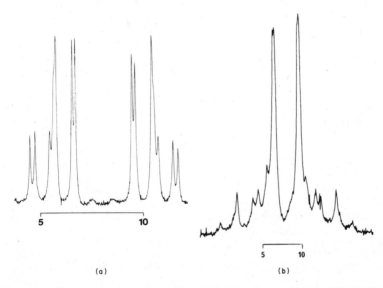

(a)   (b)

FIG. 11. The ^{31}P NMR spectra (36.7 MHz, CH_2Cl_2) of (a) 2,6-$N_4P_4(NHBu^t)_2Cl_6$ and (b) 2,4-$N_4P_4(NHBu^t)_2Cl_6$.

* It may not be easy to distinguish $AA'BB'$ and A_2B_2 spin systems when J/δ becomes relatively small.

TABLE II

^{31}P NMR DATA FOR SELECTED CYCLOPHOSPHAZENES[a]

Compound[b]	Structure	δ_{PX_2}	δ_{PCIR}	δ_{PR_2}	$^2J(P—P)$ (Hz)	Ref.
N$_3$P$_3$F$_6$	—	13.9	—	—	—	33
N$_3$P$_3$Cl$_6$	—	19.3	—	—	—	252
N$_3$P$_3$Br$_6$	—	−49.5	—	—	—	234
N$_3$P$_3$(NMe$_2$)$_2$Cl$_4$	2-trans-4	21.5	25.2	—	44.4	252
N$_3$P$_3$(NMe$_2$)$_3$Cl$_3$	2,2,4	21.7	27.3	21.7	44.8	252
N$_3$P$_3$(NMe$_2$)$_6$	—	—	—	24.6	41.2	252
N$_3$P$_3$(NEt$_2$)$_6$	—	—	—	22.5	—	252
N$_3$P$_3$(NH$_2$)$_6$	—	—	—	15.3	—	294
N$_3$P$_3$(NMe$_2$)$_2$(NHEt)$_4$	2-trans-4	—	22.8[c]	18.9[d]	42.1	252
N$_3$P$_3$(NMe$_2$)$_2$(NHEt)$_4$·HCl	2-trans-4		17.3[c]	12.3[d]	30.0	252
N$_3$P$_3$(NHEt)$_6$	—	—	—	18.0	—	252
N$_3$P$_3$(NHPr^i)$_6$	—	—	—	12.6	—	252
N$_3$P$_3$(NHPh)$_2$Cl$_4$	2,2	20.4	—	2.3	48.0	277
N$_3$P$_3$(NHBu^t)$_2$Cl$_4$	2,2	17.5	—	0.7	44.7	252
N$_3$P$_3$Cl$_2$(NHBu^t)$_4$	2,2	19.7	—	3.9	52.6	252
N$_3$P$_3$Ph$_2$Cl$_4$	2,2	17.1	—	19.5	12.1	252
N$_3$P$_3$Cl$_2$Ph$_4$	2,2	14.8	—	17.1	9.3	252
N$_3$P$_3$(SEt)$_2$Cl$_4$	2,2	17.7	—	51.7	4.8	76
N$_3$P$_3$(SEt)$_6$	—	—	—	45.7	—	76
N$_3$P$_3$Cl$_5$(OMe)	—	22.5	16.7	—	63.3	223
N$_3$P$_3$(OMe)$_6$	—	—	—	20.7	—	12
N$_3$P$_3$(OEt)$_6$	—	—	—	14.3	—	12
N$_4$P$_4$F$_8$	—	−17.0	—	—	—	294
N$_4$P$_4$Cl$_8$	—	−6.7	—	—	—	252
N$_4$P$_4$Br$_8$	—	−71.8	—	—	—	234
N$_4$P$_4$Cl$_2$(NMe$_2$)$_6$	2-trans-6	—	4.4	9.9	47.1	252
N$_4$P$_4$(NMe$_2$)$_8$	—	—	—	9.6	—	252
N$_4$P$_4$(NHEt)$_2$Cl$_6$	2-trans-6	−3.4	−4.9	—	46.0	265
N$_4$P$_4$(NHEt)$_8$	—	—	—	4.3	—	265
N$_4$P$_4$(NHBu^t)$_2$Cl$_6$	2-trans-6	−5.8	−10.6	—	38.1	373
N$_4$P$_4$(NHBu^t)$_8$	—	—	—	−3.1	—	373
N$_4$P$_4$(NHBu^t)$_8$·HCl	—	—	—	−5.0[e]	—	373

[a] The chemical shifts are with reference to 85% H$_3$PO$_4$ (external); upfield shifts are negative.

[b] Disposition of one set of substituent groups defined where necessary.

[c] ≡P(NMe$_2$)(NHEt).

[d] ≡P(NHEt)$_2$.

[e] Center of AA′BB′ multiplet (not fully analyzed).

(265), 2-trans-6-N$_4$P$_4$Cl$_2$(NMe$_2$)$_6$ (252), and 2,2,6,6-N$_4$P$_4$Cl$_4$(NMePh)$_4$ (355)] and AA′BB′ spectra [2,4-N$_4$P$_4$R$_2$Cl$_6$ (R = NMePh) (250, 355) and NHCH$_2$Ph (353)] are known. The ^{31}P NMR spectrum of the hydrochloride (37) constitutes a symmetrical AA′BB′ pattern, and a variable temperature study confirms the conclusion drawn from ^{1}H NMR data that the exchange of the acidic proton at a ring nitrogen atom(s) is slow (373).

The ^{31}P chemical shifts and phosphorus–phosphorus coupling constants, ^2J(P—N—P), for a large number of cyclophosphazene derivatives have been determined (35, 123, 184, 186, 223, 224, 252, 294, 381, 395, 399). The data for some selected compounds are shown in Table II. It has been suggested that changes in electronegativity, π-bonding, and bond angles influence ^{31}P chemical shifts (223, 284). In addition, the chemical shift of a particular phosphorus atom may be affected by substituents elsewhere on the ring. It is also possible that a bromo substituent may shield the phosphorus by a neighbor anisotropy effect (21). Empirical relationships have been proposed (186, 395) to correlate ^2J(P—N—P) with electronegativity of the substituents, but it has been noted that other factors (stereochemistry of the substituents, substituents on other phosphorus atoms) may also contribute (35, 123, 252). A comprehensive theoretical treatment of ^{31}P chemical shifts and phosphorus–phosphorus coupling constants in cyclophosphazenes has not yet emerged. Recently, four-bond phosphorus–phosphorus coupling constants have been measured for a number of phosphazenylcyclotriphosphazenes and their magnitude ($+7.5$ to -0.4 Hz) appears to be related to the conformation of the phosphazenyl group relative to the phosphazene ring (70).

3. ^{19}F NMR

The fluorine chemical shifts* observed for fluorocyclophosphazenes span a wide range from -18.0 for ≡PFBr in trans-$N_3P_3F_3Br_3$ to -71.9 for the ≡PF$_2$ unit in $N_3P_3F_6$ (21, 123). Because of strong P—F coupling [^1J(P—F) ≈ 840–1056 Hz], complex spectra are often encountered and a first-order analysis is difficult. The observed chemical shifts appear to be more useful than simple environment count or coupling constant values in determining the disposition of the substituents (22). Recently, the ^{19}F NMR spectra of a number of fluorocyclophosphazenes were recorded and analyzed by iterative computational methods. The NMR parameters (in particular the coupling constants) show marked structural dependencies. The magnitude of ^4J(F—F) is sensitive to the cis or trans orientation of the relevant P—F bonds (ca. -1.0 and ca. $+12.0$ Hz, respectively) (123).

---

* The shifts are expressed in parts per million relative to CFCl$_3$; upfield shifts are negative.

The ^{19}F NMR spectra of fluorodimethylamino derivatives of cyclotri- and cyclotetraphosphazenes confirm the structures assigned to isomeric products on the basis of ^{1}H NMR data, although a detailed analysis of the ^{19}F spectra of many derivatives could not be made (*209, 210, 213, 307, 308*). Some typical spectra are shown in Fig. 12. The structures of many chlorofluoro- (*108, 163, 223, 224*), bromofluoro- (*120*), and phenylfluorocyclophosphazenes (*32, 33, 116*) have been established mainly on the basis of ^{19}F NMR data. Also ^{19}F NMR data for a series of fluoroalkylamino- (*113, 114, 196, 197, 327*), fluoroalkyl- (*354*), and fluoroalkoxycyclophosphazenes (*441*) have been reported.

The ^{19}F NMR spectroscopic data for trifluoroethoxy derivatives, $N_3P_3R_{6-n}(OCH_2CF_3)_n$ (R = alkylamino, dialkylamino, alkoxy) can be used as an indirect method of establishing the structures of the chloro-cyclophosphazene precursors, $N_3P_3R_{6-n}Cl_n$ (*63, 139*), but this technique has received very little attention.

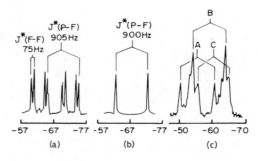

FIG. 12. The ^{19}F NMR spectra of (*a*) 2,2:4-*cis*-6:4,6-$N_3P_3F_2Cl_2(NMe_2)_2$, (*b*) 2,2:4-*trans*-6:4,6-$N_3P_3F_2Cl_2(NMe_2)_2$, and (*c*) 2-*cis*-4-*cis*-6-*trans*-8-$N_4P_4F_4(NMe_2)_4$ [$^1J^*$(P-F): A = 870 Hz, B = 880 Hz, and C = 840 Hz]. [Reproduced from Green and Sowerby (*209*) and Millington and Sowerby (*307*) by permission of The Chemical Society, London.]

B. VIBRATIONAL SPECTROSCOPY

Infrared spectroscopic studies of cyclophosphazene derivatives have been primarily used for "fingerprint" identification and differentiation of homologs (*137*). Attempts to determine the disposition of substituents solely on the basis of infrared spectroscopic data have been largely unsuccessful. The most noteworthy feature in the infrared spectra of cyclophosphazenes is a strong broad band in the region 1150–1450 cm^{-1} attributable to a degenerate ring-stretching vibration, v(P=N) (*109*). The values of v(P=N) for a large number of cyclophosphazene derivatives have been compiled (*21, 125, 203, 273, 336*), and selected data are shown in Table III. The magnitude of $v(P{=}N)$ increases with increasing electronegativity of the substituents. This effect presumably operates by contraction of phosphorus 3d orbitals permitting more effective π'-donation (Section VI) of lone pair of electrons from the ring nitrogen atom to phosphorus: the consequent strengthening of skeletal π-bonding is reflected in the increase in $v(P{=}N)$. In the ammono and amino derivatives, the electron flow from the exocyclic nitrogen atom reduces the delocalization of the lone-pair electrons on the ring nitrogen atom and weakens the skeletal π-bonding, thereby depressing the ring $P{=}N$ frequency. The extent of lowering of the ring $P{=}N$ frequency is less for aminocyclotriphosphazene derivatives than for the corresponding cyclotetraphosphazene derivatives [relative to the magnitude of $v(P{=}N)$ in $N_3P_3Cl_6$ and $N_4P_4Cl_8$, respectively] (235, 260, 355, 371).

TABLE III

VALUES OF $v(P{=}N)(cm^{-1})$
FOR SOME SELECTED CYCLOTRI-
AND CYCLOTETRAPHOSPHAZENES[a]

R	$N_3P_3R_6$	$N_4P_4R_8$
F	1300	1436
Cl	1218	1315
Br	1175	1275
Ph	1190	1213
OMe	1275	1337
NH_2	1170	1240
NHMe	1175	1215
NHEt	1180[b]	1250[c]
NHPr^n	1183	1266
NHBu^n	1197	1260
NH-n-C_5H_{11}	1190	1265
NMe_2	1195	1265

[a] From Paddock (336).
[b] From Shaw et al. (409).
[c] From Sau (371).

It may be noted that $v(P{=}N)$ for the dimethylamino derivatives, $[NP(NMe_2)_2]_{3,4}$, occur at higher frequencies than for the corresponding ethylamino derivatives. This observation is contrary to that anticipated purely on the basis of inductive effect and can be explained on the basis of greater steric hindrance to multiple bonding of the exocyclic nitrogen atom to the ring for secondary than for primary amino groups (336). The inability of the bulky substituents to assume the planar configuration required to supply electrons from exocyclic nitrogen to phosphorus is also clearly revealed by the magnitude of $v(P{=}N)$ in a series of primary alkylaminocyclotri- and cyclotetraphosphazene derivatives (Table III) (235, 260, 355, 371). A similar steric effect has been invoked to account for the greater base strengthening effect of a primary amino group compared to that of a secondary amino group in aminocyclotriphosphazenes (180). Infrared frequencies reflect ground-state properties, whereas basicity measurements pertain to perturbed state of molecules. However, it seems likely that a similar effect operates in both ground and perturbed states.

The precise assignment of $v(P{=}N)$ in amino derivatives of cyclophosphazenes is often difficult because (a) $v(P{=}N)$ invariably has a bandwidth of ca. 100 cm^{-1} and appears to become narrower with increasing degree of replacement of chlorine atoms, (b) multiple peaks can occur within the broad peak, and (c) some of the fundamental vibrations of the amino substituent occurring in the 1200–1400 cm^{-1} region may overlap considerably with $v(P{=}N)$. In some cases, no broad band attributable to $v(P{=}N)$ may be discerned at all. A careful study of the spectra of a series of aminohalogenocyclophosphazene derivatives as well as a close scrutiny of the spectra of analogous derivatives of cyclotri- and cyclotetraphosphazenes is necessary before an unambiguous assignment can be made. The infrared spectra of some t-butylamino derivatives in the region 1100–1500 cm^{-1} (Fig. 13) clearly illustrate this problem.

The feasibility of using infrared data for distinguishing positional isomers has been suggested (108, 253, 420) because the ring $P{=}N$ stretching band is more resolved in geminal than in nongeminal isomers. Unfortunately, there are many geminal derivatives that show no splitting of $v(P{=}N)$ (409) and, hence, this differentiation must be considered as unreliable. The dimethylamino derivatives of $N_3P_3Cl_6$ (420) and $N_4P_4Cl_8$ (305), which contain ${\equiv}P(NMe_2)_2$ groups, exhibit two strong and widely separated bands in the 700-cm^{-1} region and these are assigned to the symmetric and asymmetric ${\equiv}P{<}^N_N$ exocyclic stretching modes. For compounds containing only ${\equiv}PCl(NMe_2)$ groups, a single strong band at an intermediate position is observed. This approach also appears to be of limited applicability (220, 371).

Infrared spectra can be used to distinguish cyclophosphazene hydrogen halide adducts from their free bases. In general, $v(P{=}N)$ undergoes an upward shift of ca. 40–60 cm^{-1} in the hydrogen halide adduct compared to that in the free base (Table IV). This observation

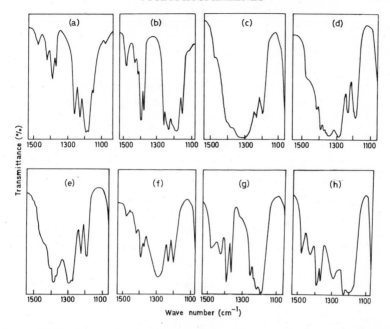

FIG. 13. Infrared spectra of (a) gem-$N_3P_3Cl_4(NHBu^t)_2$, (b) gem-$N_3P_3Cl_2(NHBu^t)_4$, (c) $N_4P_4Cl_7(NHBu^t)$, (d) 2,4-$N_4P_4(NHBu^t)_2Cl_6$, (e) 2,6-$N_4P_4(NHBu^t)_2Cl_6$, (f) 2,4,6-$N_4P_4(NHBu^t)_3Cl_5$, (g) $N_4P_4(NHBu^t)_8$, and (h) $N_4P_4(NHBu^t)_8 \cdot HCl$ in the region 1100–1500 cm^{-1} (KBr pellet).

TABLE IV

VALUES OF ν (P=N) FOR SOME SELECTED CYCLOPHOSPHAZENE
HYDROCHLORIDE ADDUCTS AND THEIR FREE BASES

Adduct	ν (P=N) (cm^{-1})	Free base	ν (P=N) (cm^{-1})	Ref.
$N_3P_3(NMe_2)_6 \cdot HCl$	1253	$N_3P_3(NMe_2)_6$	1195	*220*
$N_3P_3(NMe_2)_3(NHPh)_3 \cdot HCl$	1180, 1220	$N_3P_3(NMe_2)_3(NHPh)_3$	1165	*220*
$N_3P_3(NHEt)_6 \cdot HBr$	1230	$N_3P_3(NHEt)_6$	1180	*355*
$N_3P_3(NHPr^n)_6 \cdot HCl$	1235	$N_3P_3(NHPr^n)_6$	1183	*313*
$N_3P_3(NHBu^n)_6 \cdot HCl$	1252	$N_3P_3(NHBu^n)_6$	1197	*313*
$N_3P_3Me_6 \cdot HCl$	1170, 1230	$N_3P_3Me_6$	1180	*331, 396*
$N_4P_4(NHEt)_2(NMe_2)_6 \cdot 2HCl$ (2-*trans*-6)	1300	$N_4P_4(NHEt)_2(NMe_2)_6$ (2-*trans*-6)	1270	*372*
$N_4P_4(NHBu^t)_8 \cdot HCl$	1300	$N_4P_4(NHBu^t)_8$	1235	*371*
$N_4P_4(NHBu^t)_2(NMe_2)_6 \cdot HCl$ (2,6)	1320	$N_4P_4(NHBu^t)_2(NMe_2)_6$ (2,6)	1280	*371*
$N_4P_4Me_8 \cdot 2HCl$	1282, 1322	$N_4P_4Me_8$	1220	*331, 396*

TABLE V
X-Ray Structural Data for Some Cyclophosphazenes[a]

Compound[b]	X	Y	Ring shape	Average ring P–N bond length Å	Average P–N–P bond angle (°)	Average ring N–P–N bond angle (°)	X–P–Y bond angle (°)	Ref.
$N_3P_3F_6$	F	F	Planar	1.57	120.3	119.4	99.1	161
$N_3P_3Cl_6$	Cl	Cl	Very nearly planar	1.58	121.4	118.4	101.4	83
$N_3P_3Br_6$	Br	Br	Slightly chair	1.58	122.7	117.0	102.5	195
$N_3P_3Ph_6$	Ph	Ph	Slightly chair	1.597	122.1	117.8	103.8	3
$N_3P_3(NMe_2)_6$	NMe2	NMe2	Distorted boat	1.588	123.0	116.7	101.5	360
$N_3P_3F_3Cl_3$ (2-cis-4-cis-6)	Cl	F	Slightly boat	1.567	120.9	118.8	99.7	122
$N_3P_3Ph_2F_4$ (2,2)	F	Ph	Slightly boat	1.57[c]	120.5	118.0[d]	X–P–X = 96.9 Y–P–Y = 107.9	31
$N_3P_3Ph_2Cl_4$ (2,2)	Cl	Ph	Slightly chair	1.58[c]	121.1	118.2	X–P–X = 100.3 Y–P–Y = 104.4	290
$N_3P_3(NMe_2)_3Cl_3$ (2-trans-4-trans-6)	Cl	NMe2	Slightly boat sofa	1.573	120.6	119.0	104.9	6
$N_4P_4F_8$	F	F	Planar	1.51	147.2	122.7	99.9	302
$N_4P_4Cl_8$(K)	Cl	Cl	Boat	1.57	131.3	121.2	102.8	222
$N_4P_4Cl_8$(T)	Cl	Cl	Chair	1.56	135.6	120.5	103.1	443
$N_4P_4Br_8$(K)	Br	Br	Nearly boat	1.575	131.0	120.1	103.9	459
$N_4P_4(NMe_2)_8$	NMe2	NMe2	Puckered	1.58	133.0	120.0	104.0	82
$N_4P_4(OMe)_8$	OMe	OMe	Saddle	1.57	132.0	121.0	105.5	38
$N_4P_4Ph_8$	Ph	Ph	Hybrid boat–saddle	1.59	127.8	119.8	105.1	50

86

Compound	X	Y	Conformation					Ref.
$N_4P_4Me_8$	Me	Me	Puckered	1.60	131.6	119.8	104.0	*160*
$N_4P_4Me_2F_6$ (2,2)	F	Me	Saddle	1.52[c]	145.0	122.5	X–P–X = 93.6 Y–P–Y = 105.7	*298*
$N_4P_4(NMe_2)_4F_4$ (2-trans-4-cis-6-trans-8)	F	NMe_2	Saddle	1.557	134.6	122.3	105.3	*51*
$N_4P_4Ph_4Cl_4$ (2-cis-4-cis-6-cis-8)	Cl	Ph	Irregular crown	1.57	137.5[c]	121.0	102.8	*85*
$N_4P_4Ph_4Cl_4$ (2,2,6,6)	Cl	Ph	Saddle	1.57[c]	132.1	120.6	X–P–X = 100.8 Y–P–Y = 107.5	*90*
$N_4P_4(NMe_2)_2Cl_6$ (2-trans-6)	Cl	NMe_2	Chair	1.57[c]	134.2[c]	120.6	X–P–X = 101.7 Y–P–Y = 107.9	*87*
$N_4P_4(NMe_2)_4Cl_4$ (2-cis-4-trans-6-trans-8)	Cl	NMe_2	Hybrid crown-saddle	1.556	136.8[c]	121.1	105.1	*86*
$N_4P_4Cl_3(NMe_2)_5$ (2-cis-4-cis-6)	Cl	NMe_2	Hybrid crown-saddle	1.57[c]	135.8[c]	122.0[c]	X–P–Y = 104.2 Y–P–Y = 103.6	*46*
$N_4P_4Cl_2(NMe_2)_6$ (2-trans-6)	Cl	NMe_2	Chair	1.57[c]	135.4	120.7	X–P–Y = 103.8 Y–P–Y = 103.6	*89*
$N_4P_4Et_2Me_6·2HCl$ (2-trans-6)	Me	Et	Chair	1.665 1.572	133.2	112.2	X–P–Y = 109.1 Y–P–Y = 106.1	*99*
$N_5P_5Cl_{10}$	Cl	Cl	Very nearly planar	1.52	148.6[c]	118.4	102.0	*377*
$N_5P_5Br_{10}$	Br	Br	Puckered	1.571	135.9[c]	117.5[c]	103.3	*218*
$N_6P_6(NMe_2)_{12}$	NMe_2	NMe_2	Double-tub[d]	1.563	147.5	120.0	102.9	*444*
$N_8P_8(OMe)_{16}$	OMe	OMe	e	1.561	136.7[d]	116.7[d]	101.3	*338*

[a] Crystal structures of the following cyclophosphazenes (which are not mentioned in the table or in Sections IV,A; V,C and D) have also been reported: $N_3P_3(NCS)_6$ (173), $N_3P_3(OPh)_6$ (297), $N_3P_3FCl_5$ (333), cis-$N_3P_3F_3Br_3$ (122), $N_3P_3F_5NH_2$ (347), gem-$N_3P_3Cl_3(NMe_2)_3$ (4), cis-$N_3P_3Cl_3(NMe_2)_3$ (5), 2,2,6,6-$N_4P_4F_4Me_4$ (299), 2-cis-4-trans-6-trans-8-$N_4P_4F_4(NMe_2)_4$ (306), 2-cis-4-trans-6-trans-8-$N_4P_4Cl_4Ph_4$ (93), 2-trans-6-$N_4P_4(NMePh)_2Cl_6$ (59), and $N_5P_5F_{10}$ (219).

[b] Disposition of one set of substituent groups defined where necessary.

[c] Significant deviations from the stated mean value.

[d] Figure 15a.

[e] Figure 15b.

is consistent with protonation occurring at a ring nitrogen atom (*313*). In addition, a fairly strong band in the region 915–935 cm^{-1} and a medium to weak band at 2300–2650 cm^{-1} appear in the spectra of hydrogen chloride adducts. These have been assigned to the P—$\overset{\oplus}{N}$(H)—P linkage with no ring resonance (*413*) and the N—H stretching frequency of the group=$\overset{\oplus}{N}$H—, respectively (*220, 313, 371*).

Several attempts have been made to determine the symmetry (and hence the conformation of the phosphazene ring) of halogenocyclophosphazenes in the solid, liquid, and solution states using infrared and Raman spectroscopy (*2, 136, 249, 255, 255a, 422*). With some exceptions, there is reasonable agreement between the structures determined by diffraction methods and those predicted by vibrational spectroscopy. The calculation of force constants in $N_3P_3Cl_6$ and assignment of vibration frequencies have been discussed (*118*).

C. X-Ray Diffraction Studies

The structures of a large number of cyclophosphazene derivatives have been determined by X-ray diffraction, and the data obtained have helped to clarify the nature of the bonding in this class of compounds. Corbridge (*125*) has recently reviewed this work.

X-Ray crystallographic data for some cyclophosphazene derivatives are shown in Table V. The P—N skeletal bond lengths in cyclophosphazenes lie in the range 1.51–1.60 Å and are shorter than the accepted value (1.77 Å) for a P—N single bond (*21, 125*). This bond contraction provides good evidence for some kind of π-interaction between the skeletal phosphorus and nitrogen atoms. The presence of π-bonding between phosphorus and exocyclic groups, such as —NMe_2 or —OR, has also been inferred from X-ray crystallographic data. In homogeneously substituted cyclophosphazenes, the ring P—N bond lengths are equal and decrease with increasing electronegativity of the substituents. This attenuation is also accompanied by a decrease in the angle subtended by the exocyclic substituents at phosphorus, an increase in the ring NPN angle, and a decrease in the PNP angle. In a heterogeneously substituted compound, the ring P—N bonds are unequal. Thus, in 2,2,4,4,6,6:8,8-$N_4P_4F_6Me_2$ (*298*), there are four different P—N bond lengths with mean values 1.584, 1.470, 1.532, and 1.487 Å. Similar variations have been observed for *gem*-$N_3P_3Cl_4Ph_2$ (*290*), *gem*-$N_3P_3Cl_2Ph_4$ (*291*), *gem*-$N_3P_3F_4Ph_2$ (*31*), and 2–*cis*–4 *cis*–6:2,4,6,8,8,-$N_4P_4Cl_3(NMe_2)_5$ (*46*). These variations have been rationalized in terms of different degrees of ring π-bonding resulting from greater d-orbital contraction at the phosphorus atom bearing the most electronegative substituents. An observation that defies explanation is the presence of two different P—Cl bond lengths in gem-$N_3P_3Cl_4Ph_2$ (290) and gem-$N_3P_3Cl_2Ph_4$ (291). In cyclophosphazenes that contain nongeminal ≡PCl(NMe$_2$) groups, cooperative electron withdrawal by chlorine and electron donation by the amino nitrogen result in longer P—Cl and shorter P—N exocyclic bonds than are encountered in geminal ≡PCl$_2$ and ≡P(NMe$_2$)$_2$ groups (4, 5, 86, 87, 89).

Cyclophosphazene rings can be planar or puckered. Generally, the six-membered cyclophosphazenes are more or less planar, although in some cases small deviations from planarity have been observed. The octachloride, $N_4P_4Cl_8$, exists in two crystallographic modifications— the metastable K form having a boat conformation (222) and the T form having a centrosymmetric chair shape (443). The crystal structure analyses of a number of cyclotetraphosphazene derivatives show that the eight-membered P—N ring can adopt any of several possible conformations (planar, chair, boat, crown, saddle, or hybrid conformation). The saddle conformation of the ring in 2-trans–4-cis–6-trans-8:2,4,6,8-$N_4P_4F_4(NMe_2)_4$ is illustrated in Fig. 14 (51). Evidently, the energy differences among the various conformations are small and the attainment of a particular conformation depends on a delicate balance of a number of intra- and intermolecular factors (maximization of skeletal π-bonding, orientation of the substituents and their polar and steric nature, crystal-packing effects, and hydrogen-bonding interactions) (85, 86, 90, 161a, 306, 336).

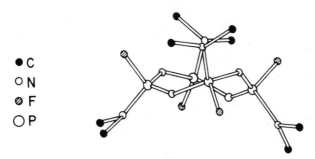

● C
○ N
◐ F
○ P

FIG. 14. Molecular configuration of 2-trans-4-cis-6-trans-8-$N_4P_4F_4(NMe_2)_4$. [Reproduced from Begley et al. (51) by permission of The Chemical Society, London.]

Higher oligomeric rings are generally puckered with the exception of $N_5P_5Cl_{10}$ (377). The ring conformations in $N_6P_6X_{12}$ (161a, 444) and $N_8P_8X_{16}$ (338, 99a) (X = OMe and NMe$_2$) are shown in Fig. 15. The sixteen-membered ring (Fig. 15b) consists of two planar and parallel segments joined by a step at two opposite nitrogen atoms. A striking

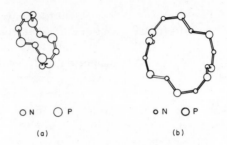

ON   OP   o N   O P

(a)   (b)

FIG. 15. Conformations of the P—N rings in (a) $N_6P_6X_{12}$ and (b) $N_8P_8X_{16}$ (X = OMe, NMe$_2$). [Adapted from Dougill and Paddock (*161a*) and Paddock *et al.* (*338*) by permission of The Chemical Society, London.]

feature of the structure of $N_8P_8(NMe_2)_{16}$ is that it exhibits both the largest angle (170.2°) at nitrogen in a cyclophosphazene and the largest mean value (156.5°) (*99a*).

The crystal structure of the bicyclic phosphazene $N_4P_4(NMe_2)_5$ (NHEt)(NEt) (**12**) shows that the original P—N heterocycle retains its phosphazene character but that P—N bonds at the bridgehead are longer and close to the accepted P—N single bond length (*100*). The bridgehead nitrogen atom has a pyramidal configuration and possesses considerable sp^3 character. An X-ray crystallographic study (*217a*) of nitrilohexaphosphonitrilic chloride, $N_7P_6Cl_9$ [discovered by Stokes (*429*) in 1897], confirms the tricyclic condensed-ring structure (**28**) suggested earlier for this compound (*263a*). The central P—N bonds (mean value 1.723 Å) are longer than for any other cyclophosphazene derivative and reflect a compromise between the requirements for optimum σ- and π-bonding (*330*). A centrosymmetric structure has been established for the two-ring assembly phosphazenes, **25** (*460*) and **32** (*385*).

It has been pointed out that most cyclophosphazene derivatives undergo protonation at a ring nitrogen atom (Section IV,A,2). X-Ray crystallographic studies of bases and their corresponding protonated species confirm that the skeletal nitrogen atom is the site of protonation and also reveal interesting bond length variations (*23, 99, 287, 292, 360, 408*). Figure 16 shows the observed bond lengths for two bases and their conjugate acids. The P—N lengths involving the protonated nitrogen atom are longer than in the free base and are phosphazane-like (*38a*). The exocyclic P—N bonds of NHPr^i and NMe$_2$ substituents on the phosphorus atoms adjacent to the protonation site are shorter in the conjugate acid than in the free base. The greater shortening observed for the —NHPr^i group (0.034 Å) than for the —NMe$_2$ group

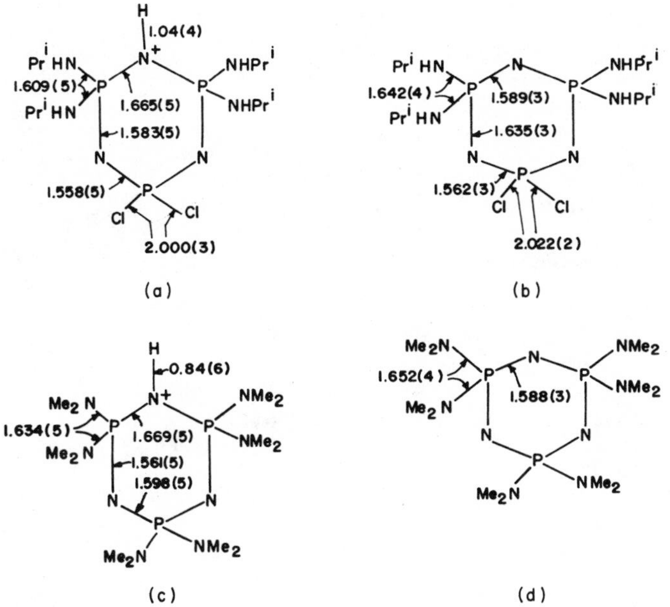

Fig. 16. Bond lengths (Å) in (a) $N_3P_3(NHPr^i)_4Cl_2 \cdot HCl$, (b) $N_3P_3(NHPr^i)_4Cl_2$, (c) $[HN_3{}^{\oplus}P_3(NMe_2)_6]_2Mo_6O_{19}{}^{2-}$, and (d) $N_3P_3(NMe_2)_6$. [Data from Shaw (408).]

(0.020 Å) is in keeping with the greater base strengthening effect of primary amino groups over that of secondary amino groups (see Sections V,B and D).

D. BASICITY MEASUREMENTS

The potentiometric titration of cyclophosphazene derivatives with perchloric acid in nitrobenzene (174) gives a relative measure of their base strength (pK_a'). Different organic substituents have a marked effect on the basicity of cyclophosphazenes (175–179) (Table VI). It has been suggested that protonation occurs at a ring nitrogen atom, and the X-ray crystal structures of protonated cyclophosphazenes substantiate this idea (Section V,C). A useful application of basicity measurements is the assignment of structures to isomeric compounds. Such assignments utilize the observation that compounds with geminal structures usually have a higher basicity than nongeminally substituted derivatives. Cis and trans isomers cannot be distinguished by this method because they have identical basicities (177).

TABLE VI

BASICITIES OF SOME CYCLOPHOSPHAZENE DERIVATIVES[a]

Compound	pK_a'	Compound	pK_a'
$N_3P_3(NHMe)_6$	8.8	$N_3P_3Ph_2Cl_2(NHMe)_2$ (2,2:4,6:4,6)	-2.8
$N_3P_3(NHEt)_6$	7.8	$N_3P_3Ph_2Cl_2(NHMe)_2$ (2,2:4,4:6,6)	-0.4
$N_3P_3(NHPr^n)_6$	7.9	$N_3P_3Cl_2(NHEt)_4$ (gem)	3.2
$N_3P_3(NHBu^t)_6$	7.9	$N_3P_3Cl_2(NHBu^t)_4$ (gem)	3.5
$N_3P_3(NMe_2)_6$	7.6	$N_3P_3Cl_5(NPPh_3)$	<-6.0
$N_3P_3(NEt_2)_6$	8.2	$N_3P_3Cl_4Ph(NPPh_3)^b$ (2,2,4,4:6:6)	-4.7
$N_3P_3(OEt)_6$	-0.2		
$N_3P_3Et_6$	6.4	$N_3P_3Cl_4(NH_2)(NPPh_3)^b$ (2,2,4,4:6:6)	-2.9
$N_3P_3Ph_6$	1.5		
$N_3P_3Cl_3(NMe_2)_3$ (trans)	-5.4	$N_3P_3Cl_4(NPPh_3)_2$ (gem)	0.4
$N_3P_3Cl_3(NMe_2)_3$ (gem)	-4.4	$N_3P_3Cl_4(NPPh_3)_2$ (nongem)	0.2

[a] Data from Refs. (177, 178, 408).

[b] Exocyclic protonation postulated; in other cases data consistent with protonation at a ring nitrogen atom.

As a result of the extensive compilation of basicity data for cyclotriphosphazenes, it has been possible to evaluate substituent constants for different groups (178–180). These constants, can be used to calculate the pK_a' value of a cyclophosphazene derivative containing one or more substituents (179, 180). Hence, a comparison of calculated and observed pK_a' values can often provide information for the complete characterization of a derivative (180). Basicity data for cyclotetraphosphazenes and higher homologs are limited (175, 321), and substituent constants have not been evaluated.

Basicity measurements for cyclophosphazenes containing a triphenylphosphazenyl substituent (—N=PPh$_3$) (323) suggest that protonation can take place either at a ring nitrogen atom (Type I) or at the nitrogen atom of the phosphazenyl substituent (Type II) (Table VI). It is believed that in Type I compounds, the N—P bond of the triphenylphosphazenyl group is more or less parallel to the local N—P—N ring segment, whereas in Type II compounds it is approximately perpendicular (408). The X-ray crystal structures of $N_3P_3Cl_4Ph(NPPh_3)$ (22) (67) and $N_3P_3Cl_5(NPPh_3)$ (23) (44) show that these compounds have Type II and Type I conformations, respectively.

Also noteworthy is the observation that in the ground state, the electron-releasing power of the $-N=PPh_3$ group is of the order of that of $-NR_2'$ and $-NHR'$ groups, whereas at the demand of a proton, the $-N=PPh_3$ group behaves as a much stronger electron donor. This behavior may be due to a major contribution from the resonance form $-\overset{\ominus}{N}-\overset{\oplus}{P}Ph_3$.

E. NUCLEAR QUADRUPOLE RESONANCE SPECTROSCOPY

Chlorocyclophosphazenes and their derivatives can be studied in the solid state by NQR spectroscopy. The hexachloride, $N_3P_3Cl_6$, has a slight "chair" configuration and the molecule is bisected by a mirror plane (83). Although the P—Cl bond lengths are almost identical, there are 4 structurally distinct chlorine atoms and these give rise to a four-line ^{35}Cl NQR spectrum (124). The ^{35}Cl NQR spectra of the K and T forms of the octachloride, $N_4P_4Cl_8$, have been reported, and the multiplicity of the signals observed has been related to the symmetry of each molecule (157, 238).

There is an approximately linear relationship between ^{35}Cl NQR frequencies and P—Cl bond lengths in closely related chlorocyclophosphazene derivatives [e.g., $N_3P_3Cl_{6-n}Ph_n$ ($n = 0, 2, 4$); $N_3P_3Cl_3$-$(NMe_2)_3$ isomers). An increase in bond length is accompanied by a decrease in the NQR frequency, and this change presumably reflects an increase in the ionic character of the bond (248).

The use of ^{35}Cl NQR spectroscopy to identify chemically distinct nuclei in chlorocyclophosphazene derivatives is limited by problems of signal sensitivity and crystal packing effects in the sample. Consequently, unambiguous structural assignments cannot always be made solely on the basis of NQR data. The ^{35}Cl NQR frequencies characteristic of $\equiv PCl_2$, $\equiv PCl(NMe_2)$, and $\equiv PPhCl$ occur in the ranges 26–29, 22–25, and 23–25 MHz, respectively, and permit a distinction of geminal and nongeminal isomers (138).

F. OTHER TECHNIQUES

The potential of dipole moment measurements for structural assignments in cyclophosphazene chemistry has been noted. The cis and trans isomers of $N_3P_3Ph_3Br_3$ (315, 316), $N_3P_3Cl_4(NMe_2)_2$, $N_3P_3Cl_3(NMe_2)_3$ (263), $N_3P_3Cl_4(NHMe)_2$ (279), and $N_3P_3F_4Ph_2$ (32) have been distinguished from dipole moment data. Dipole moments for several cis- and trans-dialkylaminocyclotriphosphazenes have also been reported (261). In view of the uncertainties in allowing for the atom polarization term and the possible deviations of the ring from planarity, dipole moment data cannot always be interpreted unambiguously (392).

The mass spectra of halogenocyclophosphazenes (79, 80, 131, 132, 390) and isothiocyanatocyclophosphazenes (445) show the inherent stability of P—N cyclic systems. The mass spectra of some aryl-substituted fluorocyclotriphosphazenes reveal differences in the major fragmentation process between geminal and nongeminal isomers (36). The fragmentation patterns of amino and aminohalogeno derivatives have not been investigated. The use of mass spectrometry for determining the disposition of substituents in cyclophosphazene derivatives is likely to be limited because the phosphazene ring does not undergo fragmentation until most substituent groups have been lost (105). The ionization potentials of several cyclophosphazene derivatives have been determined by mass spectrometry (78, 132) and photoelectron spectroscopy (78, 212).

Data obtained from other physical measurements (ultraviolet and electron-spin resonance spectroscopy, thermochemical measurements, polarography, etc.) have been compiled (249) and discussed (21). By and large, comparisons based on these data are not very instructive.

VI. Bonding and Electronic Structure

The nature of the bonding in cyclophosphazene rings has provided a considerable challenge for theoreticians. The shortness and equality of P—N bond lengths in homogeneously substituted cyclophosphazenes (Section V,C) can be interpreted as a conjugative interaction involving phosphorus and nitrogen. A precise description of this π-bonding is still a controversial point.

Craig and Paddock (134) initially suggested that the σ-bonded P—N skeleton is supplemented by π-bonding involving nitrogen p_z and phosphorus d_{xz} orbitals. Delocalization of electrons occurs through extended molecular orbitals covering all the ring nuclei. If the phosphazene ring deviates markedly from planarity, the π-bonding system can be maintained by using different combinations of d orbitals of the correct symmetry. This type of π-bonding bears a superficial resemblance to that of benzene. However, an additional type of π-bonding can be envisaged for phosphorus–nitrogen heterocycles that is not possible for aromatic carbon compounds. Overlap of the sp^2 orbital of nitrogen containing the lone-pair electrons and phosphorus d_{xy} and $d_{x^2-y^2}$ orbitals would permit "in-plane" π-bonding (designated as π'-bonding) (135). The puckering and greater PNP angles observed for cyclotetraphosphazene derivatives make conditions more favorable for significant π'-overlap. The presence of substituent groups or atoms of high electronegativity will contract the phosphorus d orbitals and, hence, facilitate donation of the nitrogen sp^2 lone-pair electrons to the π'-system. Conversely, it would be anticipated that electron-donating substituents participate in exocyclic π-bonding by means of the d_{z^2} orbital on phosphorus and, thus, allow the remaining d orbitals to expand. This backbonding would effectively localize the lone-pair electrons at ring nitrogen atoms and disrupt the π'-system. Hence, the balance of π- and π'-bonding associated with the phosphazene skeleton will be influenced by the nature of the exocyclic substituents. The orbital overlap schemes for π-bonding in cyclophosphazenes are shown in Fig. 17.

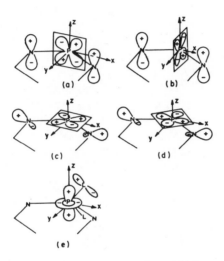

FIG. 17. Orbital overlap schemes for ring π-bonding of (a) d_{xz} and (b) d_{yz} orbitals with p_z orbitals; (c) π'-bonding of d_{xy} and (d) $d_{x^2-y^2}$ with an s-p_y hybrid, and (e) π-bonding of d_{z^2} with exocyclic substituent p orbital. [Reproduced from Corbridge (*125*) by permission of Elsevier Scientific Publ. Co., Amsterdam.]

The theory proposed by Dewar *et al.* (*154*) is also based on a σ-bonded P—N skeleton of sp^2-hybridized nitrogen and sp^3-hybridized phosphorus but postulates a pair of linear combinations of phosphorus d_{xz} and d_{yz} orbitals for overlap with an adjacent nitrogen p_z orbital. The result is a system of almost independent three-center π-bonds containing two phosphorus atoms and one nitrogen atom (Fig. 18). More detailed calculations appear to support this theoretical treatment (*167, 168, 171, 172, 318*) and indicate that conjugation beyond the three-center islands is of minor importance in cyclotriphosphazene derivatives (*167*). X-Ray crystallographic data for heterogeneously substituted cyclophosphazene derivatives (Section V,C) and evidence

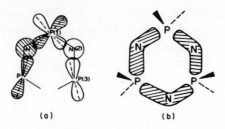

(a)   (b)

FIG. 18. (a) Orbital overlap scheme for three-center P—N—P bonds. (Shaded atomic orbitals are combined in molecular orbitals.) [Reproduced from Craig and Mitchell (*135a*) by permission of The Chemical Society, London.] (b) Three-center P—N—P islands in $N_3P_3R_6$.

from other physicochemical measurements (*81, 170, 310*) are compatible with this bonding picture. The recent work of Doggett (*158*) suggests that the delocalized model of Craig and Paddock and the three-center P—N—P approach of Dewar *et al.* are fundamentally very similar and differ only in the choice of parameters.

Recently, trans-annular phosphorus–phosphorus bonding has been postulated to account for the stability of the cyclotriphosphazene system (*40–42, 169*). In eight-membered phosphazene rings, it is believed that trans-annular bonding occurs for phosphorus atoms separated by two bonds, and trans-annular antibonding for phosphorus atoms four bonds apart (*408*).

VII. Potential Applications

The potential use of cyclophosphazene derivatives as ultrahigh-capacity fertilizers (*446*), flame retardants (*199*), pesticides (*216*), and chemosterilant insecticides (*111*) has been demonstrated. Phosphazene high polymers and, in particular, linear polyorganophosphazenes offer considerable promise as materials suitable for aerospace applications, heat-resistant coatings for electrical components, low- and high-temperature elastomers, textile impregnating agents, fire-resistant foams (*21, 24*), and body implantation plastics (*110*). By and large, phosphazenes are expensive. Prices will have to come down considerably if they are ever to gain widespread use.

ACKNOWLEDGMENTS

The authors thank the University Grants Commission, New Delhi, and the Overseas Development Ministry, U.K., for support. They are grateful to Professors R. A. Shaw and A. R. Vasudeva Murthy for their kind interest and encouragement. Thanks are also due Dr. M. N. Sudheendra Rao and Mr. K. Ramachandran for help in checking the content of this review.

REFERENCES

1. Acock, K. G., Shaw, R. A., and Wells, F. B. G., *J. Chem. Soc.* p. 121 (1964).
2. Adams, D. M., and Fernando, W. S., *J. Chem. Soc., Dalton Trans.* p. 2503 (1972).
3. Ahmed, F. R., Singh, P., and Barnes, W. H., *Acta Crystallogr., Sect. B* **25**, 316 (1969).
4. Ahmed, F. R., and Pollard, D. R., *Acta Crystallogr., Sect. B* **28**, 513 (1972).
5. Ahmed, F. R., and Pollard, D. R., *Acta Crystallogr., Sect. B* **28**, 3530 (1972).
6. Ahmed, F. R., and Gabe, E. J., *Acta Crystallogr., Sect. B* **31**, 1028 (1975).
7. Allcock, H. R., *J. Am. Chem. Soc.* **85**, 4050 (1963).
8. Allcock, H. R., *J. Am. Chem. Soc.* **86**, 2591 (1964).
9. Allcock, H. R., and Siegel, L. A., *J. Am. Chem. Soc.* **86**, 5140 (1964).
10. Allcock, H. R., and Kugel, R. L., *J. Am. Chem. Soc.* **87**, 4216 (1965).
11. Allcock, H. R., and Kugel, R. L., *Inorg. Chem.* **5**, 1016 (1966).
12. Allcock, H. R., Kugel, R. L., and Valan, K. J., *Inorg. Chem.* **5**, 1709 (1966).
13. Allcock, H. R., "Heteroatom Ring Systems and Polymers." Academic Press, New York, 1967.
14. Allcock, H. R., and Kugel, R. L., *J. Am. Chem. Soc.* **91**, 5452 (1969).
15. Allcock, H. R., Kugel, R. L., and Walsh, E. J., *J. Chem. Soc., Chem. Commun.* p. 1283 (1970).
16. Allcock, H. R., Stein, M. T., and Stanko, J. A., *J. Am. Chem. Soc.* **93**, 3173 (1971).
17. Allcock, H. R., and Walsh, E. J., *Inorg. Chem.* **10**, 1643 (1971).
18. Allcock, H. R., and Moore, G. Y., *Macromolecules* **5**, 231 (1972).
19. Allcock, H. R., and Walsh, E. J., *J. Am. Chem. Soc.* **94**, 119 (1972).
20. Allcock, H. R., and Walsh, E. J., *J. Am. Chem. Soc.* **94**, 4538 (1972).
21. Allcock, H. R., "Phosphorus-Nitrogen Compounds." Academic Press, New York, 1972.
22. Allcock, H. R., *Chem. Rev.* **72**, 315 (1972).
23. Allcock, H. R., Bissell, E. C., and Shawl, E. T., *Inorg. Chem.* **12**, 2963 (1973).
24. Allcock, H. R., *Chem. Brit.* p. 118 (1974).
25. Allcock, H. R., and Stein, M. T., *J. Am. Chem. Soc.* **96**, 49 (1974).
26. Allcock, H. R., and Cook, W. J., *Macromolecules* **7**, 284 (1974).
27. Allcock, H. R., Moore, G. Y., and Cook, W. J., *Macromolecules* **7**, 571 (1974).
28. Allcock, H. R., and Moore, G. Y., *Macromolecules* **8**, 377 (1975).
29. Allcock, H. R., Gardner, J. E., and Smeltz, K. M., *Macromolecules* **8**, 36 (1975).
30. Allcock, H. R., Kugel, R. L., and Moore, G. Y., *Inorg. Chem.* **14**, 2831 (1975).
31. Allen, C. W., Faught, J. B., Moeller, T., and Paul, I. C., *Inorg. Chem.* **8**, 1719 (1969).
32. Allen, C. W., and Moeller, T., *Inorg. Chem.* **7**, 2177 (1968).
33. Allen, C. W., Tsang, F. Y., and Moeller, T., *Inorg. Chem.* **7**, 2183 (1968).
34. Allen, C. W., *J. Chem. Soc., Chem. Commun.* p. 152 (1970).
35. Allen, C. W., *J. Magn. Reson.* **5**, 435 (1971).
36. Allen, C. W., and Toch., P. L., *J. Chem. Soc., Dalton Trans.* p. 1685 (1974).
37. Allen, G., and Mortier, R. M., *Polymer* **13**, 253 (1972).
38. Ansell, G. B., and Bullen, G. J., *J. Chem. Soc. A* p. 2498 (1971).
38a. Ansell, G. B., and Bullen, G. J., *J. Chem. Soc. A* p. 3026 (1968).
39. Appel, R., and Saleh, G., *Chem. Ber.* **106**, 3455 (1973).
40. Armstrong, D. R., Longmuir, G. H., and Perkins, P. G., *J. Chem. Soc., Chem. Commun.* p. 464 (1972).

41. Armstrong, D. R., Easdale, M. C., and Perkins, P. G., *Phosphorus* **3**, 251 (1974).
42. Armstrong, D. R., Easdale, M. C., and Perkins, P. G., *Phosphorus* **3**, 259 (1974).
43. Audrieth, L. F., Steinmann, R., and Toy, A. D. F., *Chem. Rev.* **32**, 109 (1943).
44. Babu, Y. S., Cameron, T. S., Krishnamurthy, S. S., Manohar, H., and Shaw, R. A., *Z. Naturforsch. B* **31**, 999 (1976).
45. Bailey, J. V., and Parker, R. E., *Chem. Ind. (London)* p. 1823 (1962).
46. Bamgboye, T. T., Begley, M. J., and Sowerby, D. B., *J. Chem. Soc., Dalton Trans.* p. 2617 (1975).
47. Becke-Goehring, M., John, K., and Fluck, E., *Z. Anorg. Allg. Chem.* **302**, 103 (1959).
48. Becke-Goehring, M., and Fluck, E., *Angew. Chem., Int. Ed. Engl.* **1**, 281 (1962).
49. Becke-Goehring, M., and Boppel, B., *Z. Anorg. Allg. Chem.* **322**, 239 (1963).
50. Begley, M. J., Sowerby, D. B., and Tillot, R. J., *J. Chem. Soc., Dalton Trans.* p. 2527 (1974).
51. Begley, M. J., Millington, D., King, T. J., and Sowerby, D. B., *J. Chem. Soc., Dalton Trans.* p. 1162 (1974).
52. Bergeron, C. R., and Kao, J. T. F., U. S. Patent 3,780,162 (1973); *Chem. Abstr.* **80**, 85315 (1974).
53. Bergeron, C. R., and Kao, J. T., French Patent 2,187,689 (1974); *Chem. Abstr.* **81**, 51816 (1974).
54. Bermann, M., and Utvary, K., *J. Inorg. Nucl. Chem.* **31**, 271 (1969).
55. Bermann, M., this series **14**, 1 (1972).
56. Bermann, M., and van Wazer, J. R., *Inorg. Chem.* **11**, 209 (1972).
57. Bezman, I. I., and Smalley, J. H., *Chem. Ind. (London)* p. 839 (1960).
58. Bezman, I. I., and Ford, C. T., *Chem. Ind. (London)* p. 163 (1963).
59. Bhandary, K., Manohar, H., and Babu, Y. S., *Acta Crystallogr., Sect. B* **33**, 3548 (1977).
60. Biddlestone, M., Shaw, R. A., and Taylor, D., *J. Chem. Soc., Chem. Commun.* p. 320 (1969).
61. Biddlestone, M., and Shaw, R. A., *J. Chem. Soc. A* p. 178 (1969).
62. Biddlestone, M., and Shaw, R. A., *J. Chem. Soc. A* p. 1750 (1970).
63. Biddlestone, M., and Shaw, R. A., *J. Chem. Soc. A* p. 2715 (1971).
64. Biddlestone, M., and Shaw, R. A., *Phosphorus* **3**, 95 (1973).
65. Biddlestone, M., Krishnamurthy, S. S., Shaw, R. A., Woods, M., Bullen, G. J., and Dann, P. E., *Phosphorus* **3**, 179 (1973).
66. Biddlestone, M., and Shaw, R. A., *J. Chem. Soc., Dalton Trans.* p. 2740 (1973).
67. Biddlestone, M., Bullen, G. J., Dann, P. E., and Shaw, R. A., *J. Chem. Soc., Chem. Commun.* p. 56 (1974).
68. Biddlestone, M., and Shaw, R. A., *J. Chem. Soc., Dalton Trans.* p. 2527 (1975).
69. Biddlestone, M., and Shaw, R. A., *J. Chem. Soc., Chem. Commun.* p. 407 (1968).
70. Biddlestone, M., Keat, R., Rose, H., Rycroft, D. S., and Shaw, R. A., *Z. Naturforsch. B* **31**, 1001 (1976).
71. Bik, J. D., U.S. Patent 3,794,701 (1974); *Chem. Abstr.* **80**, 122900 (1974).
72. Binder, H., *Z. Anorg. Allg. Chem.* **383**, 130 (1971).
73. Bode, H., and Bach, H., *Chem. Ber. B* **75**, 215 (1942).
74. Bode, H., Bütow, K., and Lienau, G., *Chem. Ber.* **81**, 547 (1948).
75. Bode, H., and Clausen, H., *Z. Anorg. Allg. Chem.* **258**, 99 (1949).
76. Boden, N., Emsley, J. W., Feeney, J., and Sutcliffe, L. H., *Chem. Ind. (London)* p. 1909 (1962).
77. Bond, M. R., Hewlett, C., Hills, K., and Shaw, R. A., unpublished results quoted in Keat *et al.* (*249*).

78. Branton, G. R., Brion, C. E., Frost, D. C., Mitchell, K. A. R., and Paddock, N. L., *J. Chem. Soc. A* p. 151 (1970).
79. Brion, C. E., and Paddock, N. L., *J. Chem. Soc. A* p. 388 (1968).
80. Brion, C. E., and Paddock, N. L., *J. Chem. Soc. A* p. 392 (1968).
81. Bruniquel, M. F., Faucher, J.-P., Labarre, J.-F., Hasan, M., Krishnamurthy, S. S., Shaw, R. A., and Woods, M., *Phosphorus* **3**, 83 (1973).
82. Bullen, G. J., *J. Chem. Soc.*, p. 3193 (1962).
83. Bullen, G. J., *J. Chem. Soc. A* p. 1450 (1971).
84. Bullen, G. J., and Mallinson, P. R., *J. Chem. Soc., Dalton Trans.* p. 1412 (1972).
85. Bullen, G. J., and Tucker, P. A., *J. Chem. Soc., Dalton Trans.* p. 1651 (1972).
86. Bullen, G. J., and Tucker, P. A., *J. Chem. Soc., Dalton Trans.* p. 2437 (1972).
87. Bullen, G. J., and Dann, P. E., *J. Chem. Soc., Dalton Trans.* p. 1453 (1973).
88. Bullen, G. J., Dann, P. E., Desai, V. B., Shaw, R. A., Smith, B. C., and Woods, M., *Phosphorus* **3**, 67 (1973).
89. Bullen, G. J., and Dann, P. E., *J. Chem. Soc., Dalton Trans.* p. 705 (1974).
90. Bullen, G. J., and Dann, P. E., *Acta Crystallogr., Sect. B* **30**, 2861 (1974).
91. Bullen, G. J., Dann, P. E., Evans, M. L., Hursthouse, M. B., Shaw, R. A., Wait, K., Woods, M., and Yu, Hon-Sum, *Z. Naturforsch. B* **31**, 995 (1976).
92. Burg, A. B., and Caron, A. P., *J. Am. Chem. Soc.* **81**, 836 (1959).
93. Burr, A. H., Carlisle, C. H., and Bullen, G. J., *J. Chem. Soc., Dalton Trans.* p. 1659 (1974).
94. Calhoun, H. P., Paddock, N. L., and Trotter, J., *J. Chem. Soc., Dalton Trans.* p. 2708 (1973).
95. Calhoun, H. P., and Trotter, J., *J. Chem. Soc., Dalton Trans.* p. 377 (1974).
96. Calhoun, H. P., and Trotter, J., *J. Chem. Soc., Dalton Trans.* p. 382 (1974).
96a. Calhoun, H. P., Lindstrom, R. H., Oakley, R. T., Paddock, N. L., and Todd, S. M., *J. Chem. Soc., Chem. Commun.* p. 343 (1975).
97. Calhoun, H. P., Oakley, R. T., and Paddock, N. L., *J. Chem. Soc., Chem. Commun.* p. 454 (1975).
98. Calhoun, H. P., Paddock, N. L., and Wingfield, J. N., *Can. J. Chem.* **53**, 1765 (1975).
99. Calhoun, H. P., Oakley, R. T., Paddock, N. L., and Trotter, J., *Can. J. Chem.* **53**, 2413 (1975).
99a. Calhoun, H. P., Paddock, N. L., and Trotter, J., *J. Chem. Soc., Dalton Trans.* p. 38 (1976).
100. Cameron, T. S., Mannan, Kh., Krishnamurthy, S. S., Sau, A. C., Vasudeva Murthy, A. R., Shaw, R. A., and Woods, M., *J. Chem. Soc., Chem. Commun.* p. 975 (1975).
101. Capon, B., Hills, K., and Shaw, R. A., *J. Chem. Soc.*, p. 4059 (1965).
102. Cardillo, B., Mattogno, G., and Tarli, F., *Atti Accad. Naz. Lincei, Rend. Classe Sci., Fis. Mat. Nat.* **35**, 328 (1963); *Chem. Abstr.* **61**, 4357 (1964).
103. Cardillo, B., Mattogno, G., Melera, A., and Tarli, F., *Atti Accad. Naz. Lincei, Rend. Classe Sci., Fis. Mat. Nat.* **37**, 194 (1964); *Chem. Abstr.* **62**, 13030 (1965).
104. Carroll, A. P., and Shaw, R. A., *J. Chem. Soc. A* p. 914 (1966).
105. Carroll, A. P., Shaw, R. A., and Woods, M., *J. Chem. Soc., Dalton Trans.* p. 2736 (1973).
106. Cheng, C. Y., and Shaw, R. A., *J. Chem. Soc., Perkin Trans. Sect. I* p. 1739 (1976).
107. Chapman, A. C., Paddock, N. L., Paine, D. H., Searle, H. T., and Smith, D. R., *J. Chem. Soc.* p. 3608 (1960).
108. Chapman, A. C., Paine, D. H., Searle, H. T., Smith, D. R., and White, R. F. M., *J. Chem. Soc.* p. 1768 (1961).
109. Chapman, A. C., and Paddock, N. L., *J. Chem. Soc.* p. 635 (1962).
110. *Chem. Eng. News* June 16, p. 22 (1975).

111. *Chem. Week* Dec. 14, p. 83 (1962).
112. Chivers, T., and Paddock, N. L., *J. Chem. Soc. A* p. 1687 (1969).
113. Chivers, T., and Paddock, N. L., *J. Chem. Soc., Chem. Commun.* p. 337 (1969).
114. Chivers, T., Oakley, R. T., and Paddock, N. L., *J. Chem. Soc. A* p. 2324 (1970).
115. Chivers, T., *Inorg. Nucl. Chem. Lett.* **7**, 827 (1971).
116. Chivers, T., and Paddock, N. L., *Inorg. Chem.* **11**, 848 (1972).
117. Chivers, T., and Hedgeland, R., *Can. J. Chem.* **50**, 1017 (1972).
118. Christopher, R. E., and Gans, P., *J. Chem. Soc., Dalton Trans.* p. 153 (1975).
119. Clare, P., Millington, D., and Sowerby, D. B., *J. Chem. Soc., Chem. Commun.* p. 324 (1972).
120. Clare, P., Sowerby, D. B., and Green, B., *J. Chem. Soc., Dalton Trans.* p. 2374 (1972).
121. Clare, P., and Sowerby, D. B., *J. Inorg. Nucl. Chem.* **36**, 729 (1974).
122. Clare, P., King, T. J., and Sowerby, D. B., *J. Chem. Soc., Dalton Trans.* p. 2071 (1974).
123. Clare, P., Sowerby, D. B., Harris, R. K., and Wazeer, M. I. M., *J. Chem. Soc., Dalton Trans.* p. 625 (1975).
124. Clipsham, R., Hart, R. M., and Whitehead, M. A., *Inorg. Chem.* **8**, 2431 (1969).
125. Corbridge, D. E. C., "The Structural Chemistry of Phosphorus Compounds," pp. 333–365. Elsevier, Amsterdam, 1974.
126. Cotton, F. A., and Shaver, A., *Inorg. Chem.* **10**, 2362 (1971).
127. Cotton, F. A., Rusholme, G. A., and Shaver, A., *J. Coord. Chem.* **3**, 99 (1973).
128. Coxon, G. E., Sowerby, D. B., and Tranter, G. C., *J. Chem. Soc.* p. 5697 (1965).
129. Coxon, G. E., Palmer, T. F., and Sowerby, D. B., *Inorg. Nucl. Chem. Lett.* **2**, 215 (1966).
130. Coxon, G. E., and Sowerby, D. B., *J. Chem. Soc. A* p. 1566 (1967).
131. Coxon, G. E., Palmer, T. F., and Sowerby, D. B., *J. Chem. Soc. A* p. 1568 (1967).
132. Coxon, G. E., Palmer, T. F., and Sowerby, D. B., *J. Chem. Soc. A* p. 358 (1969).
133. Coxon, G. E., and Sowerby, D. B., *J. Chem. Soc. A* p. 3012 (1969).
134. Craig, D. P., and Paddock, N. L., *Nature (London)* **181**, 1052 (1958).
135. Craig, D. P., and Paddock, N. L., *J. Chem. Soc.* p. 4118 (1962).
135a. Craig, D. P., and Mitchell, K. A. R., *J. Chem. Soc.* p. 4682 (1965).
136. Creighton, J. A., and Thomas, K. M., *Spectrochim. Acta, Sect. A* **29**, 1077 (1973).
137. Daasch, L. W., *J. Am. Chem. Soc.* **76**, 3403 (1954).
138. Dalgleish, W. H., Keat, R., Porte, A. L., Tong, D. A., Hasan, M., and Shaw, R. A., *J. Chem. Soc., Dalton Trans.* p. 309 (1975).
139. Das, R. N., Shaw, R. A., Smith, B. C., and Woods, M., *J. Chem. Soc., Dalton Trans.* p. 709 (1973).
140. Das, S. K., Keat, R., Shaw, R. A., and Smith, B. C., *J. Chem. Soc.* p. 5032 (1965).
141. Das, S. K., Shaw, R. A., and Smith, B. C., *J. Chem. Soc., Chem. Commun.* p. 176 (1965).
142. Das, S. K., Keat, R., Shaw, R. A., and Smith, B. C., *J. Chem. Soc. A* p. 1677 (1966).
143. Das, S. K., Feakins, D., Last, W. A., Nabi, S. N., Ray, S. K., Shaw, R. A., and Smith, B. C., *J. Chem. Soc. A* p. 616 (1970).
144. Das, S. K., Shaw, R. A., and Smith, B. C. *J. Chem. Soc., Dalton Trans.* p. 1883 (1973).
145. Das, S. K., Shaw, R. A., and Smith, B. C., *J. Chem. Soc., Dalton Trans.* p. 1610 (1974).
146. de Ficquelmont, A. M., *C. R. Acad. Sci., Paris* **200**, 1045 (1935).
147. Dell, D., Fitzsimmons, B. W., and Shaw, R. A., *J. Chem. Soc.* p. 4070 (1965).

148. Dell, D., Fitzsimmons, B. W., Keat, R., and Shaw, R. A., *J. Chem. Soc. A* p. 1680 (1966).
149. Denny, K., and Lanoux, S., *J. Inorg. Nucl. Chem.* **31**, 1531 (1969).
150. Derbisher, G. V., and Babaeva, A. V., *Russ. J. Inorg. Chem.* **10**, 1194 (1965).
151. Desai, V. B., Shaw, R. A., and Smith, B. C., *Angew. Chem. Int. Ed. Engl.* **7**, 887 (1968).
152. Desai, V. B., Shaw, R. A., and Smith, B. C., *J. Chem. Soc. A* p. 1977 (1969).
153. Desai, V. B., Shaw, R. A., and Smith, B. C., *J. Chem. Soc. A* p. 2023 (1970).
154. Dewar, M. J. S., Lucken, E. A. C., and Whitehead, M. A., *J. Chem. Soc.* p. 2423 (1960).
155. Dieck, R. L., and Moeller, T., *J. Inorg. Nucl. Chem.* **35**, 75 (1973).
156. Dieck, R. L., and Moeller, T., *J. Inorg. Nucl. Chem.* **35**, 737 (1973).
157. Dixon, M., Jenkins, H. D. B., Smith, J. A. S., and Tong, D. A., *Trans. Faraday Soc.* **63**, 2852 (1967).
158. Doggett, G., *J. Chem. Soc., Faraday Trans. II*, p. 2075 (1972).
159. Dostal, K., Kouril, M., and Novak, J., *Z. Chem.* **4**, 353 (1964).
160. Dougill, M. W., *J. Chem. Soc.* p. 5471 (1961).
161. Dougill, M. W., *J. Chem. Soc.* p. 3211 (1963).
161a. Dougill, M. W., and Paddock, N. L., *J. Chem. Soc., Dalton Trans.* p. 1022 (1974).
162. Emsley, J., and Udy, P. B., *J. Chem. Soc., Chem. Commun.* p. 633 (1967).
163. Emsley, J., and Paddock, N. L., *J. Chem. Soc. A* p. 2590 (1968).
164. Emsley, J., and Udy, P. B., *J. Chem. Soc. A* p. 3025 (1970).
165. Emsley, J., and Udy, P. B., *J. Chem. Soc. A* p. 768 (1971).
166. Engelhardt, G., Steger, E., and Stahlberg, R., *Z. Naturforsch. B* **21**, 586 (1966).
167. Faucher, J.-P., Devanneaux, J., Leibovici, C., and Labarre, J.-F., *J. Mol. Struct.* **10**, 439 (1971).
168. Faucher, J.-P., and Labarre, J.-F., *J. Mol. Struct.* **17**, 159 (1973).
169. Faucher, J.-P., and Labarre, J.-F., *Phosphorus* **3**, 265 (1974).
170. Faucher, J.-P., Glemser, O., Labarre, J.-F., and Shaw, R. A., *C. R. Acad. Sci., Paris Ser. C* **279**, 441 (1974).
171. Faucher, J.-P., Labarre, J.-F., and Shaw, R. A., *J. Mol. Struct.* **25**, 109 (1975).
172. Faucher, J.-P., Labarre, J.-F., and Shaw, R. A., *Z. Naturforsch. B* **31**, 680 (1976).
173. Faught, J. B., Moeller, T., and Paul, I. C., *Inorg. Chem.* **9**, 1656 (1970).
174. Feakins, D., Last, W. A., and Shaw, R. A., *J. Chem. Soc.* p. 2387 (1964).
175. Feakins, D., Last, W. A., and Shaw, R. A., *J. Chem. Soc.* p. 4464 (1964).
176. Feakins, D., Last, W. A., Neemuchwala, N., and Shaw, R. A., *J. Chem. Soc.* p. 2804 (1965).
177. Feakins, D., Last, W. A., Nabi, S. N., and Shaw, R. A., *J. Chem. Soc. A* p. 1831 (1966).
178. Feakins, D., Nabi, S. N., Shaw, R. A., and Watson, P., *J. Chem. Soc. A* p. 10 (1968).
179. Feakins, D., Last, W. A., Nabi, S. N., Shaw, R. A., and Watson, P., *J. Chem. Soc. A* p. 196 (1969).
180. Feakins, D., Shaw, R. A., Watson, P., and Nabi, S. N., *J. Chem. Soc. A* p. 2468 (1969).
181. Feistel, G. R., and Moeller, T., *J. Inorg. Nucl. Chem.* **29**, 2731 (1967).
182. Feldt, M. K., and Moeller, T., *J. Inorg. Nucl. Chem.* **30**, 2351 (1968).
183. Fields, A. T., and Allen, C. W., *J. Inorg. Nucl. Chem.* **36**, 1929 (1974).
184. Finer, E. G., *J. Mol. Spectrosc.* **23**, 104 (1967).
185. Finer, E. G., Harris, R. K., Bond, M. R., Keat, R., and Shaw, R. A., *J. Mol. Spectrosc.* **33**, 72 (1970).

186. Finer, E. G., and Harris, R. K., *Progr. Nucl. Magn. Reson. Spectrosc.* **6**, 61–118 (1971).

187. Fitzsimmons, B. W., Hewlett, C., and Shaw, R. A., *J. Chem. Soc.* p. 4459 (1964).

188. Fitzsimmons, B. W., Hewlett, C., and Shaw, R. A., *J. Chem. Soc.* p. 4799 (1965).

189. Fitzsimmons, B. W., and Shaw, R. A., *Inorg. Synth.* **8**, 77 (1966).

190. Fitzsimmons, B. W., Hewlett, C., Hills, K., and Shaw, R. A., *J. Chem. Soc. A* p. 679 (1967).

191. Ford, C. T., Dickson, F. E., and Bezmann, I. I., *Inorg. Chem.* **3**, 177 (1964).

192. Ford, C. T., Dickson, F. E., and Bezmann, I. I., *Inorg. Chem.* **4**, 890 (1965).

193. Gehlert, P., Schadow, H., Scheler, H., and Thomas, B., *Z. Anorg. Allg. Chem.* **415**, 51 (1975).

194. Gilson, I. T., and Sisler, H. H., *Inorg. Chem.* **4**, 273 (1965).

195. Giglio, E., and Puliti, R., *Acta Crystallogr.* **22**, 304 (1967).

196. Glemser, O., Niecke, E., and Roesky, H. W., *J. Chem. Soc., Chem. Commun.* p. 282 (1969).

197. Glemser, O., Niecke, E., and Thamm, H., *Z. Naturforsch. B* **25**, 754 (1970).

198. Glidewell, C., *Angew. Chem., Int. Ed. Engl.* **14**, 826 (1975).

199. Godfrey, L. E. A., and Schappel, J. W., *Ind. Eng. Chem. Prod. Res. Develop.* **9**, 426 (1970).

200. Goldin, G. S., Baturina, L. S., Novikova, A. N., and Fedorov, S. G., *Zh. Obshch. Khim.* **45**, 2566 (1975); *Chem. Abstr.* **84**, 59414 (1976).

201. Goldschmidt, J. M. E., and Weiss, J., *J. Inorg. Nucl. Chem.* **26**, 2023 (1964).

202. Goldschmidt, J. M. E., and Licht, E., *J. Chem. Soc. A* p. 2429 (1971).

203. Goldschmidt, J. M. E., *in* "Analytical Chemistry of Phosphorus Compounds" (M. Halmann, ed.), pp. 523–591. Wiley (Interscience), New York, 1972.

204. Goldschmidt, J. M. E., and Licht, E., *J. Chem. Soc., Dalton Trans.* p. 728 (1972).

205. Goldschmidt, J. M. E., and Licht, E., *J. Chem. Soc., Dalton Trans.* p. 732 (1972).

206. Graham, J. C., and Marr, D. H., *Can. J. Chem.* **50**, 3857 (1972).

207. Graham, J. C., *Tetrahedron Lett.* p. 3825 (1973).

208. Graham, J. C., German Patent 2,321,221 (1973); *Chem. Abstr.* **80**, 38922 (1974).

209. Green, B., and Sowerby, D. B., *J. Chem. Soc. A* p. 987 (1970).

210. Green, B., Sowerby, D. B., and Clare, P., *J. Chem. Soc. A* p. 3487 (1971).

211. Green, B., and Sowerby, D. B., *J. Inorg. Nucl. Chem.* **33**, 3687 (1971).

212. Green, B., Ridley, D. C., and Sherwood, P. M. A., *J. Chem. Soc., Dalton Trans.* p. 1042 (1973).

213. Green, B., *J. Chem. Soc., Dalton Trans.* p. 1113 (1974).

214. Gribova, I. A., and Ban-Yuan', U., *Russ. Chem. Rev.* (Engl. Transl.) **30**, 1 (1961).

215. Grushkin, B., Sanchez, M. G., Ernest, M. V., McClanahan, J. L., Ashby, G. E., and Rice, R. G., *Inorg. Chem.* **4**, 1538 (1965).

216. Haiduc, I., "The Chemistry of Inorganic Ring Systems," Part 2, pp. 624–761. Wiley (Interscience), New York, 1970.

217. Harrison, W., and Trotter, J., *J. Chem. Soc., Dalton Trans.* p. 61 (1973).

217a. Harrison, W., and Trotter, J., *J. Chem. Soc., Dalton Trans.* p. 623 (1972).

218. Hartsuiker, J. G., and Wagner, A. J., *J. Chem. Soc., Dalton Trans.* p. 1069 (1972).

219. Hartsuiker, J. G., and Wagner, A. J., *IUPAC Symp. Inorg. Phosphorus Comp. 2nd, 1974, Abstr. No.* **3.13**.

220. Hasan, M., Ph.D. Thesis, University of London, 1974.

221. Hasan, M., Shaw, R. A., and Woods, M., *J. Chem. Soc., Dalton Trans.* p. 2202 (1975).

222. Hazekamp, R., Migchelsen, T., and Vos, A., *Acta Crystallogr.* **15**, 539 (1962).

223. Heatley, F., and Todd, S. M., *J. Chem. Soc. A* p. 1152 (1966).
224. Heffernann, M. L., and White, R. F. M., *J. Chem. Soc.* p. 1382 (1961).
225. Herring, D. L., and Douglas, C. M., *Inorg. Chem.* **3**, 428 (1964).
226. Herring, D. L., and Douglas, C. M., *Inorg. Chem.* **4**, 1012 (1965).
227. Hota, N. K., and Harris, R. O., *J. Chem. Soc., Chem. Commun.* p. 407 (1972).
228. Hudson, R. F., "Structure and Mechanism in Organo-phosphorus Chemistry," Chapter 3 pp. 46–89. Academic Press, New York, 1965.
229. Janik, B., Ryeszutko, W., and Pel' czar, T., *Zh. Abal. Khim.* **22**, 1103 (1967).
230. Janik, B., and Zeszutko, W., *Zh. Obshch. Khim.* **43**, 274 (1973); *Chem. Abstr.* **79**, 13044 (1973).
231. Jenkins, R. W., and Lanoux, S., *J. Inorg. Nucl. Chem.* **32**, 2429 (1970).
232. Jenkins, R. W., and Lanoux, S., *J. Inorg. Nucl. Chem.* **32**, 2453 (1970).
233. John, K., Moeller, T., and Audrieth, L. F., *J. Am. Chem. Soc.* **82**, 5616 (1960).
234. John, K., and Moeller, T., *J. Inorg. Nucl. Chem.* **22**, 199 (1961).
235. John, K., Moeller, T., and Audrieth, L. F., *J. Am. Chem. Soc.* **83**, 2608 (1961).
236. John, K., and Moeller, T., *Inorg. Synth.* **7**, 76 (1963).
237. Kajiwara, M., Makihara, M., and Saito, H., *J. Inorg. Nucl. Chem.* **37**, 2562 (1975).
238. Kaplansky, M., and Whitehead, M. A., *Can. J. Chem.* **45**, 1669 (1967).
239. Keat, R., and Shaw, R. A., *Chem. Ind. (London)* p. 1232 (1964).
240. Keat, R., and Shaw, R. A., *J. Chem. Soc.* p. 2215 (1965).
241. Keat, R., Shaw, R. A., and Stratton, C., *J. Chem. Soc.* p. 2223 (1965).
242. Keat, R., and Shaw, R. A., *J. Chem. Soc.* p. 4067 (1965).
243. Keat, R., Ray, S. K., and Shaw, R. A., *J. Chem. Soc.* p. 7193 (1965).
244. Keat, R., and Shaw, R. A., *J. Chem. Soc A* p. 908 (1966).
245. Keat, R., Miller, M. C., and Shaw, R. A., *J. Chem. Soc. A* p. 1404 (1967).
246. Keat, R., and Shaw, R. A., *J. Chem. Soc. A* p. 703 (1968).
247. Keat, R., and Shaw, R. A., *Angew. Chem., Int. Ed. Engl.* **7**, 212 (1968).
248. Keat, R., Porte, A. L., Tong, D. A., and Shaw, R. A., *J. Chem. Soc., Dalton Trans.* p. 1648 (1972).
249. Keat, R., and Shaw, R. A., *in* "Organic Phosphorus Compounds" (G. M. Kosolapoff and L. Maier, eds.), Vol. 6, pp. 833–940. Wiley (Interscience), New York, 1973.
250. Keat, R., Krishnamurthy, S. S., Sau, A. C., Shaw, R. A., Rao, M. N. S., Vasudeva Murthy, A. R., and Woods, M., *Z. Naturforsch. B* **29**, 701 (1974).
251. Keat, R., *Spec. Period. Rep.: Organophosphorus Chem.* **6**, 182 (1975).
252. Keat, R., Shaw, R. A., and Woods, M., *J. Chem. Soc., Dalton Trans.* p. 1582 (1976).
253. Keat, R., and Shaw, R. A., Unpublished results quoted in Keat *et al.* (*249*).
254. Kireev, V. V., Kolesnikov, G. S., and Raigorodskii, I. M., *Russ. Chem. Rev.* (Engl. Transl.) **38**, 667 (1969).
255. Klosowski, J., and Steger, E., *Spectrochim. Acta, Sect. A* **28**, 2189 (1972).
255a. Klosowski, J., and Steger, E., *Spectrochim. Acta, Sect. A* **30**, 1889 (1974).
256. Kobayashi, E., *Nippon Kagaku Zasshi* **87**, 135 (1966); *Chem. Abstr.* **65**, 11744 (1966).
257. Kobayashi, E., *Kogyo Kagaku Zasshi* **69**, 618 (1966); *Chem. Abstr.* **69**, 108185 (1968).
258. Kobayashi, E., *Nippon Kagaku Kaishi* p. 1437 (1973); *Chem. Abstr.* **79**, 100026 (1973).
259. Kobayashi, Y., Chasin, L. A., and Clapp, L. B., *Inorg. Chem.* **2**, 212 (1963).
260. Kokalis, S. G., John, K., Moeller, T., and Audrieth, L. F., *J. Inorg. Nucl. Chem.* **19**, 191 (1961).
261. Kokoreva, I. V., Syrkin, Ya. K., Kropcheva, A. A., Kashnikova, N. M., and Mukhina, L., *Dokl. Akad. Nauk. SSSR* **166**, 155 (1966); *Chem. Abstr.* **64**, 12522 (1966).

262. Kolich, C. H., *Diss. Abstr. B* **31**, 5862 (1971).

263. Koopman, H., Spruit, F. J., Van Deursen, F., and Bakker, J., *Rec. Trav. Chim.* **84**, 341 (1965).

263a. Krause, H.-J., *Z. Elektrochem.* **59**, 1004 (1955).

264. Krishnamurthy, S. S., Shaw, R. A., and Woods M., *Curr. Sci.* **45**, 433 (1976).

265. Krishnamurthy, S. S., Sau, A. C., Vasudeva Murthy, A. R., Keat, R., Shaw, R. A., and Woods, M., *J. Chem. Soc., Dalton Trans.* p. 1405 (1976).

266. Krishnamurthy, S. S., Rao, M. N. S., Vasudeva Murthy, A. R., Shaw, R. A., and Woods, M., *Indian J. Chem., Sect. A* **14**, 823 (1976).

267. Kropacheva, A. A., Mukhina, L. E., Kashnikova, N. M., and Parshina, V. A., *J. Gen. Chem. USSR.* **31**, 957 (1961).

268. Kropacheva, A. A., and Mukhina, L. E., *J. Gen. Chem. USSR* **31**, 2274 (1961).

269. Kropacheva, A. A., and Kashnikova, N. M., *J. Gen. Chem. USSR* **32**, 512 (1962).

270. Kropacheva, A. A., and Kashnikova, N. M., *J. Gen. Chem. USSR* **32**, 645 (1962).

271. Kropacheva, A. A., and Kashnikova, N. M., *J. Gen. Chem. USSR* **35**, 1978 (1965).

272. Kropacheva, A. A., and Mukhina, L. E., *Khim. Geterotsikl. Soed.* p. 162 (1969); *Chem. Abstr.* **70**, 114966 (1969).

273. Lanoux, S., *J. Inorg. Nucl. Chem.* **33**, 279 (1971).

274. Lappert, M. F., and Srivastava, G., *J. Chem. Soc. A* p. 210 (1966).

275. Lawson, D. F., *J. Org. Chem.* **39**, 3357 (1974).

276. Lawson, D. F., Willis, J. M., and Kyker, G. S., U.S. Patent, 3,804,927 (1974); *Chem. Abstr.* **81**, 13107 (1974).

277. Lederle, H., Ottmann, G., and Kober, E., *Inorg. Chem.* **5**, 1818 (1966).

278. Lehr, W., *Z. Anorg. Allg. Chem.* **350**, 18 (1967).

279. Lehr, W., *Z. Anorg. Allg. Chem.* **352**, 27 (1967).

280. Lehr, W., *Z. Anorg. Allg. Chem.* **371**, 225 (1969).

281. Lehr, W., *Naturwissenschaften* **56**, 214 (1969).

282. Lehr, W., and Pietschmann, J., *Chem.-Ztg.* **94**, 362 (1970).

283. Lehr, W., and Rosswag, N., *Z. Anorg. Allg. Chem.* **406**, 221 (1974).

284. Letcher, J. H., and van Wazer, J. R., *Top. Phosphorus Chem.* **5**, 75–167 (1967).

285. Liebig, J., *Annalen* **11**, 139 (1834).

285a. Lingley, D. J., Shaw, R. A., Woods, M., and Krishnamurthy, S. S., *Phosphorus and Sulfur*, (In press).

286. Lund, L. G., Paddock, N. L., Proctor, J. E., and Searle, H. T., *J. Chem. Soc.* p. 2542 (1960).

287. Macdonald, A. L., and Trotter, J., *Can. J. Chem.* **52**, 734 (1974).

288. Maki, H., and Haitaka, T., Japanese Patent 7,543,097 (1975); *Chem. Abstr.* **83**, 100256 (1975).

289. Manhas, B. S., Chu, S.-K., and Moeller, T., *J. Inorg. Nucl. Chem.* **30**, 322 (1968).

290. Mani, N. V., Ahmed, F. R., and Barnes, W. H., *Acta Crystallogr.* **19**, 693 (1965).

291. Mani, N. V., Ahmed, F. R., and Barnes, W. H., *Acta Crystallogr.* **21**, 375 (1966).

292. Mani, N. V., and Wagner, A. J., *Acta Crystallogr., Sect. B* **27**, 51 (1971).

293. Mao, T. J., Dresdner, R. D., and Young, J. A., *J. Am. Chem. Soc.* **81**, 1020 (1959).

294. Mark, V., Dungan, C. H., Crutchfield, M. M., and van Wazer, J. R. *Top. Phosphorus Chem.* **5**, 227–457 (1967).

295. Markila, P. L., and Trotter, J., *Can. J. Chem.* **52**, 2197 (1974).

296. Marsh, W. C., Paddock, N. L., Stewart, C. J., and Trotter, J., *J. Chem. Soc., Chem. Commun.* p. 1190 (1970).

297. Marsh, W. C., and Trotter, J., *J. Chem. Soc. A* p. 169 (1971).

298. Marsh, W. C., and Trotter, J., *J. Chem. Soc. A* p. 573 (1971).

299. Marsh, W. C., and Trotter, J., *J. Chem. Soc. A* p. 569 (1971).

300. Marsh, W. C., and Trotter, J., *J. Chem. Soc. A* p. 1482 (1971).
301. McBee, E. T., Okuhara, K., and Morton, C. J., *Inorg. Chem.* 4, 1672 (1965).
302. McGeachin, H. M., and Tromans, F. R., *J. Chem. Soc.* p. 4777 (1961).
303. Miller, M. C., and Shaw, R. A., *J. Chem. Soc.* p. 3233 (1963).
304. Millington, D., and Sowerby, D. B., *J. Chem. Soc., Dalton Trans.* p. 2035 (1972).
305. Millington, D., and Sowerby, D. B., *Spectrochim. Acta, Sect. A* 29, 765 (1973).
306. Millington, D., King, T. J., and Sowerby, D. B., *J. Chem. Soc., Dalton Trans.* p. 396 (1973).
307. Millington, D., and Sowerby, D. B., *J. Chem. Soc., Dalton Trans.* p. 2649 (1973).
308. Millington, D., and Sowerby, D. B., *J. Chem. Soc., Dalton Trans.* p. 1070 (1974).
309. Millington, D., and Sowerby, D. B., *Phosphorus* 5, 51 (1974).
310. Mishra, S. P., and Symons, M. C. R., *J. Chem. Soc., Chem. Commun.* p. 313 (1973).
311. Moeller, T., and Tsang, F., *Chem. Ind. (London)* p. 361 (1962).
312. Moeller, T., and Lanoux, S., *J. Inorg. Nucl. Chem.* 25, 229 (1963).
313. Moeller, T., and Kokalis, S. G., *J. Inorg. Nucl. Chem.* 25, 875 (1963).
314. Moeller, T., and Kokalis, S. G., *J. Inorg. Nucl. Chem.* 25, 1397 (1963).
315. Moeller, T., and Nannelli, P., *Inorg. Chem.* 2, 659 (1963).
316. Moeller, T., and Nannelli, P., *Inorg. Chem.* 2, 896 (1963).
317. Moeller, T., and Lanoux, S., *Inorg. Chem.* 2, 1061 (1963).
318. Mracec, M., and Simon, Z., *Rev. Roum. Chim.* 16, 449 (1971).
319. Mukhina, L. E., and Kropacheva, A. A., *Zh. Obshch. Khim.* 38, 313 (1968).
320. Nabi, S. N., Shaw, R. A., and Stratton, C., *Chem. Ind. (London)* p. 166 (1969).
321. Nabi, S. N., and Shaw, R. A., *J. Chem. Soc., Dalton Trans.* p. 1618 (1974).
322. Nabi, S. N., Shaw, R. A., and Stratton, C., *J. Chem. Soc., Dalton Trans.* p. 588 (1975).
323. Nabi, S. N., Biddlestone, M., and Shaw, R. A., *J. Chem. Soc., Dalton Trans.* p. 2634 (1975).
324. Niecke, E., Glemser, O., and Roesky, H. W., *Z. Naturforsch. B* 24, 1187 (1969).
325. Niecke, E., Glemser, O., and Thamm, H., *Chem. Ber.* 103, 2864 (1970).
326. Niecke, E., Thamm, H., and Glemser, O., *Z. Naturforsch. B* 26, 366 (1971).
327. Niecke, E., Thamm, H., and Flaskerud, G., *Chem. Ber.* 104, 3729 (1971).
328. Niecke, E., Thamm, H., and Böhler, D., *Inorg. Nucl. Chem. Lett.* 8, 261 (1972).
329. Niecke, E., and Flick, W., *Angew. Chem., Int. Ed. Engl.* 13, 134 (1974).
330. Oakley, R. T., and Paddock, N. L., *Can. J. Chem.* 51, 520 (1973).
331. Oakley, R. T., and Paddock, N. L., *Can. J. Chem.* 53, 3038 (1975).
332. Okuhara, K., *Nagoya Kogyo Gijutsu Shikensho Hokuko* 16, 280 (1967).
333. Olthof, R., *Acta Crystallogr., Sect. B* 25, 2040 (1969).
334. Ottmann, G., Agahigian, H., Hooks, H., Vickers, G. D., Kober, E., and Rätz, R., *Inorg. Chem.* 3, 753 (1964).
335. Otto, R. J. A., and Audrieth, L. F., *J. Am. Chem. Soc.* 80, 3575 (1958).
336. Paddock, N. L., *Quart. Rev. Chem. Soc.* 18, 168 (1964).
337. Paddock, N. L., and Searle, H. T., this series 1, 347 (1959).
338. Paddock, N. L., Trotter, J., and Whitlow, S. H., *J. Chem. Soc. A* p. 2227 (1968).
339. Paddock, N. L., and Searle, H. T., U.S. Patent 3,407,047 (1969); *Chem. Abstr.* 70, 39397 (1969).
340. Paddock, N. L., and Searle, H. T., U.S. Patent 3,462,247 (1969); *Chem. Abstr.* 71, 82031 (1969).
341. Paddock, N. L., Ranganathan, T. N., and Todd, S. M., *Can. J. Chem.* 49, 164 (1971).
342. Paddock, N. L., Ranganathan, T. N., and Wingfield, J. N., *J. Chem. Soc., Dalton Trans.* p. 1578 (1972).
343. Paddock, N. L., and Serreqi, J., *Can. J. Chem.* 52, 2546 (1974).

344. Paddock, N. L., and Patmore, D. J., *J. Chem. Soc., Dalton Trans.* p. 1029 (1976).
345. Pashina, Yu. N., and Stepanov, B. I., *Zh. Obshch. Khim.* 44, 455 (1974); *Chem. Abstr.* 80, 115581 (1974).
346. Pashina, Yu. N., and Stepanov, B. I., *J. Gen. Chem. USSR* 44, 440 (1974).
347. Pohl, S., and Krebs, B., *Chem. Ber.* 108, 2934 (1975).
348. Pohl, S., Niecke, E., and Krebs, B., *Angew. Chem., Int. Ed. Engl.* 14, 261 (1975).
349. Prakash, H., and Sisler, H. H., *Inorg. Chem.* 11, 2258 (1972).
350. Prons, V. N., Grinblat, M. P., and Klebanskii, A. L., *Zh. Obshch. Khim.* 41, 482 (1971); *Chem. Abstr.* 75, 22102 (1971).
351. Radosavljevic, S. D., and Sasic, J. S., *Glas. Hem. Drus., Beograd* 36, 179 (1971); *Chem. Abstr.* 77, 159647 (1972).
352. Radosavljevic, S. D., and Sasic, J. S., *Glas. Hem. Drus., Beograd* 36, 189 (1971); *Chem. Abstr.* 77, 159648 (1972).
353. Ramachandran, K., Krishnamurthy, S. S., Vasudeva Murthy, A. R., Shaw, R. A., and Woods, M., unpublished results.
354. Ranganathan, T. N., Todd, S. M., and Paddock, N. L., *Inorg. Chem.* 12, 316 (1973).
355. Rao, M. N. S., Ph.D. Thesis, Indian Institute of Science, Bangalore, 1976.
356. Rapko, J. N., and Feistel, G., *Inorg. Chem.* 9, 1401 (1970).
357. Rätz, R., Kober, E., Grundmann, C., and Ottmann, G., *Inorg. Chem.* 3, 757 (1964).
358. Ray, S. K., and Shaw, R. A., *J. Chem. Soc.* p. 872 (1961).
359. Ray, S. K., Shaw, R. A., and Smith, B. C., *J. Chem. Soc.* p. 3236 (1963).
360. Rettig, S. J., and Trotter, J., *Can. J. Chem.* 51, 1295 (1973).
361. Rice, R. G., Daasch, L. W., Holden, J. R., and Kohn, E. J., *J. Inorg. Nucl. Chem.* 5, 190 (1958).
362. Roesky, H. W., and Niecke, E., *Inorg. Nucl. Chem. Lett.* 4, 463 (1968).
363. Roesky, H. W., *Z. Naturforsch. B* 25, 777 (1970).
364. Roesky, H. W., Böwing, W. G., and Niecke, E., *Chem. Ber.* 104, 653 (1971).
365. Roesky, H. W., and Wiezer, H., *Chem. Ber.* 107, 1153 (1974).
366. Roesky, H. W., and Janssen, E., *Z. Naturforsch. B* 29, 174 (1974).
367. Roesky, H. W., and Janssen, E., *Chem.-Ztg.* 98, 260 (1974).
368. Roesky, H. W., and Janssen, E., *Angew. Chem., Int. Ed. Engl.* 15, 39 (1976).
369. Rosini, G., Baccolini, G., and Cacchi, S., *J. Org. Chem.* 38, 1060 (1973).
370. Rotzsche, H., Stahlberg, R., and Steger, E., *J. Inorg. Nucl. Chem.* 28, 687 (1966).
371. Sau, A. C., Ph.D. Thesis, Indian Institute of Science, Bangalore, 1976.
372. Sau, A. C., Krishnamurthy, S. S., Vasudeva Murthy, A. R., Keat, R., Shaw, R. A., and Woods, M., *J. Chem. Res. (S)* p. 70; *(M)* p. 0860 (1977).
373. Sau, A. C., Krishnamurthy, S. S., Vasudeva Murthy, A. R., Keat, R., Shaw, R. A., and Woods, M., *J. Chem. Soc. Dalton Trans.* p. 1980 (1977).
374. Schenck, R., and Römer, G., *Chem. Ber. B* 57, 1343 (1924).
375. Schenck, R., *Chem. Ber. B* 60, 160 (1927).
376. Scherer, O. J., and Kuhn, N., *Chem. Ber.* 107, 2123 (1974).
377. Schlueter, A. W., and Jacobson, R. A., *J. Chem. Soc. A* p. 2317 (1968).
378. Schmidpeter, A., and Ebeling, J., *Angew. Chem., Int. Ed. Engl.* 7, 209 (1968).
379. Schmidpeter, A., and Weingand, C., *Z. Naturforsch. B* 24, 177 (1969).
380. Schmidpeter, A., and Weingand, C., *Angew. Chem., Int. Ed. Engl.* 8, 615 (1969).
381. Schmidpeter, A., and Schumann, K., *Z. Naturforsch. B* 25, 1364 (1970).
382. Schmidpeter, A., Ebeling, J., Stray, H., and Weingand, C., *Z. Anorg. Allgem. Chem.* 394, 171 (1972).
383. Schmidpeter, A., and Rossknecht, H., *Chem. Ber.* 107, 3146 (1974).
384. Schmidpeter, A., and Eiletz, H., *Chem. Ber.* 108, 1454 (1975).
385. Schmidpeter, A., Högel, J., and Ahmed, F. R., *Chem. Ber.* 109, 1911 (1976).

386. Schmidt, G., and Redis, V., *Stud. Univ. Babes-Bolyai, Ser. Khim.* **19**, 32 (1974); *Chem. Abstr.* **82**, 50801 (1975).
387. Schmitz-DuMont, O., and Kulkenes, H., *Z. Anorg. Allg. Chem.* **238**, 189 (1938).
388. Schmulbach, C. D., *Progr. Inorg. Chem.* **4**, 275–379 (1962).
389. Schmulbach, C. D., and Miller, V. R., *Inorg. Chem.* **5**, 1621 (1966).
390. Schmulbach, C. D., Cook, A. G., and Miller, V. R., *Inorg. Chem.* **7**, 2463 (1968).
391. Schmulbach, C. D., Derderian, C., Zeck, C., and Sahuri, S., *Inorg. Chem.* **10**, 195 (1971).
392. Schmutz, J. L., and Allcock, H. R., *Inorg. Chem.* **14**, 2433 (1975).
393. Schmutzler, R., *Z. Naturforsch. B* **19**, 1101 (1964).
394. Schmutzler, R., Moeller, T., and Tsang, F., *Inorg. Synth.* **9**, 75 (1967).
395. Schumann, K., and Schmidpeter, A., *Phosphorus* **3**, 51 (1973).
396. Searle, H. T., *Proc. Chem. Soc. (London)* p. 7 (1959).
397. Searle, H. T., Dyson, J., Ranganathan, T. N., and Paddock, N. L., *J. Chem. Soc., Dalton Trans.* p. 203 (1975).
398. Seel, F., and Langer, J., *Z. Anorg. Allg. Chem.* **295**, 316 (1958).
399. Sharov, V. N., Prons, V. N., Korolko, V. V., Klebanskii, A. L., and Kondratenkov, G. P., *Zh. Obshch. Khim.* **45**, 1953 (1975).
400. Sharts, C. M., Bilbo, A. J., and Gentry, D. R., *Inorg. Chem.* **5**, 2140 (1966).
401. Shaw, R. A., Fitzsimmons, B. W., and Smith, B. C., *Chem. Rev.* **62**, 247 (1962).
402. Shaw, R. A., Keat, R., and Hewlett, C., in "Preparative Inorganic Reactions" (W. L. Jolly, ed.), Vol. 2, pp. 1–91. Interscience, New York, 1965.
403. Shaw, R. A., *Record Chem. Progr.* **28**, 243 (1967).
404. Shaw, R. A., *Endeavour* **27**, 74 (1968).
405. Shaw, R. A., and Mukmenev, E. T., *Dokl. Akad. Nauk. SSSR* **208**, 379 (1973).
406. Shaw, R. A., and Mukmenev, E. T., *Khim. Geterotsikl. Soed.* p. 945 (1973).
407. Shaw, R. A., *Pure Appl. Chem.* **44**, 317 (1975).
408. Shaw, R. A., *Z. Naturforsch. B* **31**, 641 (1976).
409. Shaw, R. A., Woods, M., and Krishnamurthy, S. S., unpublished results.
410. Shvetsov-Shilovskii, N. I., and Pitiana, M. R., *J. Gen. Chem. USSR* **41**, 1028 (1971).
411. Siegel, L. A., and van den Hende, J. H., *J. Chem. Soc. A* p. 817 (1967).
412. Singler, R. E., Schneider, N. S., and Hagnauer, G. L., *Polymer Eng. Sci.* **15**, 321 (1975).
413. Sisler, H. H., Ahuja, H. S., and Smith, N. L., *Inorg. Chem.* **1**, 84 (1962).
414. Slawisch, A., and Pietschmann, J., *Z. Naturforsch. B* **25**, 321 (1970).
415. Sokolav, E. I., Sharov, V. N., Klebanskii, A. L., Korol'ko, V. V., and Prons, V. N., *Zh. Obshch. Khim.* **45**, 2346 (1975); *Chem. Abstr.* **84**, 59664 (1976).
416. Sorokin, M. F., and Latov, V. K., *J. Gen. Chem. USSR* **35**, 1472 (1965).
417. Sowerby, D. B., *J. Chem. Soc. (London)* p. 1396 (1965).
418. Sowerby, D. B., *MTP Int. Rev. Sci., Inorg. Chem. Ser.* **2**, 117–170 (1972).
419. Stahlberg, R., and Steger, E., *J. Inorg. Nucl. Chem.* **28**, 684 (1966).
420. Stahlberg, R., and Steger, E., *Spectrochim. Acta, Sect. A* **23**, 2005 (1967).
421. Stahlberg, R., and Steger, E., *J. Inorg. Nucl. Chem.* **29**, 961 (1967).
422. Steger, E., and Stahlberg, R., *Z. Anorg. Allg. Chem.* **326**, 243 (1964).
423. Steinman, R., Shirmer, F. B., Jr., and Audrieth, L. F., *J. Am. Chem. Soc.* **64**, 2377 (1942).
424. Stepanov, B. I., and Migachev, G. I., *Zavodsk. Lab.* **32**, 416 (1966); *Chem. Abstr.* **66**, 25854 (1967).
425. Stepanov, B. I., and Migachev, G. I., *J. Gen. Chem. USSR* **38**, 195 (1968).
426. Stokes, H. N., *Am. Chem. J.* **17**, 275 (1895).
427. Stokes, H. N., *Am. Chem. J.* **18**, 629 (1896).

428. Stokes, H. N., *Am. Chem. J.* 18, 780 (1896).
429. Stokes, H. N., *Am. Chem. J.* 19, 782 (1897).
430. Tasaka, A., and Glemser, O., *Z. Anorg. Allg. Chem.* 409, 163 (1974).
431. Tesi, G., Haber, C. P., and Douglas, C. M., *Proc. Chem. Soc., London* p. 219 (1960).
432. Tesi, G., and Slota, P. J., *Proc. Chem. Soc., London* p. 404 (1960).
433. Tesi, G., Matuszko, A. J., Zimmer-Galler, R., and Chang, M. S., *Chem. Ind.* (London) p. 623 (1964).
434. Tesi, G., and Zimmer-Galler, R., *Chem. Ind.* (London) p. 1916 (1964).
435. Tesi, G., and Douglas, C. M., *J. Am. Chem. Soc.* 84, 549 (1962).
436. Thamm, H., Lin, T. P., Glemser, O., and Niecke, E., *Z. Naturforsch. B* 27, 1431 (1972).
437. Thomas, B., Gehlert, P., Schadow, H., and Scheler, H., *Z. Anorg. Allg. Chem.* 418, 171 (1975).
438. Thomas, B., Schadow, H., and Scheler, H., *Z. Chem.* 15, 26 (1975).
439. Trotter, J., and Whitlow, S. H., *J. Chem. Soc. A* p. 455 (1970).
440. Trotter, J., and Whitlow, S. H., *J. Chem. Soc. A* p. 460 (1970).
441. Volodin, A. A., Kireev, V. V., Fomin, A. A., Edelev, M. G., and Korshak, V. V., *Dokl. Akad. Nauk SSSR* 209, 168 (1973); *Chem. Abstr.* 79, 31618 (1973).
442. Volodin, A. A., Kireev, V. V., Korshak, V. V., and Fomin, A. A., *Zh. Obshch. Khim.* 43, 2206 (1973).
443. Wagner, A. J., and Vos, A., *Acta Crystallogr., Sect. B* 24, 707 (1968).
444. Wagner, A. J., and Vos, A., *Acta Crystallogr., Sect. B* 24, 1423 (1968).
445. Wagner, A. J., and Moeller, T., *J. Chem. Soc. A* p. 596 (1971).
446. Wakefield, Z. T., Allen, S. E., McCullough, J. F., Sheridan, R. C., and Kohler, J. J., *J. Agr. Food Chem.* 19, 99 (1971).
447. Walsh, E. J., Kaluzene, S., and Jubach, T., *J. Inorg. Nucl. Chem.* 38, 397 (1976).
448. Walsh, E. J., and Derby, E., *Inorg. Chim. Acta* 14, L40 (1975).
449. Walsh, E. J., Derby, E., and Smegal, J., *Inorg. Chim. Acta* 16, L9 (1976).
450. Wende, A., and Joel, D., *Z. Chem.* 3, 467 (1963).
451. Wunsch, G., and Kiener, V., German Patent 2,302,512 (1974); *Chem. Abstr.* 81, 128758 (1974).
452. Yokohama, M., Cho, H., and Sakuma, M., *Kogyo Kagaku Zasshi* 66, 422 (1963).
453. Yu, Hon-Sum, Ph.D. Thesis, London University, 1975.
454. Zeleneva, T. P., Antonov, I. V., and Stepanov, B. I., *Zh. Obshch. Khim.* 43, 1007 (1973); *Chem. Abstr.* 79, 91164 (1973).
455. Zeleneva, T. P., and Stepanov, B. I., *Tr. Mosk. Khim.-Tekhnol. Inst. No. 70,* 133 (1972); *Chem. Abstr.* 79, 4902 (1973).
456. Zhivukhin, S. M., Tolstoguzov, V. B., Kireev, V. V., and Kuznetsova, K. G., *Russ. J. Inorg. Chem.* 10, 178 (1965).
457. Zhivukhin, S. M., Kireev, V. V., Kolesnikov, G. S., Popilin, V. P., and Fillippov, E. A., *Russ. J. Inorg. Chem.* 14, 548 (1969).
458. Zhivukhin, S. M., Kireev, V. V., Popilin, V. P., and Kolesnikov, G. S., *Russ. J. Inorg. Chem.* 15, 630 (1970).
459. Zoer, H., and Wagner, A. J., *Acta Crystallogr., Sect. B* 28, 252 (1972).
460. Zoer, H., and Wagner, A. J., *Cryst. Struct. Commun.* 1, 17 (1972).

APPENDIX

The following represents a brief account of recent papers (June 1976–July 1977) which are relevant to the theme of this Review:

A novel synthesis of phospham from red phosphorus and ammonia has been discovered (37):

$$4n\text{NH}_3 + 2n\text{P} \rightleftharpoons 2[\text{PN}_2\text{H}]_n + 5n\text{H}_2 \tag{1}$$

Preliminary experiments confirm the prediction that ammono-cyclophosphazenes may be prepared at high pressures:

$$n\text{NH}_3 + [\text{PN}_2\text{H}]_n \rightarrow [\text{NP(NH}_2)_2]_n \tag{2}$$

The reaction of $\text{N}_3\text{P}_3(\text{NH}_2)_6$ with formaldehyde (27) and the hydrolysis of $\text{N}_4\text{P}_4(\text{NH}_2)_8$ (23) have been studied. Wanek has reviewed the use of ammonophosphazenes as fertilizers (39).

The alkaline hydrolysis of $[\text{NP(OMe)}_2]_{3,4}$ gives the monohydroxy derivatives, $\text{N}_3\text{P}_3(\text{OMe})_5(\text{OH})$ and $\text{N}_4\text{P}_4(\text{OMe})_7\text{OH}$ which exist in the tautomeric oxo-form (38). The rearrangement of octamethoxycyclotetraphosphazene, $\text{N}_4\text{P}_4(\text{OMe})_8$, in the presence of methyl iodide yields two isomeric phosphazanes; the more abundant isomer has the 2-trans-4-cis-6-trans-8 structure (13).

The reactions of $\text{N}_3\text{P}_3\text{Cl}_6$ with ethylenediamine and ethanolamine give spirocyclic derivatives (25) and not ansa-compounds. Aliphatic and alicyclic diols also yield spirocyclic derivatives (26, 31). Inclusion clathrates formed by spirocyclic phosphazenes with aryldioxy substituents (2) have been studied by mass spectrometry, broadline ^{1}H NMR, and X-ray techniques (4). Some conclusions on molecular motion in these compounds have been deduced.

The replacement of chlorine atoms of $\text{N}_3\text{P}_3\text{Cl}_6$ by fluoroalkoxy groups proceeds nongeminally (21). Phosphorus–oxygen and carbon–oxygen bond cleavage have been observed in the reactions of $\text{N}_3\text{P}_3(\text{OAr})_6$ with several nucleophiles (3).

Methylcyclophosphazenes, $(\text{NPMe}_2)_n$, $n = 6\text{–}10$, are conveniently prepared by the reaction of MeMgBr with the fluorides, $(\text{NPF}_2)_n$ [18]. The crystal structures of $\text{N}_5\text{P}_5\text{Me}_{10}$ (16), $\text{N}_6\text{P}_6\text{Me}_{12}$ (29a), $\text{N}_7\text{P}_7\text{Me}_{14}$ (18) and $\text{N}_8\text{P}_8\text{Me}_{16}$ (29) have been determined. Modified syntheses of $\text{N}_3\text{P}_3\text{Me}_6$ and $\text{N}_4\text{P}_4\text{Me}_8$ have been reported. When these compounds are heated at 200–350°C, an equilibrium is established but a high polymer is not formed (5):

$$3[\text{NP(Me)}_2]_4 \rightleftharpoons 4[\text{NP(Me)}_2]_3 \tag{3}$$

Phosphazenyl cyclophosphazenes have been synthesized by a silylation method (15, 17).

Cyclophosphazene derivatives, $(\text{NPR}_2)_4$ (R = Me, NHMe) and the polymeric phosphazene, $[\text{NP(NHMe)}_2]_n$ react with K_2PtCl_4 in organic media to yield square-planar complexes, $(\text{NPR}_2)_4 \cdot \text{PtCl}_2$ and a product of composition $[NP(NHMe)_2]_n \cdot (PtCl_2)_x$ ($x:n = 1:17$), respectively (6). The salts, $[H_2N_4P_4R_8]^{2+}$ $[PtCl_4]^{2-}$ are formed in aqueous hydrochloric acid. The crystal structure of $[NP(NHMe)_2]_4 \cdot PtCl_2$ shows that the platinum atom is bonded to two antipodal ring nitrogen atoms (10). These water-soluble complexes show significant antitumor activity (6). The crystal structure of the nickel complex, $[N_3P_3Ph_4(Me)S]_2$ Ni, has been determined; the nickel atom is bonded to sulfur and two ring nitrogen atoms in a planar coordination (1). The platinum(II) and palladium(II) analogs have also been prepared (34). Infrared spectra and ^{35}Cl and ^{121,123}Sb NQR suggest that the adducts, $SbCl_5 \cdot (NPCl_2)_n$, $n = 3,4$ and $TaCl_5 \cdot N_3P_3Cl_6$, have the ionic formulas, $[N_nP_nCl_{2n-1}]$ $MCl_6]$ (24).

A more general route to spirobi(cyclotriphosphaza)pentaene derivatives [e.g., compound (4) in Fig. 1] has now been developed. The octaphenyl derivative (4) can be deprotonated to give the hitherto unknown spirobi(cyclotriphosphazene) anion, $(P_5N_6Ph_8)^-$ (32). Methylation of these anions (32) and also of the cyclotriphosphazene derivatives, $N_3P_3Ph_4X_2$($X = Me, NH_2$) (33) has been investigated.

Further reactions of P-hydridocyclotriphosphazenes (See Section IV,E) have been described. They undergo insertion reactions with aldehydes, ketones, isothiocyanates and electrophilic olefins. Addition of sulfur or oxidation with $KMnO_4$ gives thioxo-or oxo-cyclotriphosphazene derivatives which are methylated at the chalcogen to afford methylthio- and methoxy- cyclotriphosphazenes (36). Hydridocyclophosphazenes can also be oxidized to give symmetric and unsymmetric bis(cyclotriphosphazenyl) oxides (35).

^{1}H[8] and ^{13}C[9]NMR data for phenyl-substituted fluorocyclotriphosphazenes indicate that there is a strong conjugative electron release from the phenyl group into the phosphazene ring. Single crystal Raman polarization data are reported for $N_3P_3X_6$($X = Cl, Br$) (22). Conformational analysis of halogenocyclophosphazenes has been carried out (12). Cyclophosphazenes add electrons to form phosphoranyl radical anions. The ESR evidence for these anions is inconsistent with the postulate of extensive π-bonding within the P–N ring. A partial ionic structure with extensive σ-delocalization to achieve charge neutralization is proposed (28).

Details of crystal structures of the two-ring assembly phosphazene (32) (1) and the bicyclic phosphazene (12) (14) have been published. X-Ray crystallography shows that the compound, $N_4P_4F_6$(NSN) has a 2,4-fusion of rings (19) rather than the 2,6-fusion suggested earlier. The crystal structures of $N_3P_3F_4(NH_2)_2$ (30), 2-trans-4-N_4P_4(NMePh)$_2$ Cl_6 (10a) and 2-cis-4-cis-6-trans-8-$N_4P_4Cl_4$(NMe$_2$)$_4$ (11) have also been reported. Glidewell has noted that the P . . . P nonbonded contacts in many cyclophosphazenes remain constant (2.90 Å) and that this may have some bearing on the unusually large ring PNP angles in some of these compounds (20).

Recent developments in phosphazene high polymer chemistry have been summarized (7).

REFERENCES

1. Ahmed, F. R., *Acta Crystallogr., Sect. B* **32**, 3078 (1976).
2. Allcock, H. R., Stein, M. T., and Bissell, E. C., *J. Am. Chem. Soc.* **96**, 4795 (1974).
3. Allcock, H. R., and Smeltz, L. A., *J. Am. Chem. Soc.* **98**, 4143 (1976).
4. Allcock, H. R., Allen, R. W., Bissell, E. C., Smeltz, L. A., and Teeter, M., *J. Am. Chem. Soc.* **98**, 5120 (1976).
5. Allcock, H. R., and Patterson, D. B., *Inorg. Chem.* **16**, 197 (1977).
6. Allcock, H. R., Allen, R. W., and O'Brien, J. P., *J. Am. Chem. Soc.* **99**, 3984 (1977).
7. Allcock, H. R., *Angew. Chem. Inter. Edn. Eng.* **16**, 147 (1977).
8. Allen, C. W., and White, A. J., *Inorg. Chem.* **13**, 1220 (1974).
9. Allen, C. W., *J. Organomet. Chem.* **125**, 215 (1977).
10. Allen, R. W., O'Brien, J. P., and Allcock, H. R., *J. Am. Chem. Soc.* **99**, 3987 (1977).
10a. Babu, Y. S., and Manohar, H., *Cryst. Struct. Commun.* **6**, 803 (1977).
11. Begley, M. J., King, T. J., and Sowerby, D. B., *J. Chem. Soc. Dalton Trans.* P. 149 (1977).
12. Boyd, R. H., and Kesner, L., *J. Am. Chem. Soc.* **99**, 4248 (1977).
13. Bullen, G. J., Paddock, N. L., and Patmore, D. J., *Acta Crystallogr. Sect. B* **33**, 1367 (1977).
14. Cameron, T. S., and Mannan, Kh., *Acta Crystallogr. Sect. B* **33**, 443 (1977).
15. Dahmann, D., Rose, H., and Shaw, R. A., *Z. Naturforsch. B*, **32**, 236 (1977).
16. Dougill, M. W., and Sheldrick, B., *Acta Crystallogr. Sect. B* **33**, 295 (1977).
17. Du Plessis, J. A. K., Rose H., and Shaw, R. A., *Z. Naturforsch. B*, **31**, 997 (1976).
18. Gallicano, K. D., Oakley, R. T., Paddock, N. L., Rettig, S. J., and Trotter, J., *Can. J. Chem.*, **55**, 304 (1977).
19. Giern, A., Dederer, B., Roesky, H. W., and Janssen, E., *Angew. Chem. Inter. Edn. Eng.* **15**, 783 (1976).
20. Glidewell, G., *J. Inorg. Nucl. Chem.* **38**, 669 (1976).
21. Gol'din, G. S., Fedorov, S. G., Zapuskalova, S. F., and Naumov, A. D., *J. Gen. Chem. USSR*, **46**, 685 (1976).
22. Greighton, J. A., and Thomas, K. M., *Spectrochim. Acta Sect. A.* **29**, 1077 (1973).
23. Kobayashi, E., *Bull. Chem. Soc. Japan* **49**, 3524 (1976).
24. Kravchenko, E. A., Levin, B. V., Bananyarly, S. I., and Toktomatov, T. A., *Koord. Khim.* **3**, 374 (1977).
25. Krishnamurthy, S. S., Ramachandran, K., Vasudeva Murthy, A. R., Shaw, R. A., and Woods, M., *Inorg. Nucl. Chem. Letters* **13**, 407 (1977).
26. Manns, H., and Specker, H., *Z. Anorg. Allg. Chem.* **425**, 127 (1976).
27. Meznik, L., Kabela, J., Novosad, J., and Dostal, K., *Z. Chem.* **17**, 72 (1977).
28. Mishra, S. P., and Symons, M. C. R., *J. Chem. Soc. Dalton Trans.* p. 1622 (1976).
29. Oakley, R. T., Paddock, N. L. Rettig, S. J., and Trotter, J., *Can. J. Chem.* **55**, 2530 (1977).
29a. Oakley, R. T., Paddock, N. L., Rettig, S. J., and Trotter, J., *Can. J. Chem.* **55**, 3118 (1977).
30. Pohl. S., and Krebs, B., *Chem. Ber.* **109**, 2622 (1976).

31. Rose, H., and Specker, H., *Z. Anorg. Allg. Chem.* **426**, 275 (1976).

32. Schmidpeter, A., and Eiletz, H., *Phosphorus* **6**, 113 (1976).

33. Schmidpeter, A., and Eiletz, H., *Chem. Ber.* **109**, 2340 (1976).

34. Schmidpeter, A., Blanck, K., and Ahmed, F. R., *Angew. Chem. Inter. Edn. Eng.* **15**, 488 (1976).

35. Schmidpeter, A., Blanck, K., and Högel, J., *Z. Naturforsch, B*, **31**, 1466 (1976).

36. Schmidpeter, A., Blanck, K., Eiletz, H., Smetana, H., and Weingand, C., *Syn. React. Inorg. Metal-Org. Chem.* **7**, 1 (1977).

37. Sullivan, J. M., *Inorg. Chem.* **15**, 1055 (1976).

38. Vilceanu, R., and Schulz, P., *Phosphorus* **6**, 231 (1976).

39. Wanek, W., *Pure and Applied Chem.* **44**, 459 (1975).

ADVANCES IN INORGANIC CHEMISTRY AND RADIOCHEMISTRY, VOL. 21

A NEW LOOK AT STRUCTURE AND BONDING IN TRANSITION METAL COMPLEXES

JEREMY K. BURDETT*

Department of Inorganic Chemistry, The University, Newcastle upon Tyne, England

I. Introduction 113
II. Application of the Angular Overlap Method 114
  A. Octahedral Transition Metal Ions 117
  B. Tetrahedral Coordination 123
  C. Other Geometries 124
  D. Rates of Reaction of Hexaquo Ions 125
  E. Geometries of Transition Metal Complexes 127
  F. Derivation of Walsh Diagrams 134
  G. Structural Consequences in d^8 and d^9 Systems 135
  H. Ligand Site Preferences 138
  I. Intramolecular Photochemical Rearrangements 141
III. Conclusion 143
  References 143

I. Introduction

The construction of models of molecular geometry has been a popular exercise for many years. In the field of main-group stereochemistry the ideas of Sidgwick, Powell, Nyholm, and Gillespie have been successfully fused together into the VSEPR scheme (1–3), where the electron pairs around a central atom adopt the minimum energy consistent with their mutual electrostatic repulsion. The VSEPR rules have captured much of the essence of main-group stereochemistry and are unrivalled in their simplicity. In addition, Walsh diagrams (4) [quantified by the work of Gimarc and others (5–8)] and the pseudo Jahn–Teller formalism introduced by Bartell (9) into the structural field and extended by Pearson (10, 11) have provided a molecular orbital interpretation of the wide range of available structural data.

By way of contrast, fewer gas-phase structures of transition metal complexes are known: a few d^0–d^4 hexafluorides (all octahedral), d^0–d^1 tetrachlorides (tetrahedral), and some 18-electron carbonyl and phosphine structures typified by tetrahedral $Ni(CO)_4$, octahedral

---

* Present address: Chemistry Department, University of Chicago, Chicago, Illinois 60637.

$Cr(CO)_6$, and trigonal bipyramidal $Fe(CO)_5$. Many transition metal systems are polymeric in the solid state yielding little or no data concerning their ideal "gas-phase" structure. Even when monomeric structures are established, we do not generally know how strong the influence is of crystal forces on the structure and geometry. This dearth of structural data has been alleviated considerably in recent years by the synthesis in low-temperature matrices of isolated carbonyl and dinitrogenyl fragments (12), $M(CO)_n$ and $M(N_2)_n$ ($n = 1-6$), with a variety of transition metal atoms M and, hence, a spectrum of d-orbital configurations that have filled some of the gaps in the structural field. Walsh diagrams have only recently been derived for transition metal complexes (13–16) but obviously need to be prolific in order to describe as completely as possible all possible distortion modes of ML_n ($n = 1-6$) systems. A simple model with a minimum of parameters would be a significant advance in the arena of structural transition metal chemistry. In this paper, we describe a molecular orbital parametrization that is able to provide a unified pathway to many structural problems.

II. Application of the Angular Overlap Method

Our simple molecular orbital approach is based on the angular overlap model (17), which has been used mostly in the past in the interpretation of the electronic spectra of transition metal complexes. Basically, it provides the energies of the (mainly) transition metal d orbitals in an ML_n complex of given geometry in terms of just two parameters [like the Δ or Dq of the crystal field theory (CFT)]. Once these energies have been derived then many structural effects may be examined, as we shall see in the following. For example, the weighted sum of these d-orbital energies (weighted by the number of electrons in these orbitals) as a function of the molecular geometry provides the opportunity to explore the configurational potential surface and, hence, find the minimum energy geometry demanded by the particular d-orbital configuration. The weighted sum of these energies relative to those of the free ion (the total d-orbital stabilization energy) leads to a rationalization of the heats of hydration of the transition metal ions. The relative stabilities of *cis*- and *trans*-ML_4L_2' octahedral complexes may be elucidated by considering the stabilization energies of each isomer. The angular overlap model, as we shall use it here, is based on an approximation involving the interaction energy between two orbitals ϕ_i, ϕ_j on different atoms. In our case ϕ_i will represent a d orbital on the transition metal center, and ϕ_j a single-ligand orbital or

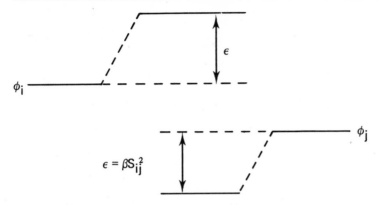

FIG. 1. Energy-level diagram showing the basis of the model. Stabilization of bonding orbital = destabilization of antibonding orbital = βS_{ij}^2.

a symmetry-adapted combination of ligand orbitals transforming under the same irreducible representation as ϕ_i.

The angular overlap approximation sets the stabilization energy of the bonding orbital proportional to the square of the overlap integral between ϕ_i and ϕ_j, $\epsilon = \beta S_{ij}^2$, where β is a constant inversely dependent on the energy separation between ϕ_i and ϕ_j. In our simple model, we shall assume in addition that the destabilization energy of the anti-bonding orbital is equal to this stabilization energy of the bonding orbital[1] (Fig. 1). The tremendous power of the model lies in the fact that the S_{ij} are, in general, dependent on simple geometric expressions as the angular metal–ligand geometry is adjusted while maintaining the same bond length. Figure 2 shows pictorially the simple function describing the overlap integral between a central atom p_z orbital and ligand σ orbital in terms of the polar coordinate θ. S_σ is a constant[2] and equal to the overlap integral between the two orbitals for $\theta = 0$.

[1] This is not strictly true (17e):

$$\Delta\epsilon_{anti} = \frac{H_{LL}^2}{H_{MM} - H_{LL}} S_{ML}^2$$

and

$$\Delta\epsilon_{bond} = \frac{H_{MM}^2}{H_{MM} - H_{LL}} S_{ML}^2$$

i.e., the antibonding orbital receives a larger destabilization than the stabilization received by the bonding orbital. Inclusion of overlap between ϕ_i and ϕ_j will also tend to push both new orbitals to higher energy and increase the disparity. (H_{LL} and H_{MM} are the ionization potentials of ligand and metal orbitals, respectively.)

[2] The notation we have chosen to use here is that of Kettle (17f).

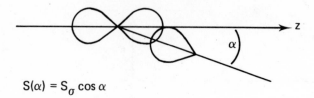

$$S(\alpha) = S_\sigma \cos \alpha$$

FIG. 2. Overlap of a ligand σ orbital with a central atom p orbital as a function of angle.

TABLE I

OVERLAP INTEGRALS OF CENTRAL ATOM
d ORBITALS AS A FUNCTION OF THE
LIGAND POSITION IN POLAR COORDINATES
(θ, ϕ) WITH A LIGAND σ ORBITAL[a,b]

d_{z^2}	$\frac{1}{2}(3\cos^2\theta - 1)S_\sigma$[c]
$d_{x^2-y^2}$	$(\sqrt{3}/2)\cos^2\phi\sin^2\theta\,S_\sigma$
d_{xz}	$(\sqrt{3}/2)\cos\phi\sin^2\theta\,S_\sigma$
d_{yz}	$(\sqrt{3}/2)\sin\phi\sin^2\theta\,S_\sigma$
d_{xy}	$(\sqrt{3}/2)\sin^2\phi\sin^2\theta\,S_\sigma$

[a] For simplicity, π-type overlap functions have been left off this table. Their values are given in Larsen and La Mar (*17g*).
[b] See Fig. 3, regarding labeling of polar coordinates and ligand.
[c] S_σ Is the overlap of a ligand σ orbital with d_{z^2} when the ligand lies along the z axis.

For the two orbitals of Fig. 2, the interaction energy as a function of angle is thus simply $\epsilon = \beta_\sigma S_\sigma{}^2 \cos^2\theta$, here we introduce β_σ as the β parameter describing interactions that are of σ type.[3] For interactions between orbitals of π type, we introduce two similar parameters β_π and S_π. Table I gives the angular dependence of the overlap integrals between the d orbitals and ligand σ-type orbitals; the ligand position is defined by the polar angles θ, ϕ. Also π-type interactions are amenable to similar treatment.[4]

We are thus in a position to be able to write the interaction energy of a pair of orbitals as a simple expression $\epsilon = kf(\theta, \phi)$, where k is a constant, and $f(\theta, \phi)$ a simple geometric function readily obtained from

[3] An interesting forerunner of the angular overlap model is given by McClure (*19*).
[4] The integrals involving s, p, and f orbitals are available in Smith and Clack (*20*).

Table I. This is the basis of our approach and is really quite a simple one to use, although obviously not as simple as the VSEPR rules (*18*) in main-group stereochemistry.

A. Octahedral Transition Metal Ions

The octahedrally coordinated molecule provides a useful introduction to the practicalities of the method. The ligand σ orbitals transform as $a_{1g} + e_g + t_{1u}$, the ligand π orbitals as $t_{1g} + t_{2g} + t_{1u} + t_{2u}$, and the metal d orbitals as e_g and t_{2g}. We are interested, therefore, in the σ interactions (of species e_g) and the π interactions of (species t_{2g}) of the ligand orbitals with the metal.

1. σ-Type Interactions

For one component of the doubly degenerate, e_g ligand combination, we may write

$$\psi_{e_g}(1) = \tfrac{1}{2}(\phi_1 - \phi_2 + \phi_3 - \phi_4) \tag{1}$$

with reference to Fig. 3. The overlap integral functions contained in Table I allow us to calculate the relevant overlap integral between

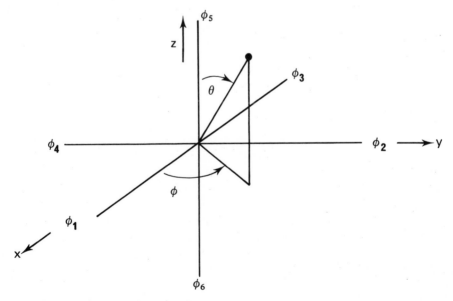

Fig. 3. Polar coordinates and ligand σ-orbital labeling in the octahedron.

$\psi_{e_g}(1)$ and $d_{x^2-y^2}$. In Table I, with reference to Fig. 3, $\theta = 90°$ and $\phi = 0°, 90°, 180°, 270°$ for the overlap integrals involving $\phi_1 \to \phi_4$. Thus we immediately find

$$S_{e_g}(1) = 1/2 \cdot 4 \frac{\sqrt{3}}{2} S_\sigma \tag{2}$$

i.e., $\epsilon(e_g) = 3\beta_\sigma S_\sigma^2$. For the octahedral geometry (and other octahedrally based structures such as the square plane), we will find Table II useful. This gives the overlap integrals of a ligand σ orbital located along x, y, or z axes with $d_{x^2-y^2}$ and d_{z^2}, and is simply a rationalized form of a part of Table I.

TABLE II

OVERLAP INTEGRALS OF A LIGAND σ ORBITAL
WITH d_{z^2} AND $d_{x^2-y^2}$

Overlap integral with	Ligand located along	
	x, y	z
d_{z^2}	$-\frac{1}{2} S_\sigma$	S_σ
$d_{x^2-y^2}$	$\pm(\sqrt{3}/2)S_\sigma$	0

We may similarly construct a ligand e_g σ-orbital combination to overlap with d_{z^2},

$$\psi_{e_g}(2) = \frac{1}{\sqrt{12}}(2\phi_5 + 2\phi_6 - \phi_1 - \phi_2 - \phi_3 - \phi_4) \tag{3}$$

and, by using Table II, we find

$$S_{e_g}(2) = \frac{1}{\sqrt{12}}(2 + 2 + \tfrac{1}{2} + \tfrac{1}{2} + \tfrac{1}{2} + \tfrac{1}{2})S_\sigma = \frac{6}{\sqrt{12}}S_\sigma \tag{4}$$

Hence the stabilization energy is $3\beta_\sigma S_\sigma^2$ as it has to be by symmetry with Eqs. (1) and (2). Figure 4 shows the molecular orbital diagram for the σ framework of the octahedral configuration using this result and our aforementioned assumption concerning stabilization in bonding and antibonding orbitals. Also shown is the well-known crystal field-splitting pattern for this geometry. We see immediately that $\Delta_{oct} = 3\beta_\sigma S_\sigma^2$.

We are now in a position to calculate the total stabilization energy of an ML_6 complex from the free ion plus ligands (21). We shall compare

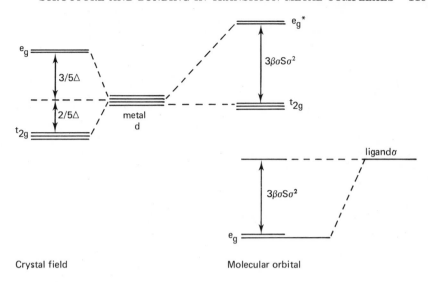

FIG. 4. Comparison of crystal field and molecular orbital methods for the octahedral environment.

the result with the familiar "double-humped" curve (22, 23) associated with ΔH^0_{hyd} for the reaction

$$M^{2+} \text{ (gas)} + 6H_2O \text{ (gas)} \rightarrow M^{+2}(H_2O)_6 \text{ (gas)} \qquad (5)$$

For all d^n complexes the e_g bonding orbitals (mainly ligand located) contain 4 electrons and these contribute (from Fig. 4) a stabilization energy of $4 \times 3\beta_\sigma S_\sigma^2 = 12\beta_\sigma S_\sigma^2$. The first 3 d electrons experience no change in energy on complex formation—they go into σ nonbonding d orbitals and, thus, $\sum(\sigma) = 12\beta_\sigma S_\sigma^2$, where we introduce symbol $\sum(\sigma)$ to describe the total σ-stabilization energy of a given electronic configuration. The fourth electron of the high-spin d^4 complex, however, enters e_g^* and is destabilized by $3\beta_\sigma S_\sigma^2$. The total σ-stabilization energy $\sum(\sigma)$ is thus equal now to only $9\beta_\sigma S_\sigma^2$. We may readily calculate the values of $\sum(\sigma)$ for all the high-spin configurations d^0-d^{10}, and these are plotted in Fig. 5. An equation we shall find useful later on in evaluating these $\sum(\sigma)$ values is

$$\sum(\sigma) = \sum_i h_i \beta_\sigma S_i^2 \qquad (6)$$

where $\beta_\sigma S_i^2$ is the interaction energy of the ith d orbital, and h_i is the number of electron holes in this orbital. Equation (6) (24) arises simply

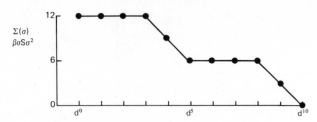

FIG. 5. Plot of d-Orbital σ-stabilization energies $[\sum(\sigma)]$ for high-spin d^n ions.

because when filling d orbitals with electrons we are filling metal–ligand *anti*bonding orbitals, and it is only the *empty* d orbitals that have filled metal–ligand bonding counterparts which contribute to the stabilization energy.

This plot of Fig. 5 is not immediately recognizable as being associated with the familiar double-humped curve describing ΔH^0_{hyd} as a function of the number of d electrons. However, we should recall that in addition to interaction with the metal nd orbitals, the ligands are attached to the transition metal ion by interactions with the $(n + 1)s$ and $(n + 1)p$ orbitals. How else would the $Zn(H_2O)_6^{2+}$ (d^{10}) complex be stabilized with all metal d-ligand antibonding orbitals occupied? Thus, the observed ΔH^0_{hyd} plot will be the sum of these s + p orbital stabilization energies and the d-orbital contribution[5] of Fig. 5.

Figure 6a shows how a sloping (s + p) contribution can be added to the d-orbital function of Fig. 5 to give the observed variation along the series. Figure 6b shows the crystal field rationalization of the effect in terms of the crystal field stabilization energy. The slope of the s + p contribution of Fig. 6a is readily rationalized. As the first-row transition metal series is traversed, the ionization potential of the metal becomes larger and shifts the entire set of s + p + d orbitals to lower energy. The s, p energy separation from the ligand σ orbitals thus decreases across the series and, correspondingly, the stabilization energy increases. This molecular orbital rationalization of the data in Fig. 6 is certainly more convincing than previous molecular orbital explanations (*23*). We have in addition produced with the minimum of effort the relative contributions of d and higher orbitals to the total stabilization energy associated with metal–ligand (in this case H_2O)

---

[5] Because we are dealing with high-spin systems, the same combination of electron repulsion terms as for the lowest-energy term of the gaseous ion will occur, in the form of the Racah parameters, in the energy description of the ground electronic state of the complex. Thus, we are perfectly justified in only considering the change in orbital energy on complex formation.

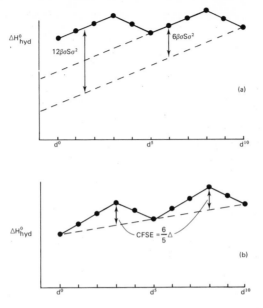

FIG. 6. Molecular orbital (a) and crystal field (b) rationalizations of the variation of ΔH^0_{hyd} with d^n.

interaction. For Ti^{2+}, the d-orbital contribution is about 50% but for Cu^{2+}, only about 15%. We must, therefore, exercise considerable care in the interpretation of structural transition metal data in d-orbital terms alone—higher orbital contributions are obviously very important in numerical terms. Having said this, we shall be surprised to find that a large amount of material can be rationalized using the d-orbital-only model.

The actual mixing of nd and $(n + 1)p$ orbitals is, of course, of crucial importance in providing a mechanism by which the Laporte forbidden d-d transitions in transition metal complexes may gain in intensity. This may occur in the static situation (e.g., tetrahedral complexes), where the p orbitals and one set of the d orbitals transform as t_2 or in the dynamic situation (as in octahedral complexes) where such mixing is only possible when the point symmetry has been reduced by an asymmetric vibration (vibronic coupling).

2. π-Bonding

The π-bonding may be treated along exactly the same lines as just shown for σ-type interactions. No new principles are needed. Structure

I shows one component of the ligand t_{2g} combination, and we may

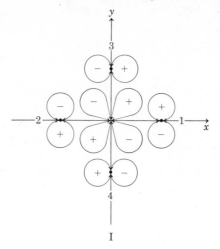

I

write

$$\psi_{t_{2g}}(1) = \tfrac{1}{2}(\pi_3 - \pi_4 + \pi_5 - \pi_6) \tag{7}$$

If the overlap of one lobe of d_{xy} with a ligand π-type orbital along the x axis is defined as S_π, the overlap of Eq. (7) with d_{xy} becomes

$$S_{t_{2g}}(1) = \tfrac{1}{2} \cdot 4 \cdot S_\pi \tag{8}$$

and, thus, the π-interaction energy $\epsilon = 4\beta_\pi S_\pi^2$. For systems containing π donors (where the ligand π orbitals lie below the d orbitals), the t_{2g} d-orbital set is raised in energy by this amount; with π acceptors (where the ligand π orbitals lie above the d orbitals), the t_{2g} d orbitals are depressed by $4\beta_\pi S_\pi^2$. The crystal field-splitting parameter Δ is then related to the angular overlap parameters by $\Delta_{\text{oct}} = 3\beta_\sigma S_\sigma^2 \pm 4\beta_\pi S_\pi^2$, the choice of sign depending on the nature of the π orbitals present. The interaction energy involving the ith d orbital may also be derived using

$$\epsilon_i = \beta \sum_j S_{ij}^2 \tag{9}$$

where the summation is over all ligand orbitals j. In the present case we have four ligand orbtials each having an overlap integral of S_π with d_{xy} and thus $\epsilon(d_{xy}) = 4 \times \beta_\pi S_\pi^2$ as before. Use of Eq. (9) gives the same results as for the σ-type interactions in the foregoing, and it is very easy quickly to derive orbital energies using this equation and Tables I and II without recourse to group theory.

B. TETRAHEDRAL COORDINATION

By the same methods as discribed in Section II, A, we may construct t_2 functions of σ symmetry with which to overlap with the d_{xz}, d_{xy}, and d_{yz} orbitals in the tetrahedral configuration. The interaction energy becomes $\epsilon(t_2) = 1.333\ \beta_\sigma S_\sigma^2$ for this geometry. Inclusion of π bonding leads to the energy levels of Fig. 7 for π donors. (For π acceptors, the π contribution is included with the opposite sign.) Immediately we notice that, whereas $\Delta_{oct} = 3\beta_\sigma S_\sigma^2 \pm 4\beta_\pi S_\pi^2$, $\Delta_{tet} = (4/3)\beta_\sigma S_\sigma^2 \pm (16/9)\beta_\pi S_\pi^2$, i.e., $\Delta_{tet} = (4/9)\Delta_{oct}$. (The equality will only hold, of course, if $\beta_\sigma S_\sigma$ is the same in both geometries, i.e., the metal–ligand distance remains constant.) One rule that we will find useful in the derivation of these figures is a sum rule concerning the orbital energies (17c).

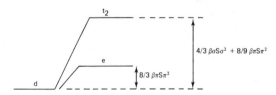

FIG. 7. Energy-level diagram for tetrahedral coordination of a σ and π donor (π contributions should be included with opposite sign for π acceptors).

In a complex ML_n, where there are $n\sigma$ orbitals directed at the central atom the sum total σ interaction energy of all the d orbitals is simply equal to $n\beta_\sigma S_\sigma^2$. Thus we know in the octahedral case that all the σ interaction is contained in the doubly degenerate e_g set of orbitals from group theory. The total available σ interaction energy ($6\beta_\sigma S_\sigma^2$) is, therefore, shared between each of the two components of the e_g set, i.e., $3\beta_\sigma S_\sigma^2$ each. In the tetrahedral molecule the total σ interaction energy ($4\beta_\sigma S_\sigma^2$) is shared between the three components of the t_2 set, i.e., $(4/3)\beta_\sigma S_\sigma^2$ each. Similarly, if each ligand has two orbitals of π symmetry the total π interaction is $2n\beta_\pi S_\pi^2$. In the octahedral case the total π interaction, $12\beta_\pi S_\pi^2$, is shared between the three components of the t_{2g} set of orbitals, $4\beta_\pi S_\pi^2$ each. In the tetrahedron, both t_2 and e sets contain π contributions, and the total interaction energy of $8\beta_\pi S_\pi^2$ is shared between the two. To calculate what fraction is located in the t_2 set and what function in the e set, we need to write down a symmetry-adapted ligand wavefunction and calculate the overlap integral as we have shown previously. However, since the sum total is $8\beta_\pi S_\pi^2$, once the π contribution to one set has been decided by this calculation, the contribution to the other set can obviously be obtained by simple subtraction. The combination of this sum rule and Eq. (9)

allows a rapid calculation of the d-orbital energies for different geometries.

C. Other Geometries

Energy-level diagrams for other geometries are now readily derived. Because inorganic chemists place greater stress on σ interactions than on π interactions, we shall emphasize the former in what follows and initially ignore the π contributions to the orbital energies. By using Eq. (9) and Table II, we can readily derive the interaction energies associated with the $d_{x^2-y^2}$ and d_{z^2} orbitals in the orthogonal geometries derived from the octahedron. For example in the square planar structure (II), the four σ ligands overlap with $d_{x^2-y^2}$ to give a total interaction energy of $4 \times \beta_\sigma(\sqrt{3/2})^2 S_\sigma^2 = 3\beta_\sigma S_\sigma^2$ and with the "collar" of d_{z^2} to give a total interaction energy of $4 \times \beta_\sigma(1/2)^2 S_\sigma^2 = \beta_\sigma S_\sigma^2$. Figure 8 shows d-orbital energy-level diagrams for some of these octahedrally based geometries and the trigonal plane and trigonal bipyramid.[6] We

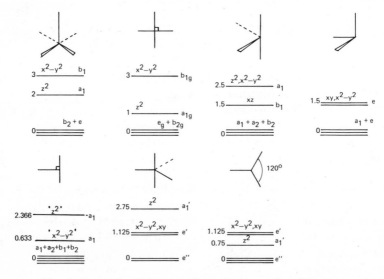

Fig. 8. Energy-level diagrams for some geometries of interest.

[6] In the low-symmetry situation of the T-shaped structure, orbitals d_{z^2} and $d_{x^2-y^2}$ must always mix together. (In the cis-divacant structure this is avoided by a suitable choice of axes.) Solution of the relevant determinant gives us $\epsilon = (1.5 \pm \frac{\sqrt{3}}{2}) \beta_\sigma S_\sigma^2$ for the two interacting orbitals [see McClure (19) for a more complete discussion]. In Burdett (24) such mixing was ignored.

remember that to lower energy, the ligand σ orbitals split apart in energy as the mirror image of this d-orbital splitting. However, we need only the d-orbital region of the energy-level diagram to calculate the total d-orbital stabilization energy of a given electronic configuration since $\sum(\sigma)$ of Eq. (6) depends only on the location of the holes in the d-orbital manifold.

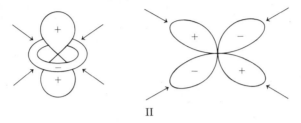

II

D. RATES OF REACTION OF HEXAQUO IONS

Knowledge of the energy-level diagram for the square pyramidal geometry allows us to rationalize immediately (22) the rates of ligand labilization in $M(H_2O)_6{}^{2+}$ molecules (k_1), determined in the classical

$$\underset{\text{octahedral}}{M(H_2O)_6{}^{2+}} \overset{k_1}{\rightleftharpoons} \underset{\text{square pyramidal}}{M(H_2O)_5{}^{2+}} + H_2O$$

researches of Eigen (25, 26). Simple subtraction of the d-orbital stabilization energies for octahedral and square pyramidal molecules gives the contribution of the metal d orbitals to the energy of activation for ligand loss shown in Fig. 9 as a function of the number of d electrons (all electronic configurations and high-spin). This is simply done by use of Eq. (6) and the relevant energy-level diagrams of Figs. 4 and 8. If a sloping contribution from s + p orbital interactions is added (cf. the heats of hydration themselves in Fig. 6a), then the curve of Fig.

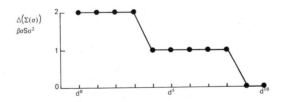

FIG. 9. Difference in d-orbital σ-stabilization energies $[\Delta(\sum(\sigma)]$ between octahedral and square pyramidal d^n ions.

JEREMY K. BURDETT

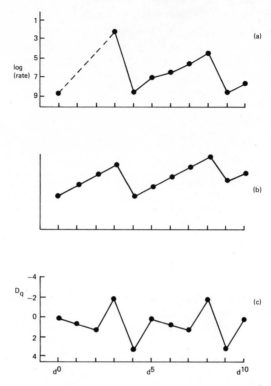

FIG. 10. Molecular orbital rationalization of the rates of reaction of hexaquo ions:
(a) experiment; (b) theory; and (c) crystal field approach.

10b results. This is to be compared with the observed rates (Fig. 10a)
and the crystal field analysis of the activation energy (Fig. 10c) obtained by simple subtraction of the CFSE's of octahedron and square
pyramid.[7]

The superiority of the molecular orbital approach is clear; the
activation energy is always positive which is not the case with the
CFT, and the molecular orbital plot is quite a faithful reproduction
of the observed log (rate) curve. The failure of the CFT is due to the
following reason. In the absence of π bonding, Fig. 8 shows that the
lowest three d orbitals are equienergetic ($b_2 + e$) for the square-based
pyramid, but the nature of the CF method removes this accidental
degeneracy considerably. In terms of Dq, the CF energies of the d

[7] For a discussion of the crystal field rationalization of this data, see Philipps and
Williams (27).

orbitals in the square pyramid are

$$
\begin{array}{ccc}
3 \quad \underline{\hspace{3cm}} & -x^2-y^2 \diagup & \underline{\quad 9.14 \quad} \\[2em]
2 \quad \underline{\hspace{3cm}} & \underline{\quad z^2 \quad} & \underline{\quad 0.86 \quad} \\[1em]
& & \underline{\quad -0.86 \quad} \\[0.5em]
& \diagup \underline{\quad xy \quad} & \\[0.5em]
0 \quad \underline{\overline{\hspace{3cm}}} & \underline{\quad xz, yz \quad} & \underline{\overline{\quad -4.57 \quad}} \\[1em]
\mathrm{MO} & & \mathrm{CF} \\
(\beta\sigma S\sigma^2) & & (Dq)
\end{array}
$$

It is the large splitting between d_{xy} and d_{xz}, d_{yz}, not reproducible on a no π-bonding molecular orbital model, which gives rise to the incorrect form of the plot of Fig. 10c. The error in the CFT will, of course, be especially apparent in those systems which are poor π bonders as is the present case with the aquo ligand. In general, then, the CFT should not be trusted to give a quantitative measure of the d-orbital splittings.

E. GEOMETRIES OF TRANSITION METAL COMPLEXES

We have seen in Section II,D the relative contributions of the metal s and p orbitals compared to d orbitals to the heats of hydration and rates of reaction of the hexaquo ions. In the field of molecular geometry, these observations suggest that here the geometry will be a compromise between that demanded by the s,p-orbital interactions and that by the d-orbital interactions with the ligands. In general, an ML_n complex will contain $n\sigma$ pairs of electrons involved in s,p (and d) interactions. Thus, on the basis of the VSEPR method (dealing exclusively with molecules involving s,p orbitals on the central atom), the geometry demanded by interactions with these higher orbitals will be (3): trigonal planar (D_{3h}) for ML_3, tetrahedral (T_d) for ML_4, trigonal bipyramidal (D_{3h}) for ML_5 and octahedral (O_h) for ML_6. We note that the VSEPR geometry in these cases is the one with minimum ligand-pair repulsions [Pauli avoidance (18)] and also the geometry containing minimum nonbonded repulsions between the ligands themselves. We will label these combined forces ligand-ligand terms. The geometries of some four-coordinate $M(CX)_4$ species with different d^n configurations, shown in Fig. 11, indicate that the d orbitals may often exert a decisive effect in modifying this VSEPR geometry.

Let us start by looking at these four-coordinate geometries. The σ stabilization energy, $\sum(\sigma)$, derived using Eq. (6) and the energy-level

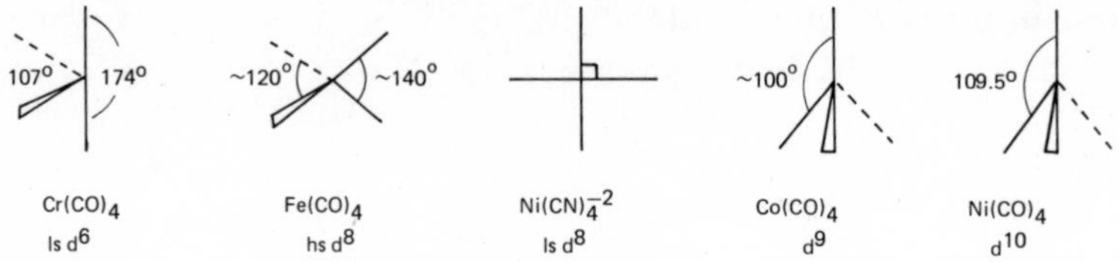

JEREMY K. BURDETT

FIG. 11. Some four-coordinate geometries (ls, low-spin; hs, high-spin).

diagrams of Fig. 8, allow ready construction of Table III for three geometries of interest, namely, the octahedral "cis-divacant" (or butterfly or disphenoidal) structure, the square plane, and the tetrahedron. For the d^{10} configuration we see that no d-orbital stabilization energy is present since all metal (d)–ligand σ-bonding and antibonding orbitals are equally occupied. The tetrahedral structure of $Ni(CO)_4$ or $ZnCl_4^{-2}$ thus results, since this is the geometry demanded by ligand-ligand interactions of the type described in the foregoing. For the d^9, low-spin d^8 and d^7 systems, the square planar geometry has the largest d-orbital stabilization energy. For the low-spin d^8 system, the square planar geometry is more favored (by $3.33\beta_\sigma S_\sigma^2$) relative to the tetrahedral (the ligand-ligand determined geometry). Low-spin d^8 systems are always square planar e.g., $Ni(CN)_4^{-2}$. The square planar geometry is similarly strongly favored for the low-spin d^7 configuration (by $3\beta_\sigma S_\sigma^2$) and all low-spin Co(II) systems are square planar.

For the d^9 configuration, square planar is still favored relative to tetrahedral but by a smaller amount ($1.66\beta_\sigma S_\sigma^2$). For Cu(II) complexes with this configuration, the geometry often depends on the crystal structure and the nature of the counterion. For example, $(NH_4)_2CuCl_4$ contains (28a) planar $CuCl_4^{-2}$ ions, but Cs_2CuCl_4 and Cs_2CuBr_4 and, in general, systems with large cations contain CuX_4^{-2} ions that are

TABLE III

TOTAL d-ORBITAL STABILIZATION ENERGIES $[\sum(\sigma)]$
FOR FOUR-COORDINATE COMPLEXES (UNITS $\beta_\sigma S_\sigma^2$)

Configuration[a]	cis-Divacant	Square planar	Tetrahedral
d^{10}	0	0	0
d^9	2.5	3	1.33
d^8(ls)	5	6	2.67
d^8(hs)	4	4	2.67
d^7(ls)	6.5	7	4
d^6(ls)	8	8	5.67

[a] Low spin (ls); high spin (hs).

squashed tetrahedra (*28b*) of the D_{2d} point group (III), i.e., halfway between the ligand-ligand (tetrahedral) and d-orbital (square planar) structures. The two geometries are close in energy for this electronic configuration, as evidenced by the fact that application of pressure (*29*) sends the squashed tetrahedron in Cs_2CuBr_4, Cs_2CuCl_4, or $[(CH_3)_2CHNH_3]_2$ $CuCl_4$ to the square planar geometry.

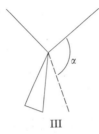

III

The process is reversible.[8] The d-orbital stabilization energy is, therefore, often only large enough to drive the structure a part of the way toward the square planar geometry. Two independent, molecular orbital calculations (*13, 15*) for a d^9 tetracarbonyl moiety predict angles α of $132°$ and $135°$ for the equilibrium geometry, i.e., halfway between tetrahedral ($110°$) and square planar ($180°$). In fact for $Co(CO)_4$ both D_{2d} and C_{3v} (angle C_{ax}-M-$C_{eq} \simeq 100°$) structures are found (*31*) for this species, trapped in a low-temperature matrix. For small distortions away from tetrahedral, there is not much to choose between these two structures for the d^9 configuration.

For two electronic configurations of Table III the square planar and cis-divacant structures are predicted to be of equal stability. In order to determine the one of lowest energy, it is necessary to extend the angular overlap approximation to the fourth power in the overlap integral S. The general result (*24*) is that the more stable structure is the one with the larger number of cis ligands—in this case, the cis-divacant structure. The fact that we need to resort to terms in S^4 means that the energy differences between the various geometries is smaller for these (e.g., low-spin d^6) systems than for low-spin d^7. We will see some consequences of this later in the stability of cis and trans octahedral isomers ML_4L_2'. The cis-divacant geometry is that observed for (*32*) matrix-isolated $Cr(CO)_4$ (Fig. 11), where the energy difference

---

[8] Ferraro and Long (*30*) have used the second-order Jahn–Teller formalism to give some clues as to why these ions should distort along this pathway linking the tetrahedron and square plane. The present author has suggested (*24*) that this is only valid for systems containing a single d-orbital hole (as in these d^9 systems), but this is not a completely general approach.

from tetrahedral is $2.33\beta_\sigma S_\sigma{}^2$. For high-spin d^8, however, the d-orbital stabilization energy difference between the two geometries is smaller $(1.33\beta_\sigma S_\sigma{}^2)$. For the four-coordinate d^9 system with a difference of $1.66\beta_\sigma S_\sigma{}^2$, we found geometries intermediate between the tetrahedron and the geometry demanded on d-orbital grounds alone. Similarly high-spin d^8 systems are usually described as distorted tetrahedral and are typified by the matrix-isolated $Fe(CO)_4$ structure (33) of Fig. 11. They are, therefore, intermediate in structure between the tetrahedral and cis-divacant arrangements.

For five-coordinate molecules, we find the stabilization energies (Table IV) for five electronic configurations of interest. For low-spin d^6, the stabilization energy difference between the square pyramid and the trigonal bipyramid (the ligand-ligand determined geometry) is largest and matrix-isolated $M(CO)_5$ (M = Cr, Mo, W) is found to have the C_{4v} geometry (34). Preliminary reports point to square pyramidal geometries for $Re(CO)_5$ (35) and $Mn(CO)_5$ (36). Ab initio molecular orbital calculations conclude a square pyramidal structure (37). However, for d^6 (intermediate spin) and low-spin d^8, the energy difference between the two geometries is very small. The species $Fe(CO)_5$ is well known in the gas phase as a trigonal bipyramidal molecule (38) [as is the $Mn(CO)_5{}^-$ in the crystal (39)], although the molecule is fluxional and rapidly exchanges axial and equatorial sites probably via the square pyramidal configuration [Berry process (40)] as intermediate. The ion $Ni(CN)_5{}^{-2}$ exists in the crystal as a mixture of trigonal bipyramidal and square pyramidal molecules, but application of pressure (41) reversibly sends all the molecules into the square pyramidal structure at 7 kbar. The photochemistry of $Cr(CO)_5$ and $Mo(CO)_4$—Pcy_3 (low-spin d^6) may be explained (42) on the basis that

TABLE IV

TOTAL d-ORBITAL STABILIZATION ENERGIES $[\sum(\sigma)]$
FOR SOME FIVE-COORDINATE COMPLEXES (UNITS $\beta_\sigma S_\sigma{}^2$)

Configuration[a]	Square pyramid	Trigonal bipyramid	Difference
d^9	3	2.75	0.25
d^8(ls)	6	5.5	0.5
d^7(ls)	8	6.625	1.375
d^6(ls)	10	7.75[b]	2.25
d^6(ls)	8	7.75[b]	0.25

[a] Low spin (ls).

[b] These two electronic configurations have identical d-orbital electronic distributions in the trigonal bipyramidal geometry.

TABLE V

TOTAL d-ORBITAL STABILIZATION ENERGIES [$\sum(\sigma)$]
FOR SOME THREE-COORDINATE COMPLEXES (UNITS $\beta_\sigma S_\sigma^2$)

Configuration[a]	Trigonal plane	T shape	C_{3v}[b]
d^{10}	0	0	0
d^9	1.125	2.366	1.5
d^8(hs)	2.25	3	3
d^8(ls)	2.25	4.732	3
d^7(ls)	3.375	5.366	4.5
d^6(ls)	4.5	6	6

[a] High spin (hs); low spin (ls).
[b] All bond angles 90°.

the first electronic excited state (intermediate-spin d^6) has a trigonal bipyramidal geometry. We return to this point later.

For three-coordinate molecules we derive the stabilization energies shown in Table V. To distinguish between the equienergetic geometries for high-spin d^8 and low-spin d^6 systems, we use the rule already developed here that the most stable geometry is the arrangement with the maximum number of cis ligands. Experimentally the d^{10} species $Ni(CO)_3$, $Ni(N_2)_3$, etc. (43) are found from matrix isolation studies to be trigonal planar (D_{3h}). The structure of $Cr(CO)_3$ (low-spin d^6) closely resembles (32) the C_{3v} structure with orthogonal ligands (IV). Here the difference in stabilization energy between C_{3v} and D_{3h} is $1.5\beta_\sigma S_\sigma^2$.

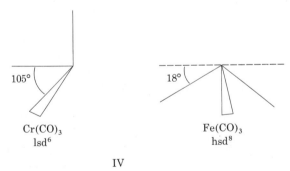

Cr(CO)₃
lsd⁶

Fe(CO)₃
hsd⁸

IV

For high-spin d^8, the stabilization energy is smaller ($0.75\beta_\sigma S_\sigma^2$) and a structure halfway between the d-orbital, and ligand–ligand determined geometries is found (44) for $Fe(CO)_3$. Perhaps this smaller energy difference between C_{3v} and D_{3h} $Cr(CO)_3$ molecules ($1.5\beta_\sigma S_\sigma^2$) than between the C_{2v} and T_d $Cr(CO)_4$ molecules ($2.33\beta_\sigma S_\sigma^2$) results in smaller deviations from the d-orbital-only geometry in the four-coordinate complex than in the three-coordinate one (Fig. 11). This behavior is certainly what would be expected under the scheme. As we go down the periodic table, however, the D_{3h} d^{10} structure is often found in distorted environments (45), e.g., in $(Ph_3P)_2AuCl$, where the M—Cl bond is long and the angle-between the two M—P bonds has opened up from 120°. This incipient two coordination is expected when, as at the bottom of the periodic table, the $(n + 1)s/nd$ orbital separation is small. Similar effects are found in five-coordinate d^{10} systems (46).

Some interesting points emerge from the preceding discussion. The molecular geometry is dominated by σ effects (we have used only σ energies in this discussion). The d orbitals involved in σ interactions are always the higher energy ones because these orbitals are metal–ligand σ antibonding. Thus, only where there are holes in the higher-energy d orbitals will there be any large differential d-orbital stabilization energy for a distorted geometry compared to the ligand-ligand determined geometry. Thus with reference to the four-coordinate geometries of Figs. 7 and 8 for d^0, d^1, d^2, low-spin d^3 and d^4 configurations, $\sum(\sigma)$ has the same value $(8\beta_\sigma S_\sigma^2)$ for all three geometries—square planar, cis-divacant, and tetrahedral. The latter geometry, the one determined by ligand-ligand interactions is, therefore, the structure adopted. This feature then rationalizes the widespread observation that the first few d electrons have little effect on the geometry of the molecule. Thus $TiCl_4$ (d^0) and VCl_4 (d^1) are both tetrahedral, but note the much larger difference in structure between $ZnCl_4^{-2}$ (d^{10}) and $CuCl_4^{-2}$ (d^9).

Molecular orbital calculations support these ideas. Small distortions from the ligand-ligand determined geometry are often found when asymmetric electron-charge distributions are present in orbitals involving π interactions only. The second- and third-row hexafluorides MF_6 (M = Mo—Rh, W—Pt) are octahedral molecules, but, where holes in the π-orbital manifold exist, small vibronic effects are observed (47, 48). This type of behavior has often been labeled the "dynamic" Jahn–Teller effect to distinguish it from the gross structural effects noted earlier. We believe that these sizably smaller effects reflect the relative importance of π interactions compared to σ interactions in determining the angular geometry. On our molecular model described in this paper, the reason for smaller π interactions compared with σ ones is the smaller value of π compared to σ overlap integrals. This discrepancy is magnified since, in the stabilization energy of Fig. 1, the overlap integral appears as the square.

The gross geometry is thus determined by the orbital occupation numbers of the three highest d orbitals. If these three orbitals are symmetrically occupied (with orbital occupation numbers 000, 111, 222), then the d-electron distribution is spherically symmetrical and the d-orbital electron configuration will not exert any distorting effect on the symmetrical ligand–ligand determined structure. Only when the distribution of electrons in these three orbitals is nonspherically symmetrical does the molecule distort as we have seen in the foregoing.

These generalizations have led to a set of rules with which to predict the lowest-energy geometry of a complex (24):

1. Neglect any electrons in the lowest two d orbitals. The *angular* geometry is determined by the occupation numbers of the three highest energy orbitals.

2. If the occupation numbers are symmetrical (000, 111, 222) then the VSEPR geometry will be observed.

3. If a hole exists in the highest-energy orbital (220, 221, 210, 110), then the structure will be that of maximum overlap with the lobes of $d_{x^2-y^2}$.[9] (A structure intermediate between this geometry and the VSEPR one may be observed for 221.)

4. If two holes exist symmetrically in the two highest energy orbital (200, 211, 100), then the structure will be that based on the octahedron containing the maximum number of cis ligands. (A structure intermediate between this geometry and the VSEPR one may be observed for 211.)

Obviously for some excited-state configurations, not contained in the scheme, e.g., 121, we need to calculate and compare the relevant stabilization energies for the various extreme geometries.

How do these conclusions fit in with the Jahn-Teller (49) theorem telling us that molecules with orbitally degenerate electronic states distort to lose this degeneracy? Some of the distortions we have seen and rationalized are allowed under the Jahn-Teller scheme, but very importantly others are not (50). For example,

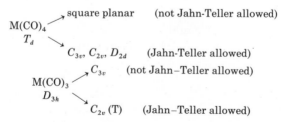

Thus, there is no reason from Jahn–Teller considerations considerations alone for $Fe(CO)_3$ (high-spin d^8) to distort from the D_{3h} geometry

[9] The first T-shaped geometry for a three coordinate is d^8 system has recently been found (73).

($^3A_2'$) to the observed C_{3v} structure ($^3A_2$), and the square planar structure for $Ni(CN_4)^{2-}$ is a very large distortion for a Jahn–Teller one. In fact, *all* the results of the "static Jahn–Teller" theorem are encompassed by the present theory, which is, as a consequence a much more powerful approach since it is able to provide energetic data on the geometry of any electronic configuration irrespective of any considerations concerning the degeneracy of the electronic state. On the present model, the smaller-energy changes involving the "dynamic Jahn–Teller" effect are considered to arise from the much smaller structural influence associated with asymmetrical charge distributions in π orbitals, as we have discussed in the preceding.

F. Derivation of Walsh Diagrams

The Walsh diagrams, showing qualitatively the energy changes associated with molecular orbitals on distortion, have a well-established place in main-group stereochemistry, but only recently have their equivalents in transition metal chemistry been comprehensively derived using the extended Hückel molecular orbital method (*13–16*). However, in some cases we may readily obtain the functional dependence on angle of distortion using the simple scheme developed here.

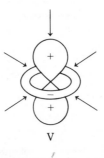

V

Consider the square pyramidal ML_5 unit with a geometry defined by the angle θ. Let us investigate how the stabilization energy associated with the d-orbital changes on distortion. d_{z^2} interacts with the axial ligand and the four equatorial ligands that overlap its collar (V). Thus, $\epsilon(d_{z^2}) = [1 + 4(1 - \frac{3}{2}\sin^2\theta)^2]\beta_\sigma S_\sigma^2$ from Table I; $d_{x^2-y^2}$ overlaps solely with the four equatorial ligands and hence $\epsilon(d_{x^2-y^2}) = 4(\sqrt{3}/2 \sin^2\theta)^2 S_\sigma^2$; d_{xz} and d_{yz}, which are σ nonbonding at $\theta = 90°$, give $\epsilon(d_{xz}, d_{yz}) = 4(\sqrt{3}/2 \sin 2\theta)^2 S_\sigma^2$. With these functions we can readily construct Fig. 12, where we see stabilization of d_{z^2} on bending away from 90°, but destabilization of d_{xz} and d_{yz}. Thus we find that whereas low-spin d^6 $Cr(CO)_6$ in a CH_4 matrix (*34a*) has a bond angle of 93°, low-spin d^7

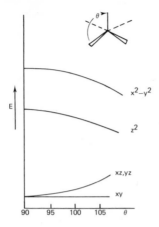

FIG. 12. Walsh diagram for distortion of a square pyramid.

$Co(CN)_5^{-3}$ has $\theta = 97°$ (51) and one-half of $Mn_2(CO)_{10}$ has $\theta = 96°$ (52). Similarly, high-spin Fe(II) in hemoglobin (53, 54) lies 0.75 Å out of the plane of the porphyrin ring ($\theta > 90°$) but, in oxyhemoglobin (low-spin d^6), the iron now lies in the plane of the ring, $\theta \sim 90°$. The change in bond angle in the low-spin d^6, d^7 pair just mentioned is in the *opposite* direction to that expected if this "d electron" was stereochemically active in the sense of being one-half of a VSEPR electron pair (VI).

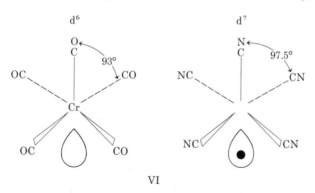

VI

G. STRUCTURAL CONSEQUENCES IN d^8 AND d^9 SYSTEMS

Three figures extracted from Tables III and IV and Fig. 5 are of immense importance in understanding some of the structural chemistry of low-spin d^8 and d^9 systems. We find that, for these two electronic configurations, the d-orbital stabilization energy of square planar, square pyramidal, and octahedral complexes are equal. This implies that the axial bond energy between ligand σ and metal d orbitals in octahedral and square pyramidal low-spin d^8 and d^9 systems is zero and that the ligands are attached purely by forces involving the (s + p) orbital interactions on the metal.

Alternatively, we can see that the axial ligands (along the z axis) may only interact with d_{z^2}. The bonding orbital d_{z^2}–ligand is occupied but, in these d^8 and d^9 systems, also its antibonding counterpart contains two electrons resulting in a zero axial d-orbital–ligand bond order. Thus, we should expect to see, in general, long axial bond lengths compared to equatorial in these systems. Table VI shows some representative bond length data. For the six-coordinate d^9 case, we immediately recall that the Jahn–Teller theorem has been used for many years to rationalize these observations (22). It is easy to see, however, that the same distortion occurs irrespective of whether the parent compound is a six-coordinate regular octahedron (e.g., $CuCl_2$), a six-coordinate octahedron but with different ligands (e.g., $CuCl_2 \cdot 2H_2O$), or a five-coordinate square pyramid. Only the first case is directly approachable by Jahn–Teller considerations since only here are $d_{x^2-y^2}$ and d_{z^2} orbitals degenerate. Yet all three cases fit within the umbrella of the present discussion.

If a soft potential surface describes the approach of the axial ligand to the transition metal center, then it is not too surprising to see some short axial metal–ligand distances in a couple of instances (as in the example of K_2CuF_4 in Table VI) where possibly, because of crystal

TABLE VI

SOME BOND LENGTHS (Å) IN d^9 COMPLEXES

Octahedral:		
$CuCl_2$	4Cl at 2.30	2Cl at 2.95
$CsCuCl_3$	4Cl at 2.30	2Cl at 2.65
$CsCl_2 \cdot 2H_2O$	2O at 2.01	2Cl at 2.31
		2Cl at 2.98
$CuBr_2$	4Br at 2.40	2Br at 3.18
CuF_2	4F at 1.93	2F at 2.27
K_2CuF_4	2F at 1.95	4F at 2.08
Square pyramidal:		
Diquo(acac)Cu(II) picrate[a]	Equatorial 1.88; axial 2.76	
N,N'-disalicylidene ethylene diamine Cu(II)[b]	Equatorial 2.01; axial 2.41	

[a] D. Hall and T. N. Wates, *J. Chem. Soc.* p. 1644 (1960).
[b] R. D. Gillard and G. Wilkinson, *J. Chem. Soc.* p. 3599 (1963).

packing, the axial ligand may be pressed into the coordination shell. Under these circumstances, we see little change in the equatorial metal–ligand distances. In the recent crystal structure determination of $K_2PbCu(NO_2)_6$, there is even a regular octahedron of nitrogen atoms about the Cu^{2+} ion (28c). Often, because of this facile approach of the fifth and sixth ligands to the square plane, it is difficult to classify solid-state Cu(II) structures as four-, five-, or six-coordinate systems.

In d^8 chemistry the structural equivalent of this well-documented structural behavior is the so-called anomalous behavior of Ni(II) systems, in which two axial ligands may be weakly bound to a square planar unit (Lifschitz's salts) producing dramatic effects in the electronic spectrum (55). The d^8 system differs slightly from Cu(II) in the sense that the octahedral geometry has an additional stabilizing force. This is the exchange energy associated with the triplet state (high spin), which is obviously lower in energy than the singlet state (low spin) when $d_{x^2-y^2}$ and d_{z^2} are equienergetic. Sometimes the Lifschitz complexes [of Ni(II) with substituted ethylenediamines] are blue and paramagentic ("octahedral," high spin), and sometimes yellow and diamagnetic (square planar, low spin). The color depends on a variety of factors, such as the particular anions present, the solvent from which the material is crystallized, the temperature, and whether the complexes are exposed to atmospheric water vapor. In several cases, four-coordinate units polymerize to give Ni(II) in five- and six-coordinate environments, e.g., nickel acetylacetonate. The position of the equilibrium is generally complex and it is clear that low-energy pathways connect square planar four-, square pyramidal five-, and octahedral six-coordinate Ni(II).

The resemblance of the low-spin d^8 systems to the "Jahn–Teller" behavior in Cu(II) is forcefully apparent. Similarly, the X-ray crystal structure of square pyramidal (low-spin) $Ni(CN)_5{}^{-3}$ shows that the axial Ni—C bond length is much longer than the equatorial ones. In high-spin d^8 systems, since d_{z^2} and $d_{x^2-y^2}$ are now equally occupied, the d-orbital contributions to both sites should be equal, and in a large number of examples there is evidence to support this. In a recent comparison (56a) between the isostructural species $Mg(Me_3AsO)_5$-$(ClO_4)_2$ (d^0) and $Ni(Me_3AsO)_5(ClO_4)_2$ (d^8), the axial MO bonds are 1.94 Å (Ni) and 1.92 Å (Mg) compared to the average equatorial bonds of 2.00 Å (Ni) and 2.03 Å (Mg). Compare these figures with those for low-spin $Ni(CN)_5{}^{-3}$ of 1.86 Å (equatorial) and 2.17 Å (axial) (41a). (In the latter molecule the axial bond length is certainly shortened to some extent by π bonding.)

Whereas the d-orbital stabilization energy difference between four and five coordination is zero for the low-spin d^8 configuration in the electronic ground state, the same is not true if an excited-state configuration is produced with a different arrangement of d electrons, as in the high-spin d^8 complexes just discussed for example. This has led to the interesting and very unusual observation of photoassociation (57) in the molecule $Ir(MeNC)_4^+$. In the first excited state with configuration 211, there is a gain in d-orbital stabilization energy on adding a fifth ligand. The photo-produced complex dissociates in the dark.

H. LIGAND SITE PREFERENCES

We may use the results of these last sections to inquire where in a mixed ML_nL_m' complex the strongest σ denors are likely to reside. In the low-spin d^8 square pyramid we saw in Section II, G that the axial bond was the weakest one. The isomer with the lowest energy will obviously be the one where the strongest σ donors (L) occupy those sites of largest latent bond strength. In the ML_4L' d^8 square pyramid, therefore, the weakest σ donor is relegated to the axial site.

Alternatively, we could have considered (58) that the most stable arrangement of the four remaining L ligands is the square planar structure, which is not disturbed by coordination of the fifth (weaker σ donor ligand). Quantitatively we may (58) use Eq. (9), define different values of $\beta_\sigma S_\sigma^2$ for each different ligand, and compare the stabilization energies for each isomer. All three approaches give us the same answer. For low-spin d^6 and high-spin d^8 systems, the stable four-coordinate geometry is the cis-divacant structure (VII). Thus the weaker σ donor here occupies an equatorial site in the square-based pyramid. However, Table III shows us that square planar and cis-divacant four-coordinate arrangements are very close in energy and, thus, the energy difference between the two isomers in these cases is small. Indeed, we see little difference in axial and equatorial bond lengths in a wide range of molecules with these electronic configurations surveyed by Orioli (59).

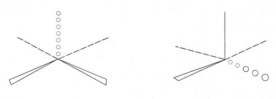

VII

For the d^8 low-spin trigonal bipyramid, we expect the stronger σ donors to occupy the axial positions from our preceding arguments. The T-shape three-coordinate structure (VIII) is much more stable

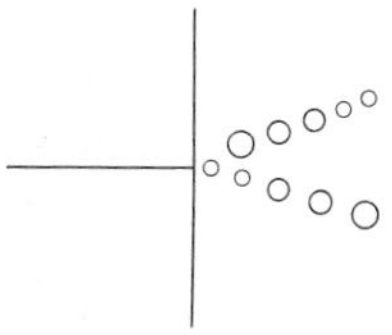

VIII

than the trigonal plane (Table V) for this electronic configuration. Alternatively, we may view the molecule in the following way (60). The total stabilization energy of the low-spin d^8 configuration arises from overlap (II) of ligand σ orbitals with d_{z^2}. (Fig. 8). Overlap of each axial ligand contributes $2\beta_\sigma S_\sigma^2$ to $\sum(\sigma)$ (from Table II), and each equatorial ligand $\frac{1}{2}\beta_\sigma S_\sigma^2$ since the latter overlap only with the collar of d_{z^2}. Thus, the axial positions contain the largest d-orbital stabilization, and it is here that the strongest σ donors should reside. With some exceptions, this is generally true. Shorter axial than equatorial bond lengths are found (Table VII) in all known low-spin d^8 trigonal bipyramidal complexes with first-row central atoms although the difference for $Fe(CO)_5$ is very small. [Recent reworking of the data suggests that there is no difference between the bond lengths of axial and equatorial groups within experimental error (61).] The situation is not quite as simple as we have suggested here, as evidenced by the disparity in bond lengths observed for $CdCl_5^3$ and $HgCl_5^{-3}$ (d^{10})

TABLE VII

SOME BOND LENGTHS (Å) IN TRIGONAL
BIPYRAMIDAL COMPLEXES

Complex	Axial	Equatorial	Ref.
$Ni(CN)_5^{-3}(d^8)$	1.838(9)	1.94[a]	41a
$Co(CNCH_3)_5^+(d^8)$	1.84(2)	1.88(2)	41c
$Fe(CO)_5(d^8)$	1.810(3)	1.833(2)	38
$CuCl_5^{-3}(d^9)$	2.2964(12)	2.3912(13)	41d
$CuBr_5^{-3}(d^9)$	2.4500(22)	2.5191(17)	41e
$CdCl_5^{-3}(d^{10})$	2.526(1)	2.561(2)	41f
$HgCl_5^{-3}(d^{10})$	2.518(4)	2.640(4)	46

[a] Average of several nonequivalent distances.

(Table VII). The tendency for the five-coordinate Hg complex to lose three ligands and become two-coordinate is reflected in the relative Hg—Cl bond lengths, and we have noted previously similar behavior in three-coordinate third-row d^{10} complexes. The usual explanation for this behavior [see Clegg (45a)] is the small d-s orbital separation at the bottom of the periodic table, i.e., heavy involvment of higher orbitals in determining the angular geometry. The trigonal bipyramidal $ZnCl_5^{-3}$ molecule is unfortunately unknown.

We may use similar arguments to conclude that six-coordinate octahedral low-spin d^6 ML_4L_2' are most stable in the cis configuration as is found to be largely the case, especially with first-row transition metals. Thus, cis-$Co(CN)_4(H_2O)_2^-$ is more stable (62) than the trans arrangement, although at 25°C in solution the two isomers are rapidly interconverting. The energy difference between them is probably small (cf. the d^6 square pyramid described in the foregoing). Similarly, for $M(CO)_4I_2$ (M = Fe, Ru, Os) the cis form is the more stable isomer (63). In a large number of FeL_4H_2 complexes (64) (L = phosphine, phosphite, CO), the cis form is of lower energy although the molecules are fluxional. For complexes with the large cage ligands P_4S_3, the cis forms are also the more stable (65) for $M(CO)_4(P_4S_3)_2$ (M = Cr, Mo). Here the trans isomer might be more likely on steric grounds. Similarly, the most stable ML_3L_3' species for low-spin d^6 should be the fac arrangement, which preserves the C_{3v} arrangement of strong-field ligands. Thus, $Co(CN)_3(H_2O)_3$ (62), $Cr(CO)_3(PH_3)_3$ (66), and $M(CO)_3(P_4S_3)_3$ (M = Cr, Mo) (65) are all found in this arrangement.

An interesting application of these ideas is the recent work on the rate of incorporation of ^{13}CO into $Mn(CO)_5Br$, $Re(CO)_5Br$, and $Cr(CO)_4$ chelate complexes (67). Thermally ^{13}CO is much more rapidly incorporated at the cis position of all three complexes than at the trans sites. By the microscopic reversibility principle, therefore, it is precisely at these sites where the ^{12}CO is preferentially labilized. We may view this in the following way. The lower energy of the two possible, square pyramidal intermediates in the case of $M(CO)_5Br$ will be the one that contains the cis-divacant arrangement of strong-field ligands as in $Cr(CO)_4$ itself. In these low-spin, d^6, five-coordinate fragments the weaker σ donor is, therefore, relegated to an equatorial position. Trans labilization, however, leaves behind a square planar strong-field ligand arrangement (IX), which is the less stable of the two. In the case of the tetracarbonyl chelate then, it is the "fac trivacant" structure which is the more stable tricarbonyl geometry (see the structure of $Cr(CO)_3$ noted above) and this is reached also by cis labilization from the six-coordinate structure.

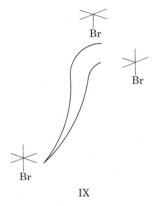

IX

I. Intramolecular Photochemical Rearrangements

The fact that the geometry of the transition metal complex depends on the precise arrangement of the electrons in the metal d orbitals means that in some cases the geometry and properties of an excited electronic state is different from that found for the electronic ground state. (We have already seen an interesting photoassociation reaction in Section II, G.) Thus, photochemical excitation is often sufficient to allow rearrangement processes to occur in otherwise stereochemically rigid systems. For example, the equilibrium geometry of the electronic ground state of the low-spin d^8 ML_2L_2' system is the square planar arrangement (cf. $Ni(CN)_4^{-2}$). The first excited state, however, with orbital occupation numbers (211) prefers the distorted tetrahedral arrangement typical of *high-spin* d^8 complexes. This rearrangement process in the excited state leads to facile (*68*) cis-trans photochemical isomerisation in solution:

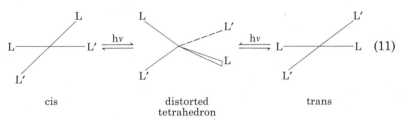

$$\qquad(11)$$

cis   distorted   trans
                       tetrahedron

The ground-state interconversion via the distorted tetrahedron is prevented by a large barrier (see in the following). Similarly, although the electronic ground state of low-spin d^6 $Cr(CO)_5$ is square pyramidal (in the low-temperature matrix, any thermal fluxional behavior is frozen out), the geometry of the first excited state (intermediate-spin

d^6) is a trigonal bipyramid from Table IV. Thus, on photoexcitation, the square pyramid may rearrange via a process much akin to the Berry process proposed for $Fe(CO)_5$. There is an exchange of axial and basal ligands,

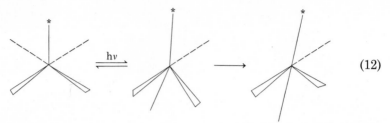

$$(12)$$

and this has been elegantly confirmed experimentally by the photochemical interconversion of the two isomers of $Cr(CO)_4CS$ in a low-temperature matrix (69).

The different spatial orientation of the right-hand product compared to starting material, indicated schematically in Eq. (12), has been followed (70) using polarized spectroscopy and photolysis for $Cr(CO)_5$ where there is no chemical "label" in the molecule. Both of these rearrangement processes are described by potential surfaces akin to the well-known cis ↔ trans ethylene interconversion (71). Figure 13 illustrates schematically for the square pyramidal/trigonal bipyramidal case an electronic ground state trapped in one of the two wells and

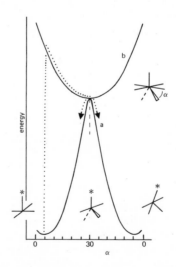

FIG. 13. Energy diagram showing interconversion of square pyramid and trigonal bipyramid for (a) d^6 low-spin ground state and (b) d^6 intermediate-spin excited state.

unable, because of the barrier between them, to interconvert thermally. These wells would equally well represent the square plane for Ni(II) L_2L_2'. The interconversion coordinate connecting the two wells is via the trigonal bipyramid (or distorted tetrahedron in the square planar examples). In the excited state, the equilibrium geometry is different from that in the ground state and the intramolecular rearrangement can proceed along the dotted pathway. (Some of the excited molecules will, of course, return to the original configuration.)

A third rearrangement process (62, 72), that of cis → trans $Co(CN)_4(H_2O)_2{}^-$ and fac → mer $Co(CN)_3(H_2O)_3$ is also approachable (58) on the scheme. In the ground electronic state (low-spin d^6), as we have just noted, the cis and fac structures are more stable than their trans and mer analogs. In the first excited electronic state with orbital occupation numbers 210 or perhaps 110, the most stable geometry of the systems are the trans and mer isomers, however. On return to the electronic ground state, it is these geometries which are frozen out and slowly revert thermally to their more stable ground-state analogs.

III. Conclusion

We have seen that the d-orbital-only model of structural transition metal chemistry is remarkably successful in rationalizing a wide spectrum of observations from structural and kinetic transition metal chemistry. It is possibly even more remarkable because of the smaller contribution to the total stabilization energy (in the $M(H_2O)_6{}^{2+}$ series at least) of the nd orbitals compared to the $(n + 1)$s and $(n + 1)$p orbitals. Its validity we suspect will be better for first-row transition metal systems compared to complexes containing the heavier elements since, as we have noted herein, higher orbitals exert a significant distorting effect in Hg(II) and Au(I) chemistry. Although we should certainly consider the influence of these higher orbitals in any detailed discussion of these structural effects (see, for example, the finely worked reasoning in Ref. 14), use of the d-orbital-only model provides a simple wide-ranging basic approach.

REFERENCES

1. Sidgwick, N. V., and Powell, H. M., *Proc. Roy. Soc. London, Ser. A* **176**, 153 (1940).
2. (a) Gillespie, R. J., and Nyholm, R. S., *Q. Rev., Chem. Soc.* **11**, 339(1957); (b) Gillespie, R. J., *J. Chem. Educ.* **40**, 295 (1963); (c) *J. Chem. Soc.* pp. 4672, 4679 (1963); *Can. J. Chem.* **38**, 818 (1960); **39**, 318 (1961).
3. Gillespie, R. J., "Molecular Geometry." Van Nostrand-Rheinhold, London, 1972.
4. Walsh, A. D., *J. Chem. Soc.* pp. 2260, 2266, 2288, 2296, 2301 (1953).
5. (a) Gimarc, B. M., *J. Am. Chem. Soc.* **92**, 266 (1970); (b) **93**, 593 (1971); (c) **93**, 815 (1971).

6. Rauk, A., Allen, L. C., and Mislow, K., *J. Am. Chem. Soc.* **94**, 3035 (1972).
7. Gavin, R. M., *J. Chem. Educ.* **46**, 413 (1969).
8. Hoffmann, R., Howell, J. M., and Muetterties, E. L., *J. Am. Chem. Soc.* **94**, 3047 (1972).
9. (a) Bartell, L. S., *J. Chem. Educ.* **45**, 754 (1968); (b) Bartell, L. S., and Gavin, R. M., *J. Chem. Phys.* **48**, 2466 (1968).
10. (a) Pearson, R. G., *J. Am. Chem. Soc.* **91**, 4947 (1969); (b) *J. Chem. Phys.* **53**, 2986 (1970).
11. (a) Pearson, R. G., *J. Am. Chem. Soc.* **91**, 1252 (1969); (b) **94**, 8287 (1972); (c) *Acc. Chem. Res.* **4**, 152 (1971).
12. See, for example, Burdett, J. K., Dubost, H., Poliakoff, M., and Turner, J. J., *in* "Advances in Infra-red and Raman Spectroscopy" (R. J. H. Clark and R. E. Hester, eds.) Vol. 11, Heyden, London, 1976.
13. Burdett, J. K., *J. Chem. Soc., Faraday 2* **70**, 1599 (1974).
14. (a) Elian, M., and Hoffmann, R., *Inorg. Chem.* **14**, 1058 (1975); (b) Rossi, A. R., and Hoffmann, R., *ibid.* **14**, 305 (1975).
15. Rosch, N., and Hoffmann, R., *Inorg. Chem.* **13**, 2656 (1974).
16. (a) Hoffmann, R., Chen, M. L., Elian, M., Rossi, A. R., and Mingos, D. M. P., *Inorg. Chem.* **13**, 2666 (1974); (b) Enemark, J. H., and Feltham, R. D., *J. Am. Chem. Soc.* **96**, 5003, 5004 (1974).
17. (a) Schaffer, C. E., *Struct. Bonding (Berlin)* **14**, 69 (1973); (b) *Proc. Roy. Soc., London A* **297**, 96 (1967); (c) Schaffer, C. E., and Jorgensen, C. K., *Mol. Phys.* **9**, 401 (1965); (d) Schaffer, C. E., *Pure Appl. Chem.* **24**, 361 (1970); (e) Gerloch, M., and Slade, R. C., "Ligand Field Parameters." Cambridge Univ. Press, London and New York, 1973; (f) Kettle, S. F. A., *J. Chem. Soc. A* 420 (1966); (g) Larsen, E., and La Mar, G. N., *J. Chem. Educ.* **51**, 633 (1974); (h) For application of the angular overlap model to main group stereochemistry, see Burdett, J. K., *Struct. Bonding (Berlin)* **31**, 67 (1976).
18. Bartell, L. S., and Plato, V., *J. Am. Chem. Soc.* **95**, 3097 (1973).
19. McClure, D. S., *in* "Advances in the Chemistry of the Coordination Compounds" (S. Kirschner, ed.) Macmillan, New York, 1961.
20. Smith, W., and Clack, D., *Rev. Roum. Chim.* **20**, 1243 (1975).
21. Burdett, J. K., *J. Chem. Soc., Dalton Trans.*, p. 1725 (1976).
22. See, for example, Cotton, F. A., and Wilkinson, G., *in* "Advanced Inorganic Chemistry." Wiley, New York, 1972.
23. Cotton, F. A., *J. Chem. Educ.* **41**, 466 (1964).
24. Burdett, J. K., *Inorg. Chem.* **14**, 375 (1975).
25. Eigen, M., *Pure Appl. Chem.* **6**, 105 (1963).
26. See also Bennett, H. D., and Caldin, B. F., *J. Chem. Soc. A* p. 2198 (1971).
27. Philipps, C. S. G., and Williams, R. J. P. "Inorganic Chemistry" Oxford Univ. Press, London and New York, 1966.
28. (a) Willett, R. D., *J. Chem. Phys.* **41**, 2243 (1964); (b) Morosin, B., and Lingafelter, *Acta Crystallogr.* **13**, 807 (1960); (c) Issacs, N. W., Kennard, C. H. L., and Wheeler, D. A., *J. Chem. Soc.A* 386 (1969).
29. (a) Willett, R. D., Ferraro, J. R., and Choca, M., *Inorg. Chem.* **13**, 2919 (1974); (b) Wang, P. J., and Drickamer, H. G., *J. Chem. Phys.* **59**, 559 (1973).
30. Ferraro, J. R., and Long, J., *Acc. Chem. Res.* **8**, 171 (1975).
31. (a) Crichton, O., Poliakoff, M., Rest, A. J., and Turner, J. J., *J. Chem. Soc., Dalton Trans.* p. 1321 (1975); (b) Ozin, G. A., *in* "Vibrational Spectroscopy of Trapped Species" (H. E. Hallam, ed.), Wiley, New York, 1973; (c) Huber, H., Hanlan, L., McGarvey, B., Kundig, E. P., and Ozin, G. A., *J. Am. Chem. Soc.* **97**, 7054 (1975).

32. Perutz, R. N., and Turner, J. J., *J. Am. Chem. Soc.* **97**, 4800 (1975).
33. (a) Poliakoff, M., and Turner, J. J., *J. Chem. Soc.* 1351 (1973); (b) 2276 (1974).
34. (a) Perutz, R. N., and Turner, J. J., *Inorg. Chem.* **14**, 262 (1965); (b) Burdett, J. K., Graham, M. A., Perutz, R. N., Poliakoff, M., Rest, A. J., Turner, J. J., and Turner, R. F., *J. Am. Chem. Soc.* **97**, 4805 (1975); (c) A trigonal bipyramidal geometry was incorrectly assigned by Kundig, E. P., and Ozin, G. A., *ibid.* **96**, 3820 (1974); (d) Perutz, R. N., and Turner, J. J., *ibid.* **97**, 4791 (1975); (e) Graham, M. A., Poliakoff, M., and Turner, J. J., *J. Chem. Soc. A* p. 2939 (1971).
35. Huber, H., Kundig, E. P., and Ozin, G. A., *J. Am. Chem. Soc.* **96**, 5585 (1974).
36. Huber, H., Kundig, E. P., Ozin, G. A., and Poe, A. J., *J. Am. Chem. Soc.* **97**, 308 (1975).
37. Veillard, A., private communication.
38. Beagley, B., Cruickshank, D. W. J., Pinder, P. M., Robiette, A. G., and Sheldrick, G. M., *Acta Crystallogr., Sect. B* **25**, 737 (1969).
39. Frenz, B. A., and Ibers, J. A., *Inorg. Chem.* **11**, 1109 (1972).
40. Berry, R. S., *J. Chem. Phys.* **32**, 933 (1960).
41. (a) Raymond, K. N., Corfield, D. W. R., and Ibers, J. A., *Inorg. Chem.* **7**, 1362 (1968); (b) Basile, L. J., Ferraro, J. R., Choca, M., and Nakamoto, K., *ibid.* **13**, 496 (1974); (c) Cotton, F. A., Dunne, T. G., and Wood, J. S., *ibid.* **4**, 318 (1965); (d) Raymond, K. N., Meek, D. W., and Ibers, J. A., *ibid.* **7**, 1111 (1968); (e) Goldfield, S. A., and Raymond, K. N., *ibid.* **10**, 2604 (1971); (f) Long, T. V., Herlinger, A. W., Epstein, E. F., and Bernal, I., *ibid.* **9**, 459 (1970).
42. Burdett, J. K., and Turner, J. J., *in* "Cryogenic Chemistry" (G. A. Ozin and M. Moskovits, eds.) Wiley, New York, 1976.
43. (a) Dekock, R. L., *Inorg. Chem.* **10**, 1205 (1971); (b) Kundig, E. P., Moskovits, M., and Ozin, G. A., *Can. J. Chem.* **51**, 2710 (1975); (c) Huber, H., Kundig, E. P., Moskovits, M., and Ozin, G. A., *J. Am. Chem. Soc.*, **95**, 332 (1973); (d) Kundig, E. P., Moskovits, M., and Ozin, G. A., *J. Mol. Struct.* **14**, 137 (1972).
44. Poliakoff, M., *J. Chem. Soc., Dalton Trans.* 210 (1974).
45. (a) Clegg, W., *Acta Crystallogr., Sect. B* **32**, 2712 (1976); (b) Guggenberger, L. J., *J. Organomet. Chem.* **81**, 271 (1974); (c) Baenziger, N. C., Dittemore, K. M., and Doyle, J. R., *Inorg. Chem.* **13**, 805 (1974); (d) Wijhoven, J. G., Bosman, W. P. J. H., and Beurskens, P. T., *J. Cryst. Mol. Struct.* **2**, 7 (1972); (e) Canty, A. J., Marker, A., and Gatehouse, B. M., *J. Organomet. Chem.* **88**, C31 (1975).
46. Clegg, W., Greenhalgh, D. A., and Straughan, B. P., *J. Chem. Soc., Dalton Trans.* 2591 (1975).
47. (a) Seip, H. M., and Seip, R., *Acta Chem. Scand.* **20**, 2698 (1966); (b) Kimura, M., Schomaker, V., Smith, D. W., and Weinstock, B., *J. Chem. Phys.* **48**, 4001 (1968); (c) Jacob, E. J., and Bartell, L. S., *ibid.* **53**, 2231 (1970).
48. Weinstock, B., and Goodman, G. L., *Advan. Chem. Phys.* **9**, 169 (1966).
49. Jahn, H. A., and Teller, E., *Proc. Roy. Soc., London, Ser. A* **161**, 220 (1937).
50. Jotham, R. W., and Kettle, S. F. A., *Inorg. Chim. Acta* **5**, 183 (1971).
51. Brown, L. D., and Raymond, K. N., *J. Chem. Soc., Chem. Commun.* **910** (1974).
52. Bennett, M., and Mason, R., *Nature (London)* **205**, 760 (1965).
53. (a) Perutz, M. F., *Nature (London)* **228**, 726 (1970); (b) **237**, 495 (1972).
54. Williams, R. J. P., *in* "Iron in Biochemistry and Medicine" (A. Jacobs and M. Worwood, eds.) Academic Press, New York, 1974.
55. (a) Higginson, W. C. E., Nyburg, S. C., and Wood, J. S., *Inorg. Chem.* **3**, 463 (1964); (b) Nyburg, S. C., and Wood, J. S., *ibid.* **3**, 468 (1964).
56. (a) Ng, Y. S., Rodley, G. A., and Robinson, W. T., *Inorg. Chem.* **15**, 303 (1976); (b) Sacconi, L., *Coord. Chem. Rev.* **8**, 351 (1972).
57. Bedford, W. M., and Rouchias, G., *J. Chem. Soc., Chem. Commun.* p. 1224 (1972).

58. Burdett, J. K., *Inorg. Chem.* **15**, 212 (1976).
59. Orioli, P. L., *Coord. Chem. Rev.* **6**, 285 (1971).
60. Burdett, J. K., *Inorg. Chem.* **14**, 931 (1975).
61. Robiette, A. G., private communication (1976).
62. (a) Viane, L., D'Olieslager, J., and De Jaegere, S., *Inorg. Chem.* **14**, 2736 (1975); (b) *J. Inorg. Nucl. Chem.* **37**, 2435 (1975).
63. (a) Pankowski, M., and Bigorgne, M., *J. Organomet. Chem.* **19**, 393 (1969); (b) Johnson, B. F. G., Lewis, J., Robinson, D. W., and Miller, R., *J. Chem. Soc. A* p. 1043 (1968).
64. Muetterties, E. L., *Acc. Chem. Res.* **3**, 266 (1970).
65. Jefferson, R., Klein, H. F., and Nixon, J. F., *J. Chem. Soc., Chem. Commun.* p. 536 (1969).
66. Fischer, E. O., Louis, E., and Kreiter, C. G., *Angew. Chem., Int. Ed. Engl.* **8**, 377 (1969).
67. (a) Attwood, J. D., and Brown, T. L., *J. Am. Chem. Soc.* **97**, 3380 (1975); (b) **98**, 3155, 3160 (1976); (c) Cohen, M. A., and Brown, T. L., *Inorg. Chem.* **15**, 1417 (1976).
68. (a) Haake, P., and Hylton, T. A., *J. Am. Chem. Soc.* **84**, 3774 (1962); (b) Balzani, V., Carassiti, V., Moggi, L., and Scandola, F., *Inorg. Chem.* **4**, 1243 (1965); (c) McGarvey, J. J., and Wilson, J., *J. Am. Chem. Soc.* **97**, 2531 (1975); (d) Eaton, D. R., *ibid.* **90**, 4272 (1968); (e) Whitesides, T. H., *ibid.* **91**, 2395 (1969).
69. Poliakoff, M., *Inorg. Chem.* **15**, 2022, 2892 (1976).
70. (a) Burdett, J. K., Perutz, R. N., Poliakoff, M., and Turner, J. J., *J. Chem. Soc., Chem. Commun.* p. 157 (1975). (b) Burdett, J. K., Gryzbowski, J. M., Perutz, R. N., Poliakoff, M., Turner, J. J., and Turner, R. F., *Inorg. Chem.* **17**, 147 (1978).
71. Salem, L., and Stohrer, W. D., *J. Chem. Soc., Chem. Commun.* p. 140 (1975).
72. Viane, L., D'Olieslager, J., and De Jaegere, S., (unpublished results).
73. Yared, Y. W., Miles, S. L., Bav, R., and Reed, C. A., *J. Am. Chem. Soc.* **99**, 7076 (1977).

ADDUCTS OF THE MIXED TRIHALIDES
OF BORON

J. STEPHEN HARTMAN and JACK M. MILLER

Department of Chemistry, Brock University, St. Catharines, Ontario, Canada

I. Introduction 147
II. Preparation, Detection, and Properties of Mixed Boron Trihalide Adducts 149
    A. Methods of Formation 149
    B. Detection. 151
    C. Properties of Specific Systems 152
    D. Comparison with BH_2X and BHX_2 Adduct Systems. 157
III. Mechanisms of Halogen Redistribution 158
    A. Dissociation of the Donor–Acceptor Bond 158
    B. Dissociation of Halide Ion 159
    C. Associated Bridging Mechanism 160
    D. Exchange Mechanisms in Tetrahaloborates 161
IV. Equilibria in Halogen Redistribution 162
V. Donor-for-Halogen Exchange: Difluoroboron Cations 166
VI. NMR Applied to Adducts: Advantages and Pitfalls 167
    A. Reliability of Mixed-Adduct NMR Data 167
    B. Complexation Shifts and Donor–Acceptor Bond Strength 168
    C. "Pairwise Interaction" NMR Parameters 171
VII. Conclusion 172
    References 172

I. Introduction

The chemistry of the boron trihalides has been extensively studied. These compounds are strong Lewis acids and form a wide range of simple 1:1 adducts that have served as model compounds for the study of Lewis acid–base interactions. Many reviews of their coordination chemistry have appeared (41, 60, 66, 120, 174, 178) and these are summarized in the more recent reviews appearing in the mid-1960s (41, 120).

In boron trifluoride adducts, fluorine displacement is relatively difficult and, thus, the widest range of adducts has been prepared for the trifluoride (66, 120). As fluorine is replaced by successively heavier halogens the displacement of halogen occurs more readily and secondary reactions of the initially formed donor–acceptor adducts

147

become more important. Whereas the initial formation of a donor–acceptor bond is a simple process, the secondary reactions of the adducts need not be simple. Many such reactions, and related catalytic activity of donor–acceptor adducts of the boron trihalides, have been reported (*41, 60, 66, 120, 174, 178*).

Uncomplexed boron trihalides undergo halogen-exchange reactions that are too rapid to allow isolation of the individual mixed halogen species formed (*113, 114, 120*). This rapid redistribution of halogens apparently deterred early investigation of adducts of the mixed boron trihalides, and these obvious but elusive compounds remained unknown until the late 1960s. The application of nuclear magnetic resonance spectroscopy during the 1960s greatly simplified the study of even very rapid redistribution reactions (*127, 128*). It became obvious that redistribution reactions are widespread and tend to occur more rapidly when the species involved are not coordinatively saturated (*31*). The nature of the substituents is also important in determining rates of redistribution, with alkyl groups tending to redistribute much more slowly than halogens (*113, 114*). Redistribution reactions in many three-coordinate boron compounds have been studied (*68, 136, 147, 168*). These can be either fast or slow depending on substituents. A dimethylamino group slows halogen redistribution, and the mixed-halogen compound $Me_2NBClBr$ can be isolated (*68*).

Redistribution of substituents tends to be especially facile for halides, hydrides, and alkyls of Groups I–III nontransition elements because these compounds are electron-deficient. Bridging groups are present in many of these compounds. Even in the boron trihalides that are not bridged, a bridged transition state making use of the empty valence shell orbitals is possible, so that redistribution can occur with a relatively low activation energy (*113*):

$$BX_3 + BY_3 \rightleftharpoons \left[X_2B \underset{Y}{\overset{X}{\diamond}} BY_2 \right] \rightleftharpoons BX_2Y + BXY_2 \qquad (1)$$

Such a low-energy reaction pathway is not available for four-coordinate boron compounds that are coordinatively saturated.

The donor–acceptor bond in many boron trihalide adducts is fairly weak (*132, 170*) and for such adducts a characteristic reaction is the rapid breaking and re-forming of this bond (*39, 62*). Nuclear magnetic resonance studies of the exchange of BX_3 among an excess of Lewis base molecules show that the dissociative mechanism of exchange predominates in a number of cases (*14, 16, 39, 55, 115, 151*). Breaking and re-forming of boron–halogen bonds is a less obvious process, although it was known to occur in the heavier tetrahaloborate anions (172) and in BF_4^- (63).

In 1968 reports of halogen redistribution in neutral boron halide adducts appeared. Chlorine–bromine exchange was detected in the formation and disproportionation of a mixed dihaloborane adduct, $PH_3 \cdot BHClBr$ (50), by observation of separate ^1H NMR peaks due to the $BHCl_2$, $BHClBr$, and $BHBr_2$ adducts at low temperatures. Fluorine exchange between boron atoms was detected in ketone–BF_3 adducts (62) as well as in the tetrafluoroborate anion (63) by ^{19}F NMR. This reaction did not give mixed-halogen species but was detectable through the collapse of the ^{10}B—^{11}B isotope shift (7, 25) in the ^{19}F spectra at higher temperatures, due to rapid and random fluorine exchange between ^{10}B and ^{11}B. In the ketone–BF_3 systems, the fluorine exchange reaction is much slower than the breaking and re-forming of the weak donor–acceptor bond (62).

This work indicated a widespread halogen-exchange process in boron trihalide adducts and the likelihood of easy formation of many adducts of the mixed trihalides of boron. Extensive work being done at the time on nucleophilic displacement of halogen in monohaloborane adducts, $D \cdot BH_2X$ (X = Cl, Br, I) (152), also suggested the susceptibility to displacement of halogens by other halogens in boron trihalide adducts. Since then, application of NMR techniques to the study of halogen redistribution in boron trihalide adducts has provided most of the available information on the mixed boron trihalide adducts, in spite of complication of the studies by other rapid exchange reactions in many of the systems.

II. Preparation, Detection, and Properties of Mixed Boron Trihalide Adducts

A. METHODS OF FORMATION

Rapid halogen redistribution at ambient temperature precludes isolation of many mixed-halogen boron trihalide adducts, and most such adducts have been observed only in solution in inert solvents such as methylene chloride. Amine adducts of the mixed boron trihalides are stable to halogen disproportionation and a few of these have been isolated (3, 10, 155) as have compounds containing mixed tetrahaloborate anions (120, 183). Other nitrogen-donor mixed boron trihalide adducts could no doubt be isolated, by analogy with nitrogen-donor adducts of BH_2X and BHX_2 (Section II,D).

The following methods have been used to prepare the mixed boron trihalide adducts $D \cdot BX_nY_{3-n}$ for in situ study by spectroscopic methods.

1. $D \cdot BX_3 + D \cdot BY_3$

If the donor–acceptor bond is weak, two different boron trihalide adducts of the same donor will exchange halogen to form the mixed-halogen adducts. The weaker the donor–acceptor bond, the more rapidly the mixed adducts form, indicating that free boron trihalide formed by dissociation is an active species in halogen exchange.

2. $D \cdot BX_3 + BY_3$

This method is a general route to mixed boron trihalide adducts. Reaction occurs readily in solution at ambient temperature even with amine adducts that do not exchange halogen in the absence of free boron trihalide (*10*). Adducts that do exchange halogen slowly in the absence of uncomplexed trihalide react much faster when uncomplexed trihalide is added (*27*). This method gives mixed $Me_3N \cdot BF_nI_{3-n}$ adducts (*10*), whereas free BF_2I and BFI_2 have never been detected except as positive ions in the mass spectrometer (*110, 111*).

3. $D + BX_nY_{3-n}$ ($n = 0\text{--}3$)

Reaction of a donor molecule with a previously equilibrated mixture of free boron trihalides gives an initial adduct mixture corresponding to the equilibrium mixture of the free boron trihalides. This method should be suitable for all mixtures except BF_3/BI_3 (as noted previously). However, equilibria in the adducts can be quite different from the corresponding equilibria in the free boron trihalides. If halogen redistribution is fast and if the mixed adducts are discriminated against, then this method does not succeed (*28*).

4. $D \cdot BX_3 + Y^-$

Halide ion Y^- in the form of soluble tetraalkylammonium salts can in some cases displace halide ion X^- forming a mixed-halogen adduct. However, the method has met with variable success. It is successful with tetrahaloborate ions (*80*) and with $Me_3N \cdot BI_3$ (*9*). A similar reaction with insoluble silver halides or pseudohalides has also given mixed adducts (*43*).

5. $BX_3 + Y^-$

Uncomplexed BX_3 interacts with triphenylmethyl halides, $(C_6H_5)_3CY$, to give the BX_3Y^- salt (*120, 183*). Halide ion Y^- can also be extracted from metal complexes by BX_3 to give species that are probably BX_3Y^- salts (*51*).

6. Reactions of Diboron Tetrafluoride Adducts

The reaction of HCl with the bis(trimethylamine) adduct of B_2F_4 gives $Me_3N \cdot BF_2Cl$ (3).

7. Other Halogen Sources

Trimethylsilyl halides will exchange halogen with $Me_3N \cdot BX_3$ adducts (9) and BF_4^- (80). Phosphorus pentahalides (9, 13), Al_2Cl_6 (9), and $AsCl_3$ (9) act similarly on amine adducts.

Tetrahaloborate ions undergo halogen exchange with methylene halides to give mixed tetrahaloborate anions (80). Because these ions are appreciably dissociated in solution (172), the reaction appears to involve halide ion attack on CH_2X_2. The complexed boron trihalide might also exchange halogen with halocarbons (64, 74).

B. DETECTION

Of the *in situ* detection methods, infrared spectroscopy was very useful in studying halogen redistribution in the uncomplexed boron trihalides (69, 92, 137). However, the mixed-halogen adducts have very low symmetry and give very complex spectra, complicated further by overlapping ligand absorptions. Thus, $Me_3N \cdot BF_2Br$ and $Me_3N \cdot BFBr_2$ gave infrared spectra that were too complex to assign by analogy with unmixed-halogen Me_3N adducts (10). An investigation of the carbonyl stretching frequency of carbonyl donor–mixed boron trihalide adducts has been carried out (87), but even here where a single strong infrared absorption can be assigned to each mixed-halogen adduct, peak overlap was serious; NMR gave more information. Raman spectroscopy has not yet been applied to the mixed adducts. Due to their higher symmetry, the mixed tetrahaloborate anions are more amenable to study by infrared and Raman spectroscopy. Such studies have been carried out on other mixed-halogen tetrahedral species, such as $AlCl_nBr_{4-n}^-$ (18) and $TiCl_nBr_{4-n}$ (35) where all members of the series ($n = 0$–4) are present together in solution, and equilibrium constants for the rapid redistribution reactions have been obtained.

Mass spectroscopy can be applied to study of the adducts only when the donor–acceptor bond is strong enough to remain intact in the gas phase under the vigorous conditions of electron impact. Thus only when the donor–acceptor bond is strong enough to allow isolation of individual mixed-halogen species does it also remain intact to a useful extent in the mass spectrometer (108); even then, halogen scrambling in the ion source is a complication (126). It has not proved possible to apply conventional mass spectroscopy to the study of the weak adducts that undergo halogen redistribution rapidly. Field desorption or field ionization mass spectroscopy might, however, be applicable.

Nuclear magnetic resonance spectroscopy has the advantage that spectra of several different nuclei may be observed in the same adduct system, providing independent verification of the species present. For adducts of organic donors, at least ^{1}H, ^{13}C, and ^{11}B nuclei are suitable for NMR studies, and ^{19}F is especially suitable for the study of fluorine-containing adducts. Resolution is generally good. Moreover, heteronuclear coupling patterns (^{1}H—^{11}B and ^{11}B—^{19}F coupling in particular) provide further confirmation of peak assignments in many cases. The importance of NMR in studying rates and kinetics of rapid reactions is well known (94). The principal experimental method in these studies has been NMR, as in extensive recent work on other rapid redistribution reactions (114, 127, 128).

C. PROPERTIES OF SPECIFIC SYSTEMS

Boron trihalide adducts of different classes of donor molecule show striking differences in both rates and equilibria of halogen redistribution. Only with phosphine adducts has this reaction been studied for a significant number of different donors of the same class, and significant variations in behavior are found even within this class. Halogen redistribution can be a complex reaction and can be strongly affected by the presence of other species in solution. As expected, the mixed boron trihalide adducts are intermediate between the corresponding unmixed adducts in properties such as NMR chemical shifts, carbonyl stretching frequencies of carbonyl-donor adducts, and donor–acceptor bond strength (where this is indicated by mass spectrometry or by equilibration studies with a limited quantity of Lewis base).

1. Ether Adducts

The first neutral mixed boron trihalide adducts were detected by NMR with dimethyl ether as donor (27, 73). Near-statistical amounts of mixed- and unmixed-halogen species are formed in the $Me_2O \cdot BF_nCl_{3-n}$ system. Equilibration is almost complete in an hour at 0°C when uncomplexed dimethyl ether is present, and is extremely rapid when excess boron trihalide is present.

Dialkyl ether adducts are typical of many boron trihalide adduct systems in that the BF_3 adducts tend to be stable, whereas the BCl_3, BBr_3, and BI_3 adducts become successively more reactive, decomposing to alkyl halide and alkoxyboron dihalide (60, 120). This instability of the boron–halogen bonds persists in the mixed-halogen adducts, limiting the study of bromine-containing mixed species and preventing study of iodine-containing mixed species. Although decomposition prevents attainment of equilibrium in bromine-containing adducts, all of the non-iodine-containing adduct species, including the ternary halogen species $Me_2O \cdot BFClBr$, have been detected.

Preliminary studies of mixed boron trihalide adducts of other ethers have been carried out (78), but many of these adducts are highly susceptible to decomposition.

2. Amine Adducts

Mixed boron trihalide adducts of trimethylamine have been extensively studied (9–11, 13, 75, 105, 108, 155). Halogen exchange among adducts does not occur spontaneously. The mixed boron trihalide adducts form readily near ambient temperature by reaction of the unmixed-halogen adducts with suitable halogenating agents such as the free boron trihalides (9–11, 75, 105), PCl_5 (9, 13) and arsenic, aluminum, and silicon chlorides (9). Lighter-halogen adducts have their halogens readily replaced by heavier halogens, but the reverse reactions are slower (9). Complexes $Me_3N \cdot BF_2Br$ and $Me_3N \cdot BFBr_2$ have been isolated from the series $Me_3N \cdot BF_nBr_{3-n}$ by vacuum sublimation and recrystallization (10) and are stable to disproportionation. The complex $Me_3N \cdot BF_2Cl$ has been obtained pure from the reaction of HCl on $Me_3N \cdot F_2BBF_2 \cdot NMe_3$ (3). All of the possible mixed boron trihalide adducts of Me_3N, including the four ternary halogen species $Me_3N \cdot BFClBr$, $Me_3N \cdot BFClI$, $Me_3N \cdot BFBrI$, and $Me_3N \cdot BClBrI$, have been detected in solution by NMR (10).

Mixed boron trihalide adducts of amines must have been present as intermediates in earlier halogen displacement studies (40, 124), but the experimental methods used led to detection of only the final unmixed-halogen product.

Boron trihalide adducts of 4-methylpyridine show halogen exchange behavior similar to that of the adducts of trimethylamine (10), and isolation of individual mixed-halogen species is feasible here too. The stability of nitrogen-donor adducts of the mixed boron trihalides parallels the stability of nitrogen-donor adducts of BH_2X and BHX_2 which have been more extensively studied (19, 153, 156).

The Me_3N adducts of mixed halogen–pseudohalogen boron compounds have a similar stability. Although three-coordinate mixed halogen–pseudohalogen boron species BX_nY_{3-n} have been known for a long time (120), adducts of these have been reported only recently, e.g., $Me_3N \cdot BBr_2CN$ (143), $Me_3N \cdot BCl_2[N(SCF_3)_2]$ (72), $R_3N \cdot BX_n(NCS)_{3-n}$

(R = Me, Et; X = Cl, Br, I; n = 1, 2) (*129*), and $Et_3N \cdot BCl_2X$ where X is $(CO)_9Co_3CO$, with the oxygen of the triply bridged carbonyl coordinated to boron (*8*). However, amine·BF_2OR adducts disproportionate above $-30°C$ in solution to give trialkoxyboron and the trialkylamine–BF_3 adduct (*176*).

Exchange of alkyl groups (R = propyl, butyl) with halogen (X = F, Cl) in $Me_3N \cdot BR_nX_{3-n}$ is slow and, like halogen exchange in $Me_3N \cdot BX_3$ adducts, requires the presence of three-coordinate boron (*37*).

3. Tetrahaloborate Anions

The tetrahaloborate anions BX_4^- constitute a special high-symmetry case of boron trihalide adducts having a halide ion as donor but with all halogens bonded in an equivalent fashion to the central boron. Synthesis and isolation of salts of a number of mixed-halogen species BX_3Y^- were reported many years ago (*120*, pp. 100–101) by methods involving hydrogen halide solvents. However, the best-characterized species, BF_3Cl^-, gave a low analysis for chlorine (*183*), suggesting that halogen redistribution may have occurred and that the salt may have contained some BF_4^- (*80*). These BX_3Y^- salts seemed in general to be difficult to handle and remained poorly characterized (*120*). More recent studies of BF_3Cl^- have indicated that disproportionation readily occurs (*26, 80, 186*). The most convincing evidence for halogen redistribution in the $BF_nCl_{4-n}^-$ (n = 0–4) system was provided by ^{19}F and ^{11}B NMR; methylene chloride solutions of BF_4^- and BCl_4^-, after standing at 60°C, gave the resonances and spin-spin splitting patterns expected for all of the $BF_nCl_{4-n}^-$ (n = 0–4) species (*80*). The redistribution process is complex and rates of redistribution are strongly affected by the presence of other species in solution (*80, 82*). Thus, impurities may account for an early solution NMR study that indicated a very rapid fluorine–chlorine exchange (*102*). Tetrahaloborate halogen redistribution is, in fact, slowest in the $BF_nCl_{4-n}^-$ system (*80*) and is sufficiently slow to be consistent with Waddington's reported isolation of individual mixed tetrahaloborate species (*120, 183*). However, temperatures well below room temperature may be required to "freeze out" all halogen exchange.

Detection of $BF_nCl_{4-n}^-$ species by NMR has been used as evidence for the formation of ionic boron trihalide adducts $D_2BF_2^+ \cdot BF_nCl_{4-n}^-$ from simple covalent adducts (*82, 83*) and as evidence for the abstraction of chloride ion from $(C_6H_5)_3PCl_2$ by BF_3 (*32*). Boron trihalides are often used as halide abstraction reagents, and when two different halogens are involved the mixed-halogen species should be formed, but often only indirect evidence has been obtained (*51*).

Further mixed tetrahaloborates detected by NMR include the $BF_nBr_{4-n}^-$, $BCl_nBr_{4-n}^-$, $BCl_nI_{4-n}^-$, and $BBr_nI_{4-n}^-$ ($n = 0-4$) series, and a number of ternary halogen species such as BF_2ClBr^- (*80*).

Boron trichloride has been reported to react with pseudohalide ions to give the mixed halogen–pseudohalogen ions $BCl_nZ_{4-n}^-$ ($Z = CN^-$, NCO^-, NCS^-, N_3^-), none of which were isolated (*107*). However, only for the $BCl_n(CN)_{4-n}^-$ series were any individual mixed-ion ^{11}B resonances reported. The BX_3OMe^- ions [$X = F$ (*117*), Cl (*54*), Br (*54*)] apparently disproportionate rapidly to give the BX_4^- ions. Early reports of fluorine-containing $BF_nX_{4-n}^-$ species ($X = $ halide, hydroxyl, etc.) have been summarized (*162*).

4. Dimethylsulfide Adducts

All possible non-fluorine-containing boron trihalide adducts of Me_2S, including the ternary halogen adduct $Me_2S\cdot BClBrI$, have been detected by ^1H and ^{11}B NMR (*28*). However, only small amounts of the mixed-halogen $Me_2S\cdot BF_nCl_{3-n}$ adducts could be detected at equilibrium and neither the BF_nBr_{3-n} nor the BF_nI_{3-n} mixed adducts could be detected. These adducts undergo especially rapid breaking and re-forming of donor–acceptor bonds and especially rapid halogen redistribution.

Contrasting behavior of dialkyl ethers and dialkyl sulfides as donors to BF_3 and BCl_3 has been discussed (*28, 189*) and is related to an earlier discussed discrepancy between BF_3 and BH_3 adducts (*65*). Sulfur-donor BF_3 bonds are anomalously weak compared to corresponding donor–acceptor bonds to BH_3 or BCl_3. There is no such anomaly in oxygen-donor bonding to BF_3 where the donor–acceptor bond strength is similar to that of the corresponding BH_3 adduct and not much less than that of the BCl_3 adduct. The peculiar halogen redistribution equilibria in Me_2S adducts appear to be related to these anomalies.

5. Phosphine Adducts

Complexes $Et_3P\cdot BCl_2Br$ and $Et_3P\cdot BClBr_2$ were the first mixed boron trihalide adducts of phosphines to be detected (*100*). Halogen redistribution is near-random in the BCl_nBr_{3-n} adducts of Et_3P (*100*) and Me_3P (*112*). The same is true for Cl—Br, Cl—I, and Br—I redistribution in $MePH_2$ and PH_3 adducts (*49*). However, phosphorus, like sulfur, is a "soft donor" (*140*) and anomalies similar to those in Me_2S adducts appear when fluorine is involved. Compound CD_3PH_2, like Me_2S, forms an anomalously weak BF_3 adduct, while PH_3 does not form a BF_3 adduct (*21, 48, 49*). As with Me_2S, an anomalously weak or nonexistent BF_3 adduct corresponds to anomalously small amounts of BF_nCl_{3-n} and BF_nBr_{3-n} mixed adducts (48, 49). Successive methylation of PH_3 gives successively more stable BF_3 adducts (21) and as the number of methyls is increased there is a corresponding tendency toward larger amounts of the BF_nX_{3-n} mixed adducts at equilibrium; these are no longer markedly discriminated against when Me_3P is the donor (30).

Certain methyldichlorophosphine–boron tribromide adducts undergo chlorine–bromine redistribution between the phosphorus and boron atoms, apparently leading to mixed boron trihalide adducts, although attempts at isolation of these species failed (104). However, spontaneous chlorine–bromine redistribution between P—Cl species and BBr_3 does not seem to occur in $aryl_nPCl_{3-n}·BBr_3$ adducts (133, 135); the mixed adducts were formed by equilibrating BCl_3 and BBr_3 prior to adduct formation (134).

6. Esters and Thioesters

Mixed boron trihalide adducts of esters (29, 78, 87) follow the pattern of oxygen donation as already described for Me_2O adducts. The carbonyl oxygen is the donor (109); halogen redistribution occurs readily to give mixed F, Cl and F, Br species. The heavier halogen adducts are prone to decomposition (60, 61). Mixed boron trihalide adducts of isomeric monothioesters have been studied (29). Replacement of the ether oxygen of methyl acetate by sulfur causes little change in adduct properties or equilibria. However, replacement of the carbonyl oxygen by sulfur causes a shift to the characteristic Me_2S adduct properties; none of the BF_nCl_{3-n} or BF_nBr_{3-n} mixed adducts can be detected. The dithioester shows similar behavior.

7. Tetramethylurea and Tetramethylthiourea

Ureas are oxygen donors in their 1:1 adducts with the boron trihalides (79, 81, 88). Mixed boron trihalide adducts have been observed in the tetramethylurea—BF_3—BCl_3 system and follow the oxygen-donor pattern of Me_2O and ester adducts (82). Thus, the mixed-halogen adducts form spontaneously from the unmixed-halogen adducts. The tetramethylurea·BF_3 adduct is among the few where spontaneous donor-for-fluorine exchange about boron has been observed, giving the ionic 1:1 adduct (tetramethylurea)$_2BF_2^+·BF_4^-$ (Section V) (82). Although in the fluorine-only system the equilibrium is strongly on the side of the simple covalent adduct, the mixed boron trihalide adduct tetramethylurea·BF_2Cl is susceptible to attack by excess tetramethylurea giving (tetramethylurea)$_2BF_2^+·BF_nCl_{4-n}^-$ ($n = 0$–4) (82).

Tetramethylthiourea·BF_3 has some typical sulfur-donor BF_3 adduct characteristics, e.g., very fast exchange between free and complexed donor, and also exchanges donor for halogen to give the ionic species (tetramethylthiourea)$_2BF_2^+·BF_4^-$ (*83*). A brief study of the tetramethylthiourea—BF_3—BCl_3 adduct system showed the presence of BF_2Cl and $BFCl_2$ adducts as well as ionic species (tetramethylthiourea)$_2BF_2^+·BF_nCl_{4-n}^-$ (*83*).

8. Ketones

Fluorine exchange in BF_3 adducts of simple methyl ketones was reported in 1968 (*62*). However the heavier halogen, boron trihalide adducts of simple ketones decompose readily (*60*), and only a preliminary NMR study of fluorine–chlorine exchange in the acetone—BF_3—BCl_3 system has been carried out; this shows that fluorine and chlorine do exchange to give BF_2Cl and $BFCl_2$ adducts of acetone (*78*). Ketones without α-hydrogens form much more stable adducts with the heavier boron trihalides (*60*), and IR and NMR studies of benzophenone and xanthone adducts show the presence of mixed BF_nCl_{3-n} and BF_nBr_{3-n} adducts (*87*).

9. Other Donors

Trimethylphosphine oxide and trimethylphosphine sulfide show a pattern of mixed-halogen adduct formation that is consistent with the oxygen-donor and sulfur-donor patterns described in the foregoing for the Me_2O and Me_2S adducts (*30*). Thus Me_3PS, unlike Me_3PO, gives only small amounts of mixed F, Cl and F, Br adducts. Binder and Fluck have reported BF_nCl_{3-n} mixed adducts of $MeN{=}PCl_3$ (*13*). The ^{11}B NMR peaks, which apparently arise from mixed BCl_nBr_{3-n} adducts of perchlorate ion, formed by halogen exchange with the chloroform solvent, were not recognized as such by Titova *et al.* (*173*). Compound $(Me_2N)_2C{=}CH_2$, which forms carbon-donor boron trihalide adducts (*88*), also forms mixed-halogen adducts (*89*). Rapid fluorine–chlorine redistribution occurs in the $CH_3CN·BF_3/CH_3CN·BCl_3$ system (*77*).

D. Comparison with BH_2X and BHX_2 Adduct Systems

Many BH_2X and BHX_2 (X = F, Cl, Br, I) adducts have been prepared from the corresponding BH_3 adducts by various halogenation procedures (*19, 44, 50, 99, 101, 153, 180, 181*) and are isolable without complications due to halogen–hydrogen redistribution. Amine-cyano-boranes have also been prepared (185). Many of these systems are stabilized by nitrogen donors and, thus, are analogous to the stable mixed boron trihalide adducts of Me_3N and 4-methylpyridine.

Weak donor–BH_nX_{3-n} adducts appear to have properties similar to those of corresponding weak donor–mixed boron trihalide adducts. Species $Et_2O \cdot BH_2Cl$ and tetrahydrofuran $\cdot BH_2Cl$, which have been used as sources of BH_2Cl for hydroboration of olefins (23), are typical weak-donor adducts. The weaker and more readily broken donor–acceptor bond in the Et_2O adduct is the cause of this adduct's greater utility over the tetrahydrofuran adduct in hydroboration. Halogen–hydrogen exchange does not seem to occur readily to give other BH_nCl_{3-n} adducts (23, 24).

Although most known BH_nX_{3-n} adducts are stable and most known BX_nY_{3-n} adducts are unstable to redistribution of substituents about boron, this need not reflect a major difference in behavior of the two types of adduct but may, instead, reflect the interests of the investigators. The BH_nX_{3-n} adduct studies have emphasized synthesis and, hence, systems that are stable to redistribution, whereas mixed boron trihalide adduct studies have emphasized redistribution reactions, which are easier to observe when the donor is weak and redistribution is rapid.

III. Mechanisms of Halogen Redistribution

Halogen exchange in the coordinatively saturated boron trihalide adducts requires either a preliminary dissociative step or an associative interaction different from the doubly halogen-bridged transition state available to three-coordinate boron compounds [Eq. (1)] (113, 114). The reactions can be classified as follows.

A. Dissociation of the Donor–Acceptor Bond

Apparently, equation

$$D \cdot BX_3 \rightleftharpoons D + BX_3 \tag{2}$$

is the rate-determining step for halogen exchange in many adducts of weak donors such as ethers (27), esters (29), ketones (62), ureas (82), phosphines (30, 142), and sulfides (28) when adduct is the only halogen source. Rates of fluorine–chlorine redistribution show an approximate inverse relationship with the donor–acceptor bond strength of the weakest adduct, i.e., the BF_3 adduct, consistent with an initial dissociation step. Thus, for $Me_2S \cdot BF_nCl_{3-n}$ [ΔH_{dissoc} of $Me_2S \cdot BF_3$ = 3.5 kcal/mole (*131*)], redistribution is very rapid at $-78°$ (*28*), whereas, for $Me_2O \cdot BF_nCl_{3-n}$ [ΔH_{dissoc} of $Me_2O \cdot BF_3$ = 13.1 kcal/mole (*132*)], equilibration takes an hour at 0°C (*27*), and, for $Me_3N \cdot BF_nCl_{3-n}$ (ΔH_{dissoc} of $Me_3N \cdot BF_3$ = 30.9 kcal/mole (*132*)], halogen exchange does not occur at room temperature in solution (*10*).

Adduct dissociation would normally be followed by recombination of BX_3 with a donor molecule, and only a small fraction of the BX_3 generated would exchange halogen first. Breaking and re-forming of donor–acceptor bonds is, indeed, a much faster process than halogen redistribution (*27, 84*). The small proportion of free boron trihalide that does not recombine unchanged with donor would probably exchange halogen by the associative bridging mechanism described in Section III,C), although the direct $BX_3 + BY_3$ reaction [Eq. (1)] would be important at high concentrations of uncomplexed BX_3. Addition of excess boron trihalide greatly increases the rate of halogen redistribution in such systems (*27*).

B. DISSOCIATION OF HALIDE ION

Halide ion dissociation

$$D \cdot BX_3 \rightleftharpoons D \cdot BX_2{}^+ + X^- \tag{3}$$

occurs in the tetrahaloborates (*172*) but has not been unambiguously established in neutral adducts. Its probability is greatest with the heaviest halogens because boron–halogen bond strengths decrease in the order F > Cl > Br > I. Krishnamurthy and Lappert (*105*) proposed adduct halogen dissociation as the mechanism of exchange of all of the $Me_3N \cdot BX_3$ adducts with BY_3, but Benton and Miller could not duplicate their results (*9, 11*). The latter workers found that the reaction of $Me_3N \cdot BI_3$ with tetraalkylammonium halides did give $Me_3N \cdot BXI_2$ (*9*) but that the lighter-halogen $Me_3N \cdot BX_3$ adducts did not react. Even with $Me_3N \cdot BI_3$, a concerted reaction has not been excluded. Blackborow (*15*) proposed dissociations involving chloride ion in his studies of exchange reactions of N,N-dimethylaniline—BCl_3 systems, but direct evidence was lacking, and his proposal does not seem consistent with the results of Benton and Miller (*9*) since halide dissociation should give spontaneous halogen redistribution in amine adducts. Benton and Miller found that traces of water increased the rate of bromine–iodine exchange between $Me_3N \cdot BI_3$ and HBr (*9*), and previous inconsistent results (*15, 105*) may arise from the presence of small amounts of HX and/or H_2O.

Kelly and co-workers have proposed that a rate-determining boron–halogen bond cleavage occurs in the hydrolysis of trimethylamine-monohaloborane adducts (*116, 184*), but they have not excluded a concerted reaction involving H_2O attack (*116*), such as seems to occur in the hydrolysis of $Me_3N{\cdot}BCl_3$ (*90, 158*). The existence of a trigonal boron cation $D{\cdot}BH_2{}^+$ has been established (*157*), and trigonal dihaloborane cations are equally plausible.

C. Associated Bridging Mechanism

Addition of free boron trihalide to an adduct gives halogen exchange in all cases studied. The initial associated species probably resembles the known (*26a, 85*) $B_2F_7{}^-$ ion in having a single halogen bridge:

$$D{\cdot}B^*X_3 + BY_3 \rightleftharpoons \qquad\qquad\qquad\qquad (4)$$

Exchange of halogen could then proceed either through a four-center transition state (which, however, requires one boron atom to be five-coordinate):

$$(5)$$

or through an ion pair such as $D{\cdot}\overset{*}{B}X_2{}^+{\cdot}BXY_3{}^-$ or

In the final dissociation to give the mixed-halogen species both boron atoms retain their initial coordination number:

$$\rightleftharpoons D{\cdot}\overset{*}{B}X_2Y + BXY_2 \qquad\qquad (6)$$

Boron-10 isotope substitution studies on $Me_3N{\cdot}BX_3 + BY_3$ systems, with all possible combinations of X and Y, confirm that the four-coordinate and three-coordinate borons retain their identities (*9*). Only above 180°C in the gas phase, where dissociation does occur (*40*), is there loss of isotopic purity (*9*).

Heavy-halogen boron trihalides rapidly replace light halogens in adducts, whereas light-halogen boron trihalides replace heavy halogens in adducts only slowly (9). This finding could be interpreted in various ways, but a clear choice cannot be made at present among the possible transition states. Boron trihalides also exchange halogen with CF_3 groups under mild conditions, and an ionic ($BFCl_3^-$) intermediate was suggested (33). However, exchange of halogen between alkyl halides and boron trihalides seems more likely to involve a four-center transition state because an ionic intermediate would involve a carbonium ion, and the isomerization of alkyl groups expected in carbonium ions was not observed (64).

Taken as a whole the evidence seems to support a doubly bridged transition state. Intermediates or transitions states of this type have been suggested in many redistribution reactions (64).

Halogen–alkyl group exchange in $Me_3N \cdot BR_n Y_{3-n}$ (R = propyl, butyl; X = F,Cl) also requires the presence of three-coordinate boron and a similar mechanism is probable (37).

D. Exchange Mechanisms in Tetrahaloborates

The dissociation mechanisms [Eqs. (2) and (3)] are equivalent in tetrahaloborate anions. Spontaneous halide ion dissociation occurs in solution in the heavier halogen species, the degree of dissociation being I > Br > Cl (172):

$$BX_4^- \rightleftharpoons BX_3 + X^- \tag{7}$$

This reaction is confirmed by attack of the halide ion X^- so generated on methylene halide solvents CH_2Y_2 to give CH_2XY (80). Possible mechanisms of formation of the mixed tetrahaloborate anions following this initial dissociation have been discussed by Hartman and Schrobilgen (80) and include (a) recombination of BX_3 with a different halide ion Y^-, (b) exchange via a singly halogen-bridged intermediate analogous to $B_2F_7^-$; (c) halide ion attack on a tetrahaloborate anion, and (d) halogen exchange between two three-coordinate boron species. Of these, mechanisms a and b are probably the most important.

The requirement for an initial dissociation step implies that halogen redistribution rates should parallel the ease of dissociation of one of the halogens, and this expectation is confirmed in fluorine–heavier halogen redistributions where the rates change from slow to very fast over the series Cl, Br, I as the B—X bond becomes weaker. Rates should also be related to stabilization of the residual boron trihalide formed on dissociation. The boron trihalide with the greatest number of fluorine atoms is the most stable, i.e., $BF_3 > BF_2X > BFX_2 > BX_3$. (The most stabilized species, BF_3, is of course the weakest Lewis acid.) It follows that mixed-halogen species $BF_nX_{4-n}^-$, and especially BF_3X^-, should be more prone to dissociation than either BF_4^- or BX_4^- because both factors favoring dissociation can be optimized. This is indeed observed (80). Fluorine redistribution in BF_4^- alone is fairly slow, the rate being quite solvent-dependent (63). Exchange of Cl—Br, Cl—I, and Br—I is slow on the NMR time scale in the tetrahaloborates except in the presence of fluorine when it becomes much faster, and this can be attributed to the especially easy dissociation of the mixed fluorine–heavier halogen tetrahaloborates that are formed (80). In such systems, BF_3X^- species exchange halogen rapidly, $BF_2X_2^-$ species at intermediate rates, and BFX_3^- species slowly. Also BF_4^- exchanges halogen rapidly under these conditions; fluoride ion dissociation is not required here as exchange can take place via fluorine-bridged $B_2F_7^-$, utilizing BF_3 from the easy dissociation of BF_3X^-.

A rate-determining initial dissociation step does account for most of the observed features of halogen exchange in the tetrahaloborates (80). An anomaly in the BF_4^-/BCl_4^- system is the late appearance of $BFCl_3^-$, which is expected to appear initially along with BF_3Cl^-. A similar anomaly was observed in an ion cyclotron resonance study of fluoride ion affinities (71) in which BF_3Cl^- and BF_4^- were observed instead of the expected $BFCl_3^-$. This is at present unexplained. There is considerable scope for more accurate kinetic analyses of halogen exchange reactions of tetrahaloborates (115). However, because carbenes can be generated from BF_4^- or F^- and a methylene halide or haloform (34), carbene reactions could further complicate such systems.

IV. Equilibria in Halogen Redistribution

Random or statistical redistribution of substituents X and Y, present in equal amounts about a central atom E in a system $E(X, Y)_n$, gives the following amounts of species (127):

$$EX_3 : EX_2Y : EXY_2 : EY_3 = 1:3:3:1 \quad \text{(for } n = 3\text{)}$$

$$EX_4 : EX_3Y : EX_2Y_2 : EXY_3 : EY_4 = 1:4:6:4:1 \quad \text{(for } n = 4\text{)}$$

In boron trihalide adducts and tetrahaloborate ions, halogen redistribution equilibria are reasonably close to this ideal random case when chlorine, bromine, and iodine are involved (27, 28, 80, 100, 112), as are equilibria in the uncomplexed heavier boron trihalides (111).

Redistribution of these halogens is near-random about a number of other central atoms as well (*97, 119*).

However, equilibria may be far from random when fluorine is involved. Table I shows the effects of the donor on disproportionation constants K_1 and K_2 (*127*) for redistribution of fluorine and chlorine:

$$K_1 = \frac{[D \cdot BF_2Cl]\,[D \cdot BCl_3]}{[D \cdot BFCl_2]^2} \tag{8}$$

$$K_2 = \frac{[D \cdot BF_3]\,[D \cdot BFCl_2]}{[D \cdot BF_2Cl]^2} \tag{9}$$

Practically all deviations from random redistribution are in the direction of less than the statistical amounts of mixed-halogen species. Deviations are especially pronounced with the sulfur-donor adducts, with uncomplexed BF_nCl_{3-n} being intermediate between the oxygen-donor and sulfur-donor systems. Constant K_1 is consistently larger

TABLE I

HALOGEN REDISTRIBUTION EQUILIBRIUM
CONSTANTS FOR $D \cdot BF_nCl_{3-n}$ [a]

Donor	K_1	K_2	Ref.
Ideal random case	0.33	0.33	*127*
Oxygen donors:			
Dimethyl ether	0.80	0.23	*27*
Methyl acetate	0.92	0.51	*29*
$CH_3C\overset{O}{\underset{SCH_3}{}}$	1.27	0.53	*29*
Uncomplexed BF_nCl_{3-n}	2.4	1.4	*111*
Sulfur donors:			
Dimethylsulfide	6.3	2.3	*28*
$CH_3C\overset{S}{\underset{OCH_3}{}}$	Large[b]	Large[b]	*29*
$CH_3C\overset{S}{\underset{SCH_3}{}}$	Large[b]	Large[b]	*29*

[a] Dichloromethane solvent was used except for the uncomplexed boron trihalides where 1,1-dichloroethane was the solvent.

[b] No mixed-halogen adducts detected.

than K_2, i.e., the $BFCl_2$ species is somewhat less favored than the BF_2Cl species.

Deviations from statistical redistribution equilibria have been studied in a variety of systems (119, 127, 128), and various explanations have been put forward. It should be kept in mind that stability differences of only a few kilocalories per mole between mixed- and unmixed-ligand complexes are sufficient to cause very large deviations from random redistribution (97). Mixed-ligand species are frequently favored over unmixed (128), and this can be explained in terms of stronger bonding in the mixed species due to decreased competition for orbitals. Thus, in $(RO)_3B$ compounds, there is strong π-bonding from oxygen to boron (170). When alkoxy ligands are mixed with non-π-bonding ligands, there are fewer RO groups to compete for boron's p_z orbital and, hence, each O—B bond can be stronger, and the mixed-ligand species are strongly favored (113, 165). Thus, dimethylamine diphenylboron (93) and aryl esters of dialkylborinic acids (22) can be synthesized by simply mixing the unmixed-ligand species in the correct ratio. By contrast, less than the statistical amounts of fluorine-containing mixed boron trihalides are present at equilibrium (111).

Deviations favoring the unmixed-ligand species are not as rare as is sometimes assumed (128). Such deviations can be rationalized using Pearson's concept of hard and soft acids and bases (140). It was pointed out some time ago that soft bases tend to cluster together on a central metal atom and, similarly, that hard bases tend to cluster together (98). Although this was originally discussed in terms of d-orbital filling in transition metal complexes (98), Pearson and Songstad extended this "symbiotic principle" to give qualitative explanations of equilibria and kinetics in a wide range of systems (141). The principle predicts that there should be extra stabilization if either several soft bases or several hard bases cluster about a single central atom. Thus, Pearson and Songstad classified BH_3 as a soft acid because H^- is soft, and BF_3 as a hard acid because F^- is hard; the symbiotic effect is maximized if a further soft base coordinates to BH_3, and if a further hard base coordinates to BF_3. Similarly, an increasing degree of soft acid character can be attributed to the other boron trihalides in the order $BCl_3 < BBr_3 < BI_3$.

In the uncomplexed boron trihalides, mixtures of fluorine with chlorine, bromine, and iodine are successively less favored (111), in accord with the symbiotic principle. This principle can rationalize halogen redistribution equilibria in adducts of the mixed boron trihalides as well. Incompatibility of fluorine with the heavier halogens is greatest with the softest donors. In the extreme case of PH_3 as donor, BF_3 is unable to form an adduct at all (21), and this can be attributed to the very unfavorable combination of a soft PH_3 and 3 hard fluorines about boron. The combination of Me_2S or CH_3PH_2 with 3 fluorines about boron is not prevented, but the BF_3 adducts of these donors are anomalously weak (21, 131). All three of these donors discriminate against mixed fluorine–heavier halogen adducts: the adducts are either not observed at all or are present in abnormally small amounts (28, 48). Some irregularities exist, however. The symbiotic principle predicts that F,Br mixing should be less favorable than F,Cl mixing in the adducts, and this is borne out in the Me_2S adduct system (28), but not in the PH_3 adduct system where the mixed F,Br adducts, but not the mixed F,Cl adducts, have been detected (48). [Both F,Cl and F,Br mixed adducts of CD_3PH_2 have been detected (48).] The symbiotic principle does help to rationalize results but is clearly not quantitative in its predictions.

The near-random redistribution of fluorine and chlorine in oxygen-donor boron trihalide adducts (27, 29) might be rationalized using this principle since oxygen is between fluorine and chlorine on the hard–soft scale (140), and there is less discontinuity of ligands in terms of hard–soft properties. However, amine adducts of BHF_2 do tend to disproportionate to give the BF_3 and BH_3 adducts (130).

A similar tendency for fluorine to segregate itself from the heavier halogens has been noted in redistribution reactions about many other central atoms, including phosphorus (47, 144, 145), silicon (139, 146, 163), germanium (139), arsenic (139), tungsten (56), niobium (36), and tantalum (36). However, relative stabilities of mixed and unmixed species involving fluorine vary greatly in systems with other central atoms, as they do in boron systems. The mixed fluorine-containing tetrahaloborate ions are formed in large amounts (80), as are mixed-halogen adducts in the TiF_nCl_{4-n} system (17).

Because hardness and softness are qualitative concepts that no doubt originate from a variety of factors (140), ambiguities in interpretation may arise (96). Recently, more quantitative treatments of the bonding in series of mixed-ligand trigonal and tetrahedral boron compounds have been reported. Application of the virial partitioning method by Runtz and Bader (150) indicates that Lewis acidity of BH_3 and BF_3 is determined more by the properties of the H and F fragments than by the net charge and energy of the boron fragment. However, these calculations do not show an incompatibility of H and F ligands in $BH_nF_{4-n}^-$. These authors discuss destabilizing repulsions between fluorine and neighboring atoms that may account for anomalies in the chemistry of fluorine but show that in specific cases other factors may compensate. An *ab initio* molecular orbital study of the $BF_n(OH)_{3-n}$ ($n = 0-3$) species indicates that replacement of a fluorine of BF_3 by a hydroxyl group is energetically unfavored, and the authors relate this to known stability of BF_4^- with respect to BF_3OH^- (*1*). Unfortunately, discussions become more qualitative when the heavier halogens are involved. Comparisons of BF_3 and BCl_3 adducts have concentrated on the donor–acceptor rather than on the boron–halogen bonds (*148, 171*). The uncomplexed heavier boron trihalides have been investigated in much more detail than their adducts (*5, 6, 12, 103*).

V. Donor-for-Halogen Exchange: Difluoroboron Cations

Because both donor–acceptor and boron–halogen bonds are labile in many adducts, donor-for-halogen exchange is possible giving ionic adducts of 1:1 stoichiometry:

$$2D \cdot BX_3 \rightleftharpoons D_2BX_2^+ + BX_4^- \tag{10}$$

Some BCl_3 and BBr_3 adducts have been reported to rearrange spontaneously in solution according to Eq. (10) to give varying amounts of the ionic adducts (*120, 160, 161*). Reaction of either excess base or excess acid with an initially formed covalent adduct of a secondary amine can also give ionic adducts $(R_2NH)_2BX_2^+ \cdot BX_4^-$ (X = Cl, Br) (*154*). The heavier the halogen, the more easily it is displaced from boron. Thus one, two, or three bromines of $Me_3N \cdot BBr_3$ can be displaced by pyridines to give boron cations of charge $+1$, $+2$, and $+3$, respectively, and the formation of the $+3$ ions occurs even more readily with the BI_3 adduct (*152*). Highly charged ions have been postulated to form spontaneously in amine·BCl_3 systems as well, but direct evidence is lacking (*15*).

Fluorine is the least readily displaced halogen, and few BF_3 adducts have been reported to rearrange to the ionic form. Ionic species do form when chelating donors favor $D_2BF_2^+$ over $D \cdot BF_3$ (*4, 187*), although these chelates form less readily than other boron chelates due to the difficulty of displacing fluorine (*177*). Other chelating ligands with BF_3 form neutral BF_2 species by loss of HF (*25, 122, 175*). With similar but nonchelating ligands, i.e., mono- rather than diketones, fluorine redistribution does occur but ionic species have not been detected (*62*), indicating that the equilibrium of Eq. (10) is far to the left. A slow rearrangement of vinyl BF_3^- to (vinyl)$_2BF_2^-$ and BF_4^- has been reported to occur in aqueous solution (*167*).

The tetramethylurea–BF_3 adduct in methylene chloride is in equilibrium with a very small amount of the corresponding ionic adduct, (tetramethylurea)$_2$BF$_2$$^+$·BF$_4$$^-$ (*82*), and similar ionic-covalent equilibria occur in tetramethylthiourea–BF$_3$ and tetramethylselenourea–BF$_3$ adducts (*83*). Ionic species can be favored by excess BF$_3$ which gives D$_2$BF$_2$$^+$·B$_2F_7$$^-$ (*83*). Mixed boron trihalide adducts can also favor the formation of fluoroboron cations. Excess tetramethylurea displaces Cl$^-$ from the tetramethylurea·BF$_2$Cl adduct to form large amounts of (tetramethylurea)$_2$BF$_2$$^+$ and mixed chlorofluoroborate anions; mixed donor cations, DD'·BF$_2$$^+$, can also be formed in this way (*82*). It should be possible to prepare many fluoroboron cations by preferential displacement of heavier halogens from mixed fluorine— heavier halogen boron trihalide adducts. The reaction is analogous to the formation of dihydroboron cations DD'·BH$_2$$^+$ (*138, 152, 164*) through halide ion displacement from D·BH$_2$Cl and D·BH$_2$Br by D' (*152*).

Johnson and Shore have discussed factors favoring symmetrical vs. unsymmetrical cleavage of B$_2$H$_6$, giving D·BH$_3$ or D$_2$BH$_2$$^+$·BH$_4$$^-$, respectively (*95*). Similar stability factors may be applicable to reaction (10), i.e., bulky donors may be prevented from forming D$_2$BX$_2$$^+$. However, the type of initial adduct that forms from B$_2$H$_6$ is related to the preferred course of bridge cleavage reactions (*95*), whereas the monomeric boron trihalides should form an initial D·BX$_3$ adduct and any ionic species should form by subsequent rearrangement. Thus ionic boron trihalide adducts have not been observed as frequently as ionic borane adducts. The presence of fluorine and a heavier halogen does favor the formation of ionic boron trihalide adducts by displacement of the heavier halogen.

VI. NMR Applied to Adducts: Advantages and Pitfalls

A. Reliability of Mixed-Adduct NMR Data

Identification of most mixed boron trihalide adducts, and all information on rates and equilibria of halogen redistribution in these systems, depends on correct NMR peak assignments. The reliability of these is assured by a combination of two or more of the following: distinctive heteronuclear NMR coupling patterns; systematic variation of peak areas as the ratio of two halogens is changed; identification and consistent behavior of peaks due to the same species in spectra of two or more of ^{1}H, ^{19}F, ^{11}B, and ^{13}C nuclei; agreement of observed parameters with values calculated from "pairwise interaction" parameters (*76, 123*). Fine points of structure can also be determined by

NMR; thus, ^{19}F splittings due to ^{11}B—^{19}F coupling, and their disappearance at lower temperatures due to quadrupole relaxation of ^{11}B, give a measure of the electric field gradient about boron and, hence, of the adduct structure (27, 28).

B. COMPLEXATION SHIFTS
AND DONOR–ACCEPTOR BOND STRENGTH

Conclusions about Lewis acidity should not be based solely on shifts of donor ^{1}H resonances to low field on adduct formation (45, 159). The assumption that this shift is a measure of Lewis acid strength neglects other factors, such as bond reorganization of the Lewis acid prior to adduct formation, that can affect the heat of formation of the adduct (106, 125). Thus Kristoff and Shriver have pointed out that spectroscopic probes in general measure the relative strength of the donor–acceptor bond but not the Lewis acidity (106). Lewis acid strengths usually parallel acidities because the donor–acceptor bond strength usually dominates the heat of formation of the adduct. Even the use of complexation shifts as a measure of donor–acceptor bond strengths neglects factors such as changes in long-range shielding and hybridization that can occur on adduct formation. A major change in donor hybridization on adduct formation does cause quite different donor ^{1}H chemical shifts in the adduct (88). In a few cases, donor ^{1}H resonances can even shift to high field on adduct formation (67, 88). An anisotropic solvent, such as benzene, can change the direction of a normally "well-behaved" complexation shift (30, Table I).

When the base is changed, ^{1}H complexation shifts bear little relationship to the heat of dissociation of the donor–acceptor bond, as illustrated in Table II. The same is true even when the donor atom stays the same, as in a series of amine–BF_3 complexes (179). A number of factors affect NMR parameters in adducts, and trends in chemical shifts in closely related series of adducts (70).

However, when boron trihalide adducts of the same base are studied and only the halogens are changed, and there is negligible hybridization change in the donor on complexation, the ^{1}H complexation shifts do correlate with the heats of formation of the complexes, as in $Me_3N\cdot BX_3$ and pyridine·BX_3 (X = F, Cl, Br, I) (125). In such cases, ^{1}H complexation shifts should give a measure of donor–acceptor bond strength for the mixed-halogen adducts as well (10). For trimethylamine adducts, this order is also consistent with mass spectrometric data for the mixed boron trihalide adducts (108). In $Me_2O\cdot BF_nCl_{3-n}$ (n = 1–3) adducts, ^{1}H complexation shifts do correlate with Lewis

TABLE II

ADDUCT COMPLEXATION SHIFTS AND HEATS OF DISSOCIATION OF THE
DONOR–ACCEPTOR BOND

Adduct	ΔH_{dissoc} (kcal/mole)	Ref.	Complexation shifts (ppm)[a]			Ref.
			^1H	^{19}F	^{11}B	
$Me_2S \cdot BF_3$	3.5	131	0.31	15.3	8.6	28
$Me_2S \cdot BCl_3$	12.1	131	0.45	—	40.5	28
$Me_2O \cdot BF_3$	13.1	132	0.62	34.2	11.5	27
$Me_3P \cdot BF_3$	18.9	2, 21	0.45	13.4	10.7	30,189
$Me_3N \cdot BF_3$	30.9	65	0.40	40.1	11.1	10,189
	26.6	121				
$Me_3N \cdot BCl_3$	30.5	121	0.78	—	37.8	10,189

[a] ^1H, to low field; ^{19}F, to high field; free BF_3 absorbs at 124.0 ppm to high field of $CFCl_3$; ^{11}B, to high field of free BF_3 and free BCl_3 [6.6 ppm to high field, and 29.3 ppm to low field, respectively, from external $(MeO)_3B$.]

acidity as determined by relative proportions of the different BF_nCl_{3-n} species complexed with a limited quantity of Me_2O (27).

When a donor such as methyl acetate, with more than one proton absorption, is complexed to a series of mixed and unmixed boron trihalides, the different proton environments give similar but not identical orderings of complexation shift for the series of Lewis acids (29). This confirms that the relationship between proton complexation shift and acceptor strength can be only approximate and that the complexation shift cannot arise solely from an inductive shift of electron density toward the donor site. Changes in long-range shielding are likely to be important in causing irregularities in series of complexation shifts, especially when heavy halogens can be in close proximity to some of the protons being observed.

Use of the complexation shift as a measure of donor–acceptor interaction is especially treacherous with nuclei other than protons, because chemical shifts of these nuclei are more dependent on the paramagnetic than on the diamagnetic term of the screening tensor (52, 53). Both ^{19}F and ^{11}B resonances of the boron trihalides do shift to high field on complexation, as expected if the complexation shift were due to the increase in electron density on the boron trihalide, and early work indicated that ^{19}F complexation shifts of BF_3 could be correlated with enthalpies of formation of the complexes. Although this is true for BF_3 adducts of some series of closely related donors (42, 91, 151), such correlations do not occur in other series (169). Table II illustrates that, although there is a tendency for the strongest

BF_3 adducts to have ^{19}F resonances at the highest field, there are inversions in this order.

Much of the ^{19}F complexation shift appears to be related to the change of coordination geometry about boron from trigonal planar to tetrahedral; this component is related to donor strength only to the extent that weak donors may be unable to cause this change to be carried through to fully tetrahedral boron (20). The buildup of negative charge on BF_3 is only one of the factors determining ^{19}F complexation shifts in BF_3 adducts. Similar considerations apply to mixed fluorine–heavier halogen adducts, and it would be unwise to attempt to deduce relative donor–acceptor bond strengths of BF_3, BF_2X, and BFX_2 adducts from ^{19}F complexation shifts.

Mooney and co-workers considered ^{11}B complexation shifts of boron Lewis acids to be a measure of the acceptor strengths of the Lewis acids and of the donor abilities of various ligands (59). However, conclusions based on these assumptions are not valid. Tetrahedral boron shifts do not follow the same pattern as trigonal boron shifts, and boron trihalide complexes with quite different heats of formation can have very similar ^{11}B complexation shifts, e.g., the BF_3 adducts of Table II. Also $Me_2S\cdot BCl_3$ and $Me_2O\cdot BF_3$, with similar heats of formation, have very different complexation shifts (Table II). The ^{11}B chemical shifts and complexation shifts are determined by the halogens present as well as by the donor (76), as discussed in the following. In spite of this, some workers continue to use ^{11}B complexation shifts as a measure of donor–acceptor bond strength (134).

Also, ^{13}C complexation shifts cannot be directly related to donor–acceptor bond strength. In ether and ketone complexes with BF_3, different carbons shift to high and to low field in the same donor molecule (57, 58, 86). Similar irregularities occur in the only series of mixed boron trihalide adducts for which ^{13}C NMR data are available. The ^{13}C complexation shift of $Me_3N\cdot BF_3$ is to high field, but, in all other mixed- and unmixed-halogen adducts of Me_3N, the shift is to low field (123). Trends in chemical shift are regular in all series of adducts $Me_3N\cdot BX_nY_{3-n}$, from $n = 0$ to $n = 3$, and are consistent with the ordering of ^1H shifts, i.e., the lowest-field resonance appears to correspond to the strongest donor–acceptor bond, but the complexation shifts are not related quantitatively to donor–acceptor bond strength in any of these systems.

Correlations of NMR parameters to electronic structure have been carried out for a wide range of three-coordinate boron compounds (12, 188). However, the extensive NMR data that exist for boron trihalide adducts (70) are less easily interpreted. Among the factors leading to difficulties in correlating NMR parameters with structure is the lack of a σ-π separation of orbitals in tetrahedral molecules, such as occurs in trigonal planar molecules. Interpretations of NMR parameters of the mixed tetrahaloborate anions by various methods, including CNDO calculation of chemical shifts, allows a "fluorine anomaly" in [11]B chemical shifts to be interpreted in various ways (80). Correlations of [11]B and ^{19}F chemical shifts in fluorine-containing adducts can be made in terms of survival of varying amounts of fluorine-to-boron multiple bonding in the adducts (20). However, interpretation has lagged behind the accumulation of NMR data.

Similar difficulties have occurred in attempts to find a general correlation between coupling constants across the donor–acceptor bond in borane adducts and the strength and nature of the donor–acceptor bond (38, 149).

Mixed boron trihalide adducts should facilitate interpretation of NMR parameters because properties can be changed gradually by changing a single halogen rather than all three. Studies of mixed boron trihalide adducts have already led to application of pairwise interaction NMR parameters, which are discussed next. Although these are not appreciably closer to an explanation of NMR parameters in terms of electronic structure and bonding, they do have the virtue of greater simplicity.

C. "Pairwise Interaction" NMR Parameters

Nuclear spin-spin coupling constants and chemical shifts of various nuclei can be pairwise additive with respect to the substituent groups (182), i.e., the chemical shift (or spin- spin coupling) can be expressed as

$$\delta = \sum \eta_{i,j} \tag{11}$$

where $\eta_{i,j}$ is a parameter associated with substituents i and j and independent of all other substituents. The sum is taken over all substituents about a central atom, excluding the nucleus observed in the NMR experiment. In the case of $BF_2Cl_2{}^-$

$$\delta_{11_B} = \eta_{F,F} + 4\eta_{F,Cl} + \eta_{Cl,Cl} \tag{12}$$

$$\delta_{19_F} = 2\eta'_{F,Cl} + \eta'_{Cl,Cl}$$

where $\eta_{i,j}$ is different for each nucleus observed.

Theoretical justification of the pairwise additivity rule for chemical shifts and nuclear spin-spin couplings can be found in the work of Vladimiroff and Malinowski (182). Chemical shifts of [27]Al (118), [13]C

(*123*, *166*), ^{11}B (*76*, *80*, *166*), and ^{19}F (*76*, *80*) as well as of a number of other nuclei are pairwise additive, as are many coupling constants. Application of the pairwise approach to the mixed boron trihalide adducts eliminates some apparent anomalies in chemical shifts and shows that ^{11}B complexation shifts cannot correlate to donor–acceptor bond strengths in the general case (*76*). The halogen–halogen pairwise interaction terms, rather than the donor–halogen terms, frequently dominate in determining complexation shifts.

Pairwise interaction parameters can perhaps be interpreted in chemical bonding terms (*76*, *80*, *123*, *182*). Pairwise terms involving fluorine are often anomalous and this may be related to some surviving multiple bonding in the boron–fluorine bonds in the adducts (*20*). Pairwise interaction parameters of the Me_3N mixed trihalide adduct system are reported to correlate consistently for ^{1}H, ^{13}C, ^{19}F, and ^{11}B chemical shifts and for ^{11}B—^{19}F coupling constants (*123*). It is perhaps surprising that even the proton shifts can be correlated in this way.

VII. Conclusion

The mixed boron trihalide adducts hold few surprises in terms of their donor–acceptor bond behavior, but provide striking examples of dependence of halogen redistribution behavior on the nonhalogen substituent. The simplicity and accessibility of these systems suggests their use as model compounds in the study of ligand redistribution reactions. Many of the features complicating ligand redistribution in, for example, metal carbonyl systems (*46*) are simplified or absent here.

ACKNOWLEDGEMENTS

We thank our colleagues for helpful discussions and the National Research Council of Canada for financial support of our work in this field.

REFERENCES

1. Armstrong, D. R., and Perkins, P. G., *Inorg. Chim. Acta* 10, 77 (1974).
2. Arnett, E. M., *Progr. Phys. Org. Chem.* 1, 223 (1963).
3. Ashcroft, B. W. C., and Holliday, A. K., *J. Chem. Soc. A* p. 2581 (1971).
4. Axtell, D. D., Campbell, A. C., Keller, P. C., and Rund, J. V., *J. Coord. Chem.* 5, 129 (1976).
5. Barker, G. K., Lappert, M. F., Pedley, J. B., Sharp, G. J., and Westwood, N. P. C., *J. Chem. Soc., Dalton Trans.* p. 1765 (1975).
6. Bassett, P. J., and Lloyd, D. R., *J. Chem. Soc. A* p. 1551 (1971).

7. Batiz-Hernandez, H., and Bernheim, R. A., *Progr. Nucl. Magn. Reson. Spectrosc.* **3**, 63 (1968).
8. Bätzel, V., Müller, U., and Allmann, R., *J. Organomet. Chem.* **102**, 109 (1975).
9. Benton, B. W., and Miller, J. M., *Can. J. Chem.* **52**, 2866 (1974).
10. Benton-Jones, B., Davidson, M. E. A., Hartman, J. S., Klassen, J. J., and Miller, J. M., *J. Chem. Soc., Dalton Trans.* p. 2603 (1972).
11. Benton-Jones, B., and Miller, J. M., *Inorg. Nucl. Chem. Lett.* **8**, 485 (1972).
12. Berger, H.-O., Kroner, J., and Nöth, H., *Chem. Ber.* **109**, 2266 (1976).
13. Binder, H., and Fluck, E., *Z. Anorg. Allg. Chem.* **381**, 116, 123 (1971).
14. Blackborow, J. R., and Lockhart, J. C., *Chem. Commun.* p. 726 (1968).
15. Blackborow, J. R., *J. Chem. Soc., Dalton Trans.* p. 2139 (1973).
16. Blackborow, J. R., and Lockhart, J. C., *J. Chem. Soc. A* p. 3015 (1968).
17. Borden, R. S., Loeffler, P. A., and Dyer, D. S., *Inorg. Chem.* **11**, 2781 (1972).
18. Bradley, R. H., Brier, P. N., and Jones, D. E. H., *J. Chem. Soc. A* p. 1397 (1971).
19. Bratt, P. J., Brown, M. P., and Seddon, K. R., *J. Chem. Soc., Dalton Trans.* p. 2161 (1974).
20. Brown, D. G., Drago, R. S., and Bolles, T. F., *J. Am. Chem. Soc.* **90**, 5706 (1968).
21. Brown, H. C., *J. Chem. Soc.* p. 1248 (1956).
22. Brown, H. C., and Gupta, S. K., *J. Am. Chem. Soc.* **93**, 2802 (1971).
23. Brown, H. C., and Ravindran, N., *J. Am. Chem. Soc.* **94**, 2113 (1972), and references contained therein.
24. Brown, H. C., and Ravindran, N., *J. Am. Chem. Soc.* **98**, 1785 (1976).
25. Brown, N. M. D., and Bladon, P., *J. Chem. Soc. A* p. 526 (1969).
26. Brownstein, S., *Can. J. Chem.* **45**, 2403 (1967).
26a. Brownstein, S., and Paasivirta, J., *Can. J. Chem.* **43**, 1645 (1965).
27. Bula, M. J., Hamilton, D. E., and Hartman, J. S., *J. Chem. Soc., Dalton Trans.* p. 1405 (1972).
28. Bula, M. J., and Hartman, J. S., *J. Chem. Soc., Dalton Trans.* p. 1047 (1973).
29. Bula, M. J., Hartman, J. S., and Raman, C. V., *J. Chem. Soc., Dalton Trans.* p. 725 (1974).
30. Bula, M. J., Hartman, J. S., and Raman, C. V., *Can. J. Chem.* **53**, 326 (1975).
31. Burke, J. J., and Lauterbur, P. C., *J. Am. Chem. Soc.* **83**, 326 (1961).
32. Burton, D. J., and Koppes, W. M., *J. Org. Chem.* **40**, 3026 (1975).
33. Chivers, T., *Can. J. Chem.* **48**, 3856 (1970).
34. Clark, J. H., and Miller, J. M., unpublished observations.
35. Clark, R. J. H., and Willis, C. J., *Inorg. Chem.* **10**, 1118 (1971).
36. Colditz, L. and Calov, V., *Z. Anorg. Allg. Chem.* **376**, 1 (1970).
37. Costes, J.-P., Cros, G., and Laurent, J.-P., *J. Chim. Phys.* **76**, 16 (1976).
38. Cowley, A. H., and Damasco, M. C., *J. Am. Chem. Soc.* **93**, 6815 (1971).
39. Cowley, A. H., and Mills, J. L., *J. Am. Chem. Soc.* **91**, 2911 (1969).
40. Coyle, T. D., *Proc. Chem. Soc.* p. 172 (1963).
41. Coyle, T. D., and Stone, F. G. A., *Progr. Boron Chem.* **1**, 83 (1964).
42. Craig, R. A., and Richards, R. E., *Trans. Faraday Soc.* **59**, 1962 (1963).
43. Dazord, J., Mongeot, H., Atchekzai, H., and Tuchagues, J. P., *Can. J. Chem.* **54**, 2135 (1976).
44. Denniston, M. C., Chiusano, M. A., and Martin, D. R., *J. Inorg. Nucl. Chem.* **38**, 979 (1976).
45. Deters, J. F., McCusker, P. A., and Pilger, R. C., Jr., *J. Am. Chem. Soc.* **90**, 458 (1968).
46. Dobson, G. R., *Acc. Chem. Res.* **9**, 300 (1976).
47. Drake, J. E., and Goddard, N., *J. Chem. Soc. A* p. 2587 (1970).
48. Drake, J. E., and Rapp, B., *J. Chem. Soc., Dalton Trans.* p. 2341 (1972).

49. Drake, J. E., and Rapp, B., *J. Inorg. Nucl. Chem.* **36**, 2613 (1974).
50. Drake, J. E., and Simpson, J., *J. Chem. Soc. A* p. 974 (1968).
51. Druce, P. M., Lappert, M. F., and Riley, P. N. K., *J. Chem. Soc., Dalton Trans.* p. 438 (1972).
52. Ebraheem, K. A. K., Webb, G. A., and Witanowski, M., *Org. Magn. Reson.* **8**, 317 (1976).
53. Emsley, J. W., Feeney, J., and Sutcliffe, L. H., "High Resolution Nuclear Magnetic Resonance Spectroscopy." Pergamon, London, 1965.
54. Fischer, E. O., and Richter, K., *Angew. Chem., Int. Ed., Engl.* **14**, 34 (1975).
55. Fogelman, J., and Miller, J. M., *Can. J. Chem.* **50**, 1262 (1972).
56. Fraser, G. W., Gibbs, C. J. W., and Peacock, R. D., *J. Chem. Soc. A* p. 1708 (1970).
57. Fratiello, A., Kubo, R., and Chow, S., *J. Chem. Soc., Perkin Trans. 2* p. 1205 (1976).
58. Fratiello, A., Kubo, R., Liu, D., and Vidulich, G., *J. Chem. Soc., Perkin Trans. 2* p. 1415 (1975).
59. Gates, P. N., McLaughlan, E. J., and Mooney, E. F., *Spectrochim. Acta* **21**, 1445 (1965).
60. Gerrard, W., and Lappert, M. F., *Chem. Rev.* **58**, 1081 (1958).
61. Gerrard, W., and Wheelans, M. P., *J. Chem. Soc.* p. 4296 (1956).
62. Gillespie, R. J., and Hartman, J. S., *Can. J. Chem.* **46**, 2147 (1968).
63. Gillespie, R. J., Hartman, J. S., and Parekh, M., *Can. J. Chem.* **46**, 1601 (1968).
64. Goldstein, M., Haines, L. I. B., and Hemmings, J. A. G., *J. Chem. Soc., Dalton Trans.* p. 2260 (1972).
65. Graham, W. A. G., and Stone, F. G. A., *J. Inorg. Nucl. Chem.* **3**, 164 (1956).
66. Greenwood, N. N., and Martin, R. L., *Q. Rev., Chem. Soc.* **8**, 1 (1954).
67. Greenwood, N. N., and Storr, A., *J. Chem. Soc.* p. 3426 (1965).
68. Greenwood, N. N., and Walker, A., *Inorg. Nucl. Chem. Lett.* **1**, 65 (1965).
69. Gunn, R. S., and Sanborn, R. H., *J. Chem. Phys.* **33**, 955 (1960).
70. Gur'yanova, E. N., Gol'dshtein, I. P., and Romm, I. P., "Donor-Acceptor Bond," Isr. Progr. Sci. Transl., Jerusalem, Chapter III. Wiley, New York, 1975.
71. Haartz, J. C., and McDaniel, D. H., *J. Am. Chem. Soc.* **95**, 8562 (1973).
72. Haas, A., Haberlein, M., and Krüger, C., *Chem. Ber.* **109**, 1769 (1976).
73. Hamilton, D. E., Hartman, J. S., and Miller, J. M., *Chem. Commun.* p. 1417, (1969).
74. Hanna, Z., and Miller, J. M., unpublished observations.
75. Hartman, J. S., and Miller, J. M., *Inorg. Nucl. Chem. Lett.* **5**, 831 (1969).
76. Hartman, J. S., and Miller, J. M., *Inorg. Chem.* **13**, 1467 (1974).
77. Hartman, J. S., and Miller, J. M., unpublished observations.
78. Hartman, J. S., and Raman, C. V., unpublished observations.
79. Hartman, J. S., and Schrobilgen, G. J., *Can. J. Chem.* **50**, 713 (1972).
80. Hartman, J. S., and Schrobilgen, G. J., *Inorg. Chem.* **11**, 940 (1972).
81. Hartman, J. S., and Schrobilgen, G. J., *Can. J. Chem.* **51**, 99 (1973).
82. Hartman, J. S., and Schrobilgen, G. J., *Inorg. Chem.* **13**, 874 (1974).
83. Hartman, J. S., Schrobilgen, G. J., and Stilbs, P., *Can. J. Chem.* **54**, 1121 (1976).
85. Hartman, J. S., and Stilbs, P., *J. Chem. Soc., Chem. Commun.* p. 566 (1975).
86. Hartman, J. S., Stilbs, P., and Forsén, S., *Tetrahedron Lett.* p. 3497 (1975).
87. Hartman, J. S., and Yetman, R. R., *Can. J. Spectrosc.* **19**, 1 (1974).
88. Hartman, J. S., and Yetman, R. R., *Can. J. Chem.* **54**, 1130 (1976).
89. Hartman, J. S., and Yetman, R. R., unpublished observations.
90. Heaton, G. S., and Riley, P. N. K., *J. Chem. Soc. A* p. 952 (1966).
91. Heitsch, C. W., *Inorg. Chem.* **4**, 1019 (1965).
92. Higgins, T. H. S., Leisegang, E. C., Raw, C. J. G., and Rossouw, A. J., *J. Chem. Phys.* **23**, 1544 (1955).

93. Hofmeister, I. H. K., and Van Wazer, J. R., *J. Inorg. Nucl. Chem.* **26**, 1209 (1964).
94. Jackman, L. M., and Cotton, F. A., eds., "Dynamic Nuclear Magnetic Resonance Spectroscopy." Academic Press, New York, 1975.
95. Johnson, H. D., and Shore, S. G., *Fortschr. Chem. Forsch.* **15**, 87 (1970).
96. Johnson, M. P., Shriver, D. F., and Shriver, S. A., *J. Am. Chem. Soc.* **88**, 1588 (1966).
97. Jones, D. E. H., *J. Chem. Soc., Dalton Trans.* p. 567 (1972).
98. Jorgensen, C. K., *Inorg. Chem.* **3**, 1201 (1964).
99. Jugie, G., and Laussac, J.-P., *C. R. Acad. Sci., Ser. C* **274**, 1668 (1972).
100. Jugie, G., Laussac, J.-P., and Laurent, J.-P., *Bull. Soc. Chim. Fr.* pp 2542, 4238 (1970).
101. Jugie, G., Laussac, J.-P., and Laurent, J.-P., *J. Inorg. Nucl. Chem.* **32**, 3455 (1970).
102. Kemmitt, R. D. W., Milner, R. S., and Sharp, D. W. A., *J. Chem. Soc.* p. 111 (1963).
103. King, G. H., Krishnamurthy, S. S., Lappert, M. F., and Pedley, J. B., *Faraday Disc. Chem. Soc.* **54**, 70 (1972).
104. Kren, R. M., Mathur, M. A., and Sisler, H. H., *Inorg. Chem.* **13**, 174 (1974).
105. Krishnamurthy, S. S., and Lappert, M. F., *Inorg. Nucl. Chem. Lett.* **7**, 919 (1971).
106. Kristoff, J. S., and Shriver, D. F., *Inorg. Chem.* **12**, 1788 (1973).
107. Landesman, H. L., and Williams, R. E., *J. Am. Chem. Soc.* **83**, 2663 (1961).
108. Lanthier, G. F., and Miller, J. M., *J. Chem. Soc. A* p. 346 (1971).
109. Lappert, M. F., *J. Chem. Soc.* p. 817 (1961), p. 542 (1962).
110. Lappert, M. F., Litzow, M. R., Pedley, J. B., Riley, P. N. K., and Tweedale, A., *J. Chem. Soc. A* p. 3105 (1968).
111. Lappert, M. F., Litzow, M. R., Pedley, J. B., Riley, P. N. K., and Nöth, H., *J. Chem. Soc. A* p. 383 (1971).
112. Laussac, J.-P., Jugie, G., Laurent, J.-P., and Gallais, F., *C. R. Acad. Sci., Ser. C* **276**, 1497 (1973).
113. Lockhart, J. C., *Chem. Rev.* **65**, 131 (1965).
114. Lockhart, J. C., "Redistribution Reactions." Academic Press, New York, 1970.
115. Lockhart, J. C., *Int. Rev. Sci., Inorg. Chem., Ser. Two*, 9, 1–20 (1974).
116. Lowe, J. R., Uppal, S. S., Weidig, C., and Kelly, H. C., *Inorg. Chem.* **9**, 1423 (1970).
117. Ludman, C. J., and Waddington, T. C., *J. Chem. Soc. A* p. 1816 (1966).
118. Malinowski, E. R., *J. Am. Chem. Soc.* **91**, 4701 (1969).
119. Marcus, Y., and Eliezer, L., *Coord. Chem. Rev.* **4**, 273 (1969).
120. Massey, A. G., this series **10**, 1 (1967).
121. McCoy, R. E., and Bauer, S. H., *J. Am. Chem. Soc.* **78**, 2061 (1956).
122. Medvedeva, V. G., Skoldinov, A. P., and Shapet'ko, N. N., *Zh. Obsch. Khim.* **39**, 460 (1969); Engl. Transl.: *J. Gen. Chem. USSR* **39**, 432 (1969).
123. Miller, J. M., and Jones, T. R. B., *Inorg. Chem.* **15**, 284 (1976).
124. Miller, J. M., and Onyszchuk, M., *Can. J. Chem.* **41**, 2898 (1963).
125. Miller, J. M., and Onyszchuk, M., *Can. J. Chem.* **42**, 1518 (1964); **44**, 899 (1966).
126. Miller, J. M., and Wilson, G. L., this series **18**, 229 (1976).
127. Moedritzer, K., *Adv. Organomet. Chem.* **6**, 171 (1968).
128. Moedritzer, K., *Organomet. React.* **2**, 1 (1971).
129. Mongeot, H., Dazord, J., Atchekzai, H. R., and Tuchagues, J. P., *Syn. React. Inorg. Metal-Org. Chem.* **6**, 191 (1976).
130. Mongeot, H., Dazord, J., and Tuchagues, J. P., *J. Inorg. Nucl. Chem.* **34**, 825 (1972).
131. Morris, H. L., Kulevsky, N. I., Tamres, M., and Searles, S., *Inorg. Chem.* **5**, 124 (1966).
132. Mortimer, C. T., "Reaction Heats and Bond Strengths," Chapter 6. Pergamon Press, London, 1962.
133. Muylle, E., and van der Kelen, G. P., *Spectrochim. Acta, Part A* **31**, 1045 (1975).

134. Muylle, E., van der Kelen, G. P., and Claeys, E. G., *Spectrochim. Acta, Part A* **32** 1149 (1976).

135. Muylle, E., van der Kelen, G. P., and Eeckhaut, Z., *Spectrochim Acta, Part A* **31** 1039 (1975).

136. Niedenzu, K., *Organometal. Rev.* **1**, 305 (1966).

137. Nightingale, R. E., and Crawford, B., *J. Chem. Phys.* **22**, 1468 (1954).

138. Nöth, H., *Progr. Boron Chem.* **3**, 211 (1970).

139. Pace, S. C., and Riess, J. G., *J. Organomet. Chem.* **76**, 325 (1974).

140. Pearson, R. G., *J. Am. Chem. Soc.* **85**, 3533 (1963); *Science* **151**, 172 (1966); *J. Chem. Educ.* **45**, 581, 643 (1968).

141. Pearson, R. G., and Songstad, J., *J. Am. Chem. Soc.* **89**, 1827 (1967).

142. Rapp, B., and Drake, J. E., *Inorg. Chem.* **12**, 2868 (1973).

143. Repasky, J. E., Weidig, G., and Kelly, H. C., *Syn. React. Inorg. Metal-Org. Chem.* **5**, 337 (1975).

144. Riess, J. G., and Bender, R., *Bull. Soc. Chim. Fr.* p. 3700 (1972).

145. Riess, J. G., Elkaim, J.-C., and Pace, S. C., *Inorg. Chem.* **12**, 2874 (1973).

146. Riess, J. G., and Pace, S. C., *Inorg. Chim. Acta* **9**, 61 (1974).

147. Ritter, J. J., and Coyle, T. D., *J. Chem. Soc. A* p. 1303 (1970).

148. Ronan, R. J., Gilje, J. W., and Biallas, M. J., *J. Am. Chem. Soc.* **93**, 6811 (1971).

149. Rudolph, R. W., and Schultz, C. W., *J. Am. Chem. Soc.* **93**, 6821 (1971).

150. Runtz, G. R., and Bader, R. F. W., *Mol. Phys.* **30**, 129 (1975).

151. Rutenberg, A. C., and Palko, A. A., *J. Phys. Chem.* **69**, 527 (1965).

152. Ryschkewitsch, G. E., *in* "Boron Hydride Chemistry" (E. L. Muetterties, ed.), Chapter 6 Academic Press, New York, 1975.

153. Ryschkewitsch, G. E., and Miller, V. R., *J. Am. Chem. Soc.* **95**, 2836 (1973).

154. Ryschkewitsch, G. E., and Myers, W. H., *Syn. React. Inorg. Metal-Org. Chem.* **5**, 123 (1975).

155. Ryschkewitsch, G. E., and Rademaker, W. J., *J. Magn. Reson.* **1**, 584 (1969).

156. Ryschkewitsch, G. E., and Wiggins, J. W., *Inorg. Chim. Acta* **4**, 33 (1970).

157. Ryschkewitsch, G. E., and Wiggins, J. W., *J. Am. Chem. Soc.* **92**, 1790 (1970).

158. Ryss, I. G., and Mahonin, U. D., *Ukr. Khim. Zh.* **37**, 9, 863 (1971).

159. Satchell, R. S., Bukka, K., and Payne, C. J., *J. Chem. Soc., Perkin Trans. II* p. 541 (1975).

160. Schmulbach, C. D., and Ahmad, I. Y., *Inorg. Chem.* **8**, 1414 (1969).

161. Schmulbach, C. D., and Ahmad, I. Y., *Inorg. Chem.* **11**, 228 (1972).

162. Sharp, D. W. A., *Adv. Fluorine Chem.* **1**, 68 (1960).

163. Sharp, K. G., and Bald, J. F., Jr., *Inorg. Chem.* **14**, 2553 (1975).

164. Shitov, O. P., Ioffe, S. L., Tartakovskii, V. A., and Novikov, S. S., *Russ. Chem. Rev.* **40**, 905 (1970).

165. Skinner, H. A., and Smith, N. B., *J. Chem. Soc.* p. 3930 (1954).

166. Spielvogel, B. F., and Purser, J. M., *J. Am. Chem. Soc.* **93**, 4418 (1971).

167. Stafford, S. L., *Can. J. Chem.* **41**, 807 (1963).

168. Steinberg, H., "Organoboron Chemistry," Vol. 1. Wiley, New York, 1964.

169. Stephens, R. S., Lessley, S. D., and Ragsdale, R. O., *Inorg. Chem.* **10**, 1610 (1971).

170. Stone, F. G. A., *Chem. Rev.* **58**, 101 (1958).

171. Swanson, B., Shriver, D. F., and Ibers, J. A., *Inorg. Chem.* **8**, 2182 (1969).

172. Thompson, R. J., and Davis, J. C., Jr., *Inorg. Chem.* **4**, 1464 (1965).

173. Titova, K. V., Malov, Y. I., Kolmakova, E. I., and Rosolovskii, V. Y., *Bull. Acad. Sci. USSR, Div. Chem.* **24**, 664 (1975).

174. Topchiev, A. V., Zavgorodnii, S. V., and Paushkin, Ya. M., "Boron Fluoride and Its Compounds as Catalysts in Organic Chemistry." Pergamon, London, 1959.

175. Tuchagues, J.-P., Castan, P., Commenges, G., and Laurent, J.-P., *Syn. React. Inorg. Metal-Org. Chem.* **5**, 279 (1975).
176. Tuchagues, J.-P., and Laurent, J.-P., *J. Inorg. Nucl. Chem.* **36**, 1469 (1974).
177. Umland, F., Hohaus, E., and Brodte, K., *Chem. Ber.* **106**, 2427 (1973).
178. Urry, G., *in* "The Chemistry of Boron and Its Compounds" (E. L. Muetterties, ed.), Chapter 6. Wiley, New York, 1967.
179. Vandrish, G., and Onyszchuk, M., *Proc. 14th International Conference on Coordination Chemistry*, p. 434 (1972).
180. van Paasschen, J. M., and Geanangel, R. A., *J. Am. Chem. Soc.* **94**, 2680 (1972).
181. van Paasschen, J. M., Hu, M. G., Peacock, L. A., and Geanangel, R. A., *Syn. React. Inorg. Metal-Org. Chem.* **4**, 11 (1974).
182. Vladimiroff, T., and Malinowski, E. R., *J. Chem. Phys.* **46**, 1830 (1967).
183. Waddington, T. C., and Klanberg, F., *J. Chem. Soc.* p. 2332 (1960).
184. Weidig, C., Lakovits, J. M., and Kelly, H. C., *Inorg. Chem.* **15**, 1783 (1976).
185. Weidig, C. Uppal, S. S., and Kelly, H. C., *Inorg. Chem.* **13**, 1763 (1974).
186. Wharf, I., and Shriver, D. F., *J. Inorg. Nucl. Chem.* **32**, 1831 (1970).
187. Wiberg, N., and Buchler, J. W., *Chem. Ber.* **96**, 3000 (1963).
188. Wrackmeyer, B., and Nöth, H., *Chem. Ber.* **109**, 1075 (1976).
189. Young, D. E., McAchran, G. E., and Shore, S. G., *J. Am. Chem. Soc.* **88**, 4390 (1966).

ADVANCES IN INORGANIC CHEMISTRY AND RADIOCHEMISTRY, VOL. 21

REORGANIZATION ENERGIES OF OPTICAL ELECTRON TRANSFER PROCESSES

R.D. CANNON

School of Chemical Sciences, University of East Anglia, Norwich, England

I. Introduction 179
    A. Definitions 180
    B. Valency States 182
    C. Inner- and Outer-Sphere Energy Terms 184
II. Review of Data 185
    A. Single-Ion Processes (Photoemission) 185
    B. Donor–Acceptor Pairs 189
    C. Ion Pairs 190
    D. Linked Pairs 194
    E. Bridged Binuclear Complexes 195
    F. Directly Bonded Complexes 202
    G. Further Single-Ion Processes 207
III. Theory 211
    A. Ionic Solvation 211
    B. Continuum Theory for Electron Transfer Processes 213
IV. Conclusions 217
    A. Single-Ion Processes 217
    B. Donor–Acceptor Processes 221
    References 225

I. Introduction

Electron transfer reactions and spectroscopic charge-transfer transitions have been extensively studied, and it has been shown that both processes can be described with a similar theoretical formalism. The activation energy of the thermal process and the transition energy of the optical process are each determined by two factors: one due to the difference in electron affinity of the donor and acceptor sites, and the other arising from the fact that the electronically excited state is a nonequilibrium state with respect to atomic motion (Franck–Condon principle). Theories of electron transfer have been concerned with predicting the magnitude of the Franck–Condon barrier; but, in the field of thermal electron transfer kinetics, direct comparisons between theory and experimental data have been possible only to a limited extent. One difficulty is that in kinetic studies it is generally difficult to separate the electron transfer process from the complex formation

179

and rearrangement processes that precede it; another is that quite small energy differences can wholly change the course of a reaction, so that errors of, say ± 10 kcal in the theoretical estimate of an activation energy can lead to entirely wrong predictions. So far, therefore, theory has been used mainly to rationalize comparisons between closely related reactions, and not to predict reaction rates absolutely.

The purpose of this review is to assemble the data in the area of spectroscopy for comparisons between theory and experiment. It seems at first sight that the data are simpler to evaluate than are those of kinetics, and certainly the energies involved are larger, which is an advantage where absolute predictions are required. The main obstacle to progress hitherto has been the lack of thermodynamic electron-affinity data for the systems of interest. There is, however, a growing number of systems for which both the spectroscopic and the thermodynamic data are sufficiently accurate to encourage attempts to calculate the Franck–Condon energy.

In Section II, we shall collect the relevant data and calculate values of the Franck–Condon energies; and we shall attempt to rationalize them qualitatively by making suitable comparisons between related systems. In Section III we shall briefly review the electrostatic theories of solvation and reorganization, and in Section IV we shall make some preliminary comparisons between theory and experiment.

A. DEFINITIONS

A thermal electron transfer reaction between complex ions in solution may be written schematically as

$$(A^+ \cdots B)(env) \to (A \cdots B^+)(env) \qquad (1)$$
$$\quad p \qquad\qquad\qquad s$$

in which A and B denote atoms each of which possesses two valence states differing by one electron, and the ($+$) sign denotes the higher of the two valences. This notation will be used throughout for the sake of simplicity, regardless of the actual valences and ionic charges involved. Systems ($A^+ \cdots B$) and ($A \cdots B^+$) may be binuclear complexes of more-or-less fixed geometry, or they may be reactive pairs in systems undergoing a bimolecular reaction. We shall refer to them in general as the *precursor* (p) and *successor* (s) states. The state symbol (env) denotes the total ligand and solvent environment of the reacting ions, at thermal equilibrium.

The optically excited electron transfer process may then be written

$$(A^+ \cdots B)(env) \xrightarrow{h\nu_{CT}} (A \cdots B^+)(env^*) \qquad (2)$$
$$\quad p \qquad\qquad\qquad\qquad s^*$$

in which the product s^* has an electronic configuration similar to that of s, but, by the Franck–Condon principle, the ligand and solvent molecules are in the same positions as in p.

Without attempting any justification, we shall ascribe thermodynamic properties to the state s^*, so that a thermodynamic cycle may be written

$$\tag{3}$$

giving

$$\Delta Y_{CT} = \Delta Y_{PS} + \Delta Y_{FC} \tag{4}$$

where symbol Y denotes energy (U), enthalpy (H), Gibbs free energy (G), or entropy (S). Thus ΔG_{PS} is the standard free-energy change associated with the thermal process [Eq. (1)]; and ΔG_{FC} is the reorganizational free energy, i.e., the work done in reversibly distorting the atomic configuration of the successor state into a configuration typical of the precursor state. We shall, moreover, assume* that the entropy change ΔS_{CT} associated with the optical absorption process is zero (unless there is a change in multiplicity, which however we shall ignore). The absorption frequency ν_{CT} then measures the molar energy, enthalpy, and free energy of process Eq. (2):

$$Lh\nu_{CT} = \Delta U_{CT} = \Delta H_{CT} \cong \Delta G_{CT} \tag{5}$$

(the quantities ΔY are molar quantities, L being the Avogadro constant). We note that, by our assumption,

$$\Delta S_{FC} = -\Delta S_{PS} \tag{6}$$

For the reverse electron transfer process, we may write

$$(A \cdots B^+)\text{env} \xrightarrow{\;\Delta Y_{CT}'\;} (A^+ \cdots B)(\text{env}^*) \tag{7}$$
$$s \qquad\qquad\qquad p^*$$

giving the cycle

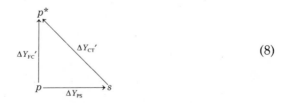

$$\tag{8}$$

<hr />

* This assumption was first made explicit by Treinin (127f).

with

$$\Delta Y_{CT}' = -\Delta Y_{PS} + \Delta Y_{FC}' \tag{9}$$

which, on combining with Eq. (4) yields

$$\Delta Y_{CT} - \Delta Y_{CT}' = 2\Delta Y_{PS} + (\Delta Y_{FC} - \Delta Y_{FC}') \tag{10}$$

$$\Delta Y_{CT} + \Delta Y_{CT}' = \Delta Y_{FC} + \Delta Y_{FC}' \tag{11}$$

whereas from Eq. (6) we have

$$\Delta S_{FC} = -\Delta S_{FC}' \tag{12}$$

It will be seen in the following that under certain conditions the free-energy terms in parentheses in Eq. (10) cancel, i.e., $\Delta G_{FC} = \Delta G_{FC}'$ [Eq. (74), Section III,B].

B. Valency States

The foregoing discussion is restricted to cases in which the valencies in the binuclear complexes are localized, so that, for example, the ground state of p can properly be described as $(A^+ \cdots B)$ without appreciable mixing of the state $(A \cdots B^+)$. The range of possible variations, from zero to complete mixing, is shown in Fig. 1, following a discussion by Mayoh and Day (96). Ordinate U is the energy of the wave function of the transferring electron (assumed separable from the wave functions of all other electrons in the system), and abscissa x is a "reaction coordinate" describing the course of the reaction $p \to s$.

Figure 1a shows the case of a complex that is symmetrical as regards the component atoms (i.e., $A \equiv B$) but not necessarily symmetrical in its electronic configuration. If states $(A^+ \cdots A)$ and $(A \cdots A^+)$ do not mix, the energies are given by the zero-order curves (0,0). (According to the theories outlined in Section III, these curves are parabolas.) States p and s at the two minima are mixed-valence complexes of the Robin and Day "Class I" (119a). The excitation $p \to s^*$ has a probability of zero. For a weak interaction, the curves (1,1) differ significantly only at the crossing point. The complexes have localized valences, but are designated Class II, because the $p \to s^*$ transition produces an observable "intervalence transfer" band in the spectrum. The extent of interaction and, thus, the resonance energy H at the crossing point, can be estimated from the transition dipole moment (96). With stronger interactions, the ground states also can have appreciable mixing of

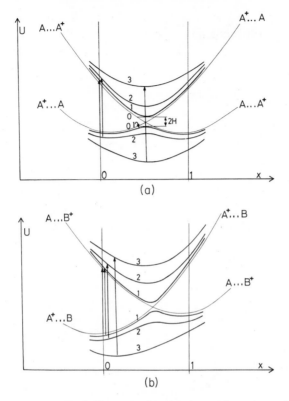

FIG. 1. Idealized energy-level diagrams showing intervalence-type charge-transfer transitions in mixed-valence binuclear complexes. (a) Symmetrical case ($\Delta G_E = 0$), after Mayoh and Day (*96*). The fine lines are parabolas representing the energies of zero-order wave functions for the electronic configurations $A^+ \cdots A$ and $A \cdots A^+$. The bold lines 1, 1; 2, 2; 3, 3 show ground and excited electronic states of the mixed-valence complexes with (1) weak interaction, (2) moderate interaction, (3) strong interaction. The vertical transitions shown by the arrow have intervalence charge-transfer character ($p \to s^*$) in cases 1 and 2 but not in case 3. (b) Unsymmetrical case ($\Delta G_E > 0$). The labeling of the curves corresponds to that of the symmetrical case (a), but it is seen that even with strong interaction the ground- and excited-state curves (3, 3) have minima at difference abscissas x, hence the vertical transitions retain some charge-transfer character.

the ($A^+ \cdots A$) and ($A \cdots A^+$) configurations (curve 2,2) and, finally, if the resonance energy exceeds a certain critical value (*11, 67*) states p and s coalesce into a single minimum (curves 3,3) giving a wholly delocalized Class III complex, which should now be termed an "average-valence" rather than a "mixed-valence" complex. There is still a characteristic electronic transition, but it no longer has intervalence charge-transfer character (*11*).

Figure 1b shows the corresponding situations for unsymmetrical complexes. Similar comments apply, except for the fact that even in the case of strong interaction, when the ground state has only one minimum, the optical transition retains some charge-transfer character.

There are now well-attested examples of all the degrees of interaction represented in Fig. 1, some of which will be mentioned in the following. On the other hand, by no means all complexes showing charge transfer have yet been classified. In this review we shall concentrate on complexes with weak interaction, but others will also be considered with appropriate caution, as discussed especially in Section II,F.

It must be stressed that diagrams of the type of Fig. 1 contain other major assumptions, which, however, we shall not discuss here. One assumption is that a single reaction coordinate can be used to characterize the energy changes in the ground and excited states. In fact, the vibrational modes associated with the thermal activation processes are almost wholly unknown, and the interactions between vibrational and electronic excitations are only beginning to be explored, for example by the resonance Raman technique (22). Another assumption is that the excited states produced by charge transfer from the ground state s do, in fact, correlate with the precursor p. There are many other possibilities, as has been made clear from studies of photochemical reaction mechanisms (6). A classification of mechanisms based on the mixing or nonmixing of reactants and products' ground- and excited-state wave functions has been proposed by Endicott (45). In his terminology, only Class I and Class I' processes are relevant to the present study; but we shall discuss one of the alternative kinds of process in Section II,C.

C. INNER- AND OUTER-SPHERE ENERGY TERMS

The total reorganization energy ΔG_{FC} will in general be made up of three components, which we may call the bonding, the ligand, and the solvent terms:

$$\Delta G_{FC} = \Delta G_{FC}^{bond} + \Delta G_{FC}^{ligand} + \Delta G_{FC}^{solvent} \tag{13}$$

The bonding term arises whenever the donor and acceptor atoms are close enough to interact directly, and may be positive or negative depending on whether the excited state s^* is subject to a repulsive or an attractive internuclear force. The ligand term is due to differences in the equilibrium positions of the ligands directly bonded to the two atoms. Experience from kinetic studies suggests that this contribution is likely to be significant for highly ionic complexes with ligands such as H_2O or NH_3, but insignificant for strongly π-bonded complexes with ligands such as CN^- or $2:2'$-bipyridyl. The solvent term is the free energy of reorganization of solvent molecules. In this review we shall use the commonly accepted terminology, grouping the bond and ligand terms as the *inner-sphere* component and calling the solvent term the *outer-sphere* component:

$$\Delta G_{FC} = \Delta G_{FC}^{in} + \Delta G_{FC}^{out} \qquad (14)$$

The same subdivision is used in the qualitative discussion of ionic solvation energies, and it is found that the outer-sphere part is satisfactorily predicted by electrostatic continuum theories (as reviewed in Section III), whereas the inner-sphere part is best obtained by considering the distortion of individual bonds, using force constants from vibrational spectroscopy.

II. Review of Data

A. SINGLE-ION PROCESSES (PHOTOEMISSION)

The process of ionization of an atom or molecule A in the gas phase may be written

$$A(g) \rightarrow A^+(g) + e^-(g) \qquad (15)$$

where the state symbol (g) denotes a free species *in vacuo*, having zero translational kinetic energy. An analogous equation (*111*) may be written for a process in solution:

$$A(env) \longrightarrow A^+(env) + \{e^-\}_{env} \qquad (16)$$

where the state symbol (env) denotes the total solvent environment at thermal equilibrium with the species in question, and the symbol $\{e^-\}_{env}$ denotes an electron delocalized throughout the bulk of the solvent (and, hence, not specifically interacting with any chemical species present), but having zero kinetic energy.* Such an electron has been called "quasi-free" (*36*) or "dry" (*40*); in quantum-mechanical language, it is at the lowest continuum energy level of the system.

---

* Notation $\{e^-\}_{aq}$, $\{-\}_{aq}$ is based on that of Noyes, whose paper (*111*) should be consulted for a clear discussion of the problems of definining transfer energies of charged species.

If, however, we specify that the ionization process is instantaneous, so that all atoms and solvent molecules remain in the positions characteristic of the ground state, we may write

$$A(env) \xrightarrow{\Delta U_\phi} A^+(env^*) + \{e^-\}_{env} \tag{17}$$

by analogy with Eq. (2) This type of ionization was first explicitly discussed by Platzman and Franck in 1952 (116); but it did not become accessible to direct measurement until some 20 years later. Photoemission spectra have now been reported for a variety of chemically reducing species in solution, notably the ferrocyanide ion (4, 7), the solvated electron (8, 40), and a number of other organic and inorganic species (109). The mechanism of emission has been analyzed theoretically (35) in terms of two consecutive processes: generation of electrons in unbound states and the escape of electrons into the gas phase. Sometimes the energy ΔU_ϕ is close to a peak in the photoemission spectra, but equations have been given from which ΔU_ϕ can be calculated when this is not the case (35a).

Combining equation (17) with expressions for the standard reduction of $A^+(env)$ and for the formation and hydration of the proton, we obtain the Franck–Condon parameters as follows:

$$
\begin{array}{lll}
A(env) \rightarrow A^+(env^*) + \{e^-\}_{env} & \Delta Y_\phi(A) & (17) \\
A^+(env) + \tfrac{1}{2}H_2(g) \rightarrow A(env) + H^+(env) & \Delta Y_E(A^+) & (18) \\
H^+(env) + \{e^-\}_{env} \rightarrow H^+(g) + e^-(g) & & (19) \\
H^+(g) + e^-(g) \rightarrow H(g) & \left.\begin{array}{c}\\\\\\\end{array}\right\} \; \Delta Y_H & (20) \\
H(g) \quad\;\;\; \rightarrow \tfrac{1}{2}H_2(g) & & (21) \\
\hline
A^+(env) \rightarrow A^+(env^*) & \Delta Y_{FC}(A^+) & (22)
\end{array}
$$

It will be noted that Eq. (19) is written[†] so as to avoid transferring any net charge across the liquid–vacuum interface (111). Values of ΔG_H, ΔH_H, and ΔS_H for aqueous medium are calculated in Table I.

The reverse process, of electron transfer from the continuum to the oxidized ion,

$$A^+(env) + \{e^-\}_{aq} \rightarrow A(env)^* \qquad \Delta Y_\phi(A^+ + e^-) \tag{23}$$

has apparently not been studied. By a similar argument, it would

---

[†] Throughout this review, we use the units cal = 4.184 J; kK = 10^3 cm^{-1}. Electric potentials and equilibrium constants reported in the literature have been converted to standard free-energy changes by means of the expressions $-\Delta G^\ominus = -zFE^\ominus = 2.303\,RT \log K$, where $F = 23.060$ kcal mole^{-1} volt^{-1}, $2.303\,RT = 1.359$ kcal mole^{-1}, at 25°C. Wave numbers of optical transitions reported in the literature have been converted to energy changes by means of the expression $\Delta U = Lhc^{-1}v^{-1}$, where $Lhc^{-1} = 2.862$ kcal mole^{-1} kK^{-1}.

TABLE I

CALCULATION OF THERMODYNAMIC PARAMETERS[a] FOR THE ION $H^+(aq)$

Reaction	ΔH (kcal mole^{-1})	ΔG (kcal mole^{-1})	$T\Delta S$ (kcal mole^{-1})	ΔS (cal K^{-1} mole^{-1})
$H^+(aq) + Cl^-(aq) \rightarrow H^+(g) + Cl^-(g)$	348.8[b]	333.6[c]	—	51.0[b]
$\{e^-\}_{aq} + Cl^-(g) \rightarrow Cl^-(aq) + e^-(g)$	−88.7[d]	−84.2[d]	—	−15.0[d]
Subtotal:				
$\{e^-\}_{aq} + H^+(aq) \rightarrow H^+(g) + e^-(g)$	260.1[e,f]	249.4[e]	10.74	36.0[e]
$H^+(g) + e^-(g) \rightarrow H(g)$	−315.0[b]	−315.0[c]	0.00[g]	0.0[g]
$H(g) \rightarrow \frac{1}{2}H_2(g)$	−52.1[b]	−48.6[b]	−3.57[b]	−11.8[b]
Total:				
$\{e^-\}_{aq} + H^+(aq) \rightarrow \frac{1}{2}H_2(g)$	−107.0	−114.2	7.14	24.2

[a] All temperature-dependent parameters relate to $T = 298$ K.
[b] Calculated from data in Ref. 108.
[c] Calculated from the other parameters using $\Delta G = \Delta H - T\Delta S$.
[d] W. M. Latimer et al. (83).
[e] Sum of the above two parameters.
[f] A value 260.7 kcal mole^{-1} has been obtained (58) using somewhat different assumptions to those of Latimer et al. (83).
[g] Assumed.

yield the Franck–Condon parameters for the process,

$$A(env) \rightarrow A(env)^* \qquad \Delta Y_{FC}(A) \qquad\qquad (24)$$

where $A(env^*)$ denotes a species having the electronic configuration of $A(env)$ but the atomic configuration of $A^+(env)$. If Eq. (74) (see Section III,B) is valid, we have

$$\Delta G_{FC}(A) = \Delta G_{FC}(A^+) \qquad\qquad (25)$$

Calculations of the Franck–Condon energy ΔG_{FC} for three species are shown in Table II. One complication in the use of the data at this early stage is the variety of solvent systems used; another is the fact that, in the case of $Fe(CN)_6^{4-}$ ion, different salts have been used, and in the less polar media the nature of the cation may be expected to be significant. It is worth noting, however, that, again in the case of $Fe(CN)_6^{4-}$, the photoemission data of Ballard and Griffith (4),

TABLE II

SINGLE IONS: FREE-ENERGY PARAMETERS ΔG_ϕ FOR THE PHOTOEMISSION PROCESS
$A(env) \rightarrow A^+(env^*) + \{e^-\}_{aq}$

Emitting complex (A):	$Fe(CN)_6^{4-}$	I^-	$Fe(C_5H_5)_2$
Solvent (env):	aq^a	Tetraglyme	Tetraglyme
Temperature (°C):	1.2	25	25
Radius of complex A^+ (10^{-10} m)	4.3^h	1.33^i	3.8^j
ΔG (kcal mole^{-1}):			
$A(env) \qquad\qquad \rightarrow A^+(env^*) + \{e^-\}_{env}$	137^b	143^e	143^e
$A^+(aq) + \frac{1}{2}H_2(g) \rightarrow A(aq) + H^+(aq)$	-8^c	-31.1^f	-8.0^g
$\{e^-\}_{aq} + H^+(aq) \rightarrow \frac{1}{2}H_2(g)$	-114^d	-114.2^d	-114.2^d
Totals:	15	-2	21

[a] A value for $0.25M$ $Fe(CN)_6^{4-}$ in water (cation not specified).

[b] Delahay (35a).

[c] Using ΔG ($T = 298$ K) values from Hanania (59). ΔH is calculated from the following data (ΔH in units of kcal mole^{-1}): $Fe(CN)_6^{4-}(aq) + \frac{1}{2}Br_2(l) \rightarrow Fe(CN)_6^{3-}(aq) + Br^- \cdot aq(-4.2)$, $\frac{1}{2}Br_2(aq) \rightarrow \frac{1}{2}Br_2(l)$ (0.6); $Br^-(aq) + H^+(aq) \rightarrow \frac{1}{2}Br_2(aq) + \frac{1}{2}H_2(g)$ (28.8) (Refs. 61, 108, 107, respectively).

[d] Table I.

[e] Delahay (37).

[f] Calculated from the following data (ΔG in units of kcal mole^{-1}, at 298 K): $\frac{1}{2}I_2(s) + \frac{1}{2}H_2(g) \rightarrow I^-(aq) + H^+(aq)$ (-12.3); $\frac{1}{2}I_2(aq) \rightarrow \frac{1}{2}I_2(s)$ (-2.0); $I(aq) \rightarrow \frac{1}{2}I_2(aq)$ (assumed same as for gas phase, -16.8) (Refs, 126, 134, 108, respectively).

[g] Kolthoff and Thomas (77).

[h] Estimated.

[i] Phillips and Williams (114, p. 109).

[j] Average of maximum and minimum estimated radii of $Fe(C_5H_5)_2$ (128).

obtained for glycerol media, are evidently continuous with those of Delahay *et al.* (7), for aqueous media.

It is interesting to note that the two complexes $Fe(CN)_6^{4-}$ and $Fe(C_5H_5)_2$, of somewhat similar dimensions, give similar values of ΔG_{FC}. On the other hand, the negative value found for I^- is clearly incorrect. The error may be due to the use of data from different solvent systems, and it might be expected that the I^- ion, with its smaller radius, would be more sensitive to changes of solvent than the other two ions. But it is difficult to discuss this without more comparative data.

Ballard (4) points out that, below the energy of the band maximum, the photoemission current varies exponentially with photon energy, which is evidence that the band width is due to thermal occupation of vibrational levels in the electronic ground state, $Fe(CN)_6^{4-}$(env). The threshold energy was found to be approximately 119 kcal mole^{-1} (4b) [118.3 kcal mole^{-1} by another method (7b)], i.e. about 18 kcal mole^{-1} lower than ΔU_ϕ. It seems likely that the low-energy threshold of the spectrum represents emission from those $Fe(CN)_6^{4-}$ ions which happen to have solvent environments similar to those that characterize the ground state of $Fe(CN)_6^{3-}$.

B. Donor–Acceptor Pairs

Transfer of charge between localized reducing and oxidizing sites is well known. In a general case, with oxidant A^+ and reductant B, situated at infinite distance apart, we may write

$$A^+(env) + B(env) \xrightarrow{h\nu} A(env)^* + B^+(env)^* \tag{26}$$

and the thermodynamic parameters will be given by

$$\Delta Y_{CT} = \Delta Y_{FC}(A) + \Delta Y_E(A^+) + \Delta Y_{FC}(B^+) - \Delta Y_E(B^+) \tag{27}$$

Thus, the Franck–Condon barrier for the oxidant–reductant pair is given by

$$\Delta Y_{FC}^{pair} = \Delta Y_{FC}(A) + \Delta Y_{FC}(B^+) = \Delta Y_{CT} - \{\Delta Y_E(A^+) - \Delta Y_E(B^+)\}$$
$$= \Delta Y_{CT} - \Delta Y_0 \tag{28}$$

where ΔY_0 denotes the energy, enthalpy, etc., change associated with the equilibrium:

$$A^+(env) + B(env) \rightleftharpoons A(env) + B^+(env) \tag{29}$$

In principle, the converse charge-transfer process* should also be observable:

$$A(env) + B^+(env) \xrightarrow{h\nu'} A^+(env)* + B(env)*$$ (30)

with

$$\Delta Y_{CT}' = \Delta Y_{FC}(A^+) - \Delta Y_E(A^+) + \Delta Y_{FC}(B) + \Delta Y_E(B^+)$$ (31)

If, as predicted by Eq. (74) (see Section III,B), $\Delta G_{FC}(A^+) = \Delta G_{FC}(A)$ and $\Delta G_{FC}(B^+) = \Delta G_{FC}(B)$, then, as before, we obtain

$$\Delta G_{CT} - \Delta G_{CT}' = 2\{\Delta G_E(A^+) - \Delta G_E(B^+)\} = 2\Delta G_0$$ (32)

As yet there are no data for charge transfer over long distances, defined by Eq. (26), although there are some results for intramolecular electron transfer in systems where the donor and acceptor are, if not independent, at least only weakly coupled. Nor have we any data for the variation in charge-transfer energy as a function of distance, except in systems where the coupling between donor and acceptor is evidently strong. The possibility that thermal electron transfer between like-charged ions in solution may take place over large distances has been discussed several times, and some of the quantitative theories involve integration of the rate of reaction over all possible distances (93); other theories, however, assume that the bulk of electron transfer takes place between ions in contact (93), as discussed in the following.

C. Ion Pairs

When the oxidant and reductant occur in complexes of opposite charge, the assumption that charge transfer occurs mainly between contact pairs becomes reasonable; and, in addition, the free energy of formation of the contact pair can often be estimated. Sometimes, indeed, the study of the charge-transfer spectrum has proved to be the most useful method of obtaining the free energy of formation. In the reaction scheme,

$$
\begin{array}{ccccc}
A^+(env) + B(env) & \xrightarrow{\Delta Y_W} & A^+ \cdot B(env) & \xrightarrow{\Delta Y_{FC}'(A^+ \cdot B)} & A^+ \cdot B(env*) \\
 & & p & & p* \\
\Big\downarrow \Delta Y_0 & & \Big\downarrow \Delta Y_{PS} & \Delta Y_{CT} \diagdown \diagup \Delta Y_{CT}' & \\
A(env) + B^+(env) & \xrightarrow{\Delta Y_W'} & A \cdot B^+(env) & \xrightarrow{\Delta Y_{FC}(A \cdot B^+)} & A \cdot B^+(env*) \\
 & & s & & s*
\end{array}
$$ (33)

* Reversible, optical electron transfer has been observed in the solid state, between the ions Eu^{2+} and Sm^{3+} doped into crystalline CaF_2 (143a).

formula $A^+ \cdot B$(env) denotes the precursor complex p in which A^+ and B have the same inner-sphere ligands as A^+(env) and B(env), respectively, and the remaining solvent molecules are appropriately rearranged; $A \cdot B^+$(env) is the successor complex s similarly related to A(env) and B^+(env). Thermal electron transfer reactions of the type $p \to s$ have been extensively discussed (93), and in a few instances, directly measured (50). One of the current goals of research in this field is to measure the reaction rate and the total free-energy change ΔG_{PS} for the same system. So far this is still to be achieved.

From Eq. (33) we have

$$\Delta Y_{FC}(A \cdot B^+) = \Delta Y_{CT} - \Delta Y_{PS}$$
$$= \Delta Y_{CT} + (\Delta Y_W - \Delta Y_W') - \Delta Y_0 \qquad (34)$$

where ΔY_W and $\Delta Y_W'$ are "work terms" (93) for the formation of the precursor and successor complexes, and ΔY_0 is the same as before. At the present time there are no cases in which ΔY_W and $\Delta Y_W'$ are known for the same system, but, in some cases, at least the free-energy terms may be estimated from data on analogous complexes, since it is evident that these do not vary greatly between outer-sphere systems of similar charge type (12).

If the algebraic difference in charge between the complexes A^+ and B happens to be $+1$ [as in the pairs Fe^{3+}, Cr^{2+}; $Fe(C_5H_5)_2^+$, $Fe(C_5H_5)_2$; $Fe(CN)_6^{3-}$, $Fe(CN)_6^{4-}$], the two work terms may be expected to cancel; but no spectrum has yet been measured for such a pair (cf. Table IV).

If the redox potentials are not too different, it should be possible to measure spectra for both complexes, $A^+ \cdot B$ and $A \cdot B^+$; however, this again has still to be achieved. From Eq. (33), we find

$$\Delta Y_{FC}'(A^+ \cdot B) = \Delta Y_{CT}' + \Delta Y_{PS} \qquad (35)$$

and the energies of the two spectra are related by

$$\Delta Y_{CT} + \Delta Y_{CT}' = \Delta Y_{FC}(A \cdot B^+) + \Delta Y_{FC}'(A^+ \cdot B) \qquad (36)$$

$$\Delta Y_{CT} - \Delta Y_{CT}' = 2\Delta Y_{PS} + \{\Delta Y_{FC}(A \cdot B^+) - \Delta Y_{FC}'(A^+ \cdot B)\} \qquad (37)$$

On the basis of the electrostatic theory outlined in the following, the braced terms in Eq. (37) will cancel provided that the Groups A and B in the complex have similar geometry. (See Section III,B).

Calculations of ΔG_{FC} based on Eq. (34) are shown in Table III. The values of ΔG_W and $\Delta G_W'$ used are only approximate, because they are based on data for various ionic strengths. But this is not a serious error since they are all small—certainly small enough to justify the assumptions made in the foregoing. Data for cobalt(III) complexes have been included in the Table III, but they cannot be used for calculations of ΔG_{FC} because the cobalt(III) product of the optical redox

TABLE III

CONTACT ION PAIRS: ENERGY PARAMETERS FOR THE CHARGE-TRANSFER PROCESS
$(A^+ \cdot B) \rightarrow (A \cdot B^+)$

A^+(env)	B(env)	ΔG_{CT}	$-\Delta G_E(A^+)$	$\Delta G_E(B)$	ΔG_W	$-\Delta G_W'$	$\Delta G_{FC}(A \cdot B^+)$
$Ru(NH_3)_6^{3+}$	Cl^-	97.3	2.3^h	-57.4^j	-1.6^a	0.0^m	40.6
$Ru(NH_3)_6^{3+}$	Br^-	88.6	2.3^h	-45.0^j	-1.4^a	0.0^m	44.5
$Ru(NH_3)_6^{3+}$	I^-	71.2	2.3^h	-31.5^j	-1.4^a	0.0^m	40.6
$Ru(NH_3)_6^{3+}$	NCS^-	88.2	2.3^h	-39.6^k	-1.4^m	0.0^m	49.5
$Ru(NH_3)_6^{3+}$	$S_2O_3^{2-}$	71.0					
$Ru(en)_3^{3+}$	I^-	63.6^c	4.9^h	-31.5^j	-0.9^n	0.0^m	36.1
$Co(NH_3)_6^{3+}$	I^-	106.8^d	2.3^i	-31.5^j	-2.3^o	0.0^m	—
$Co(en)_3^{3+}$	I^-	100.2^e	-2.6^i	-31.5^j	-1.8^p	0.0^m	—
$Co(en)_3^{3+}$	NCS^-	100.2^f	-2.6^i	-39.6^k	-1.8^m	0.0^m	—
$Co(en)_3^{3+}$	$S_2O_3^{2-}$	109.9^f					
$Co(en)_3^{3+}$	$Fe(CN)_6^{4-}$	64.9^g	-2.6^i	-8.2^l	-4.3^q	3.8^r	—

[a] Waysbort *et al.* (*143*).

[b] Armor (*3a*).

[c] Elsbernd and Beattie (*43*).

[d] Yokoyama and Yamatera (*152*).

[e] Schmidtke (*122*).

[f] Yoneda (*153*).

[g] Larsson (*81*).

[h] Meyer and Taube (*98a*).

[i] Martell and Sillén (*94*).

[j] Berdnikov and Bazhin (*14*).

[k] Wilmarth (*147*).

[l] Hanania *et al.* (*59*).

[m] Assumed.

[n] Estimated from $Ru(NH_3)_6^{3+} + I^-$, by analogy with pairs of $Co(NH_3)_6^{3+}$ and $Co(en)_3^{3+}$ complexes (*12*). On the other hand, Elsbernd and Beattie (*43*) found $\Delta G_W \geq +1.8$ kcal mole^{-1}.

[o] Estimated by analogy with $Co(NH_3)_6^{3+} + X^-$ (X = Cl, Br) (*12*).

[p] Estimated from $Co(NH_3)_6^{3+} + I^-$ by analogy with other pairs of $Co(NH_3)_6^{3+}$ and $Co(en)_3^{3+}$ complexes (*12*).

[q] Assumed to be same as for $Co(NH_3)_5OH_2^{3+} + Fe(CN)_6^{4-}$ (*50*), by analogy with the pair of complexes $Co(NH_3)_5OH_2^{3+} \cdot SO_4^{2-}$ and $Co(en)_3^{3+} \cdot SO_4^{2-}$.

[r] Assumed by analogy with various $M^{2+} \cdot Fe(CN)_6^{3-}$ complexes (*94*).

process is believed to be the low-spin state, whereas the thermodynamic data relate to the high-spin state. Thus the optical process for the complex $Co(NH_3)_6^{3+} \cdot I^-$ is

$$Co(NH_3)_6^{3+}(aq) \cdot I^-(aq) \xrightarrow{\Delta G_{CT}} Co[t_{2g}^6 e_g^1](NH_3^*)_6^{2+}(aq^*) \cdot I(aq^*) \tag{38}$$

where the asterisks are used to show that, in the excited state, both the NH_3 molecules and the solvent molecules are in positions characteristic of the cobalt(III) complex.

If we define an excitation energy ΔG_{ex} for the process,

$$Co[t_{2g}^5 e_g^2](NH_3^*)_6^{2+}(aq^*) \cdot I(aq^*) \xrightarrow{\Delta G_{ex}} Co[t_{2g}^6 e_g^1](NH_3^*)_6^{2+}(aq^*) \cdot I(aq^*) \tag{39}$$

then Eq. (34) need only be modified by the addition of $-\Delta G_{ex}$ on the right-hand side. Provided ΔG_{ex} is the same for complexes with different anions, it is possible to make valid comparisons between the data.

Examining first the optical transition energies ΔG_{CT}, we see that, for $Ru(NH_3)_6^{3+}$ complexes, these decrease in the order $Cl^- > Br^- > NCS^- > I^- \approx S_2O_3^{2-}$; and for $Co(en)_3^{3+}$ the order is similar, $NCS^- > I^- = S_2O_3^{2-} > Fe(CN)_6^{4-}$. The differences in ΔG_{CT} for the hexammine- and *tris*ethylenediamine metal complexes, with I^- as anion, are similar for cobalt(III) and ruthenium(III). The difference between $Ru(NH_3)_6^{3+}$ and $Co(en)_3^{3+}$ is similar for iodide and thiosulfate, but somewhat less for thiocyanate. Except for this last observation, which may have a bearing on the mode of bonding of NCS^- in the outer sphere, these comparisons agree with the additive model of Eq. (34) and provide some support for it. The difference in ΔG_{FC} between the pairs $Co(en)_3^{3+} \cdot Fe(CN)_6^{4-}$ and $Co(en)_3^{3+} \cdot I^-$ can be calculated from the data of Table III provided we assume that the excitation energies ΔG_{ex} cancel. The result, 11 kcal mole^{-1}, may be compared with the apparent difference between ΔG_{FC} for the single species $Fe(CN)_6^{3-}$ and I, namely 17 kcal mole^{-1} (Table II). Furthermore the value of ΔG_{ex} itself can also be roughly estimated. From the data for the $Co(NH_3)_6^{3+} \cdot I^-$ pair (Table III), we may deduce $\Delta G = 64.3$ kcal mole^{-1} for the process

$$Co(NH_3)_6^{2+}[t_{2g}^5 e_g^2](aq) \to Co(NH_3^*)_6^{2+}[t_{2g}^6 e_g^1](aq^*) \tag{40}$$

and if the contribution due to Franck–Condon factors, such as bond compression and solvent reorganization, is the same as for $Ru(NH_3)_6^{2+} \cdot (aq) \to Ru(NH_3)_6^{2+}(aq^*)$, then we have $\Delta G_{ex} = 34.7$ kcal mole^{-1}. It is interesting to compare this with the energy of the lowest d-d band in the spectrum of $Co(NH_3)_6^{2+}$ ion, which may be ascribed to the transition

$$Co(NH_3)_6^{2+}[t_{2g}^5 e_g^2](aq) \to Co(NH_3)_6^{2+}[t_{2g}^4 e_g^3](aq) \tag{41}$$

i.e., $\Delta G = 25.8$ kcal mole^{-1} (5). The corresponding figures for the *tris*ethylenediamine complexes are 28 and 26.9 kcal mole^{-1}. It should be noted that the equilibrium configuration of low-spin cobalt(II) would be expected to be strongly distorted toward a tetragonal structure, whereas these calculations apply to octahedral systems. Hence the equilibrium

$$Co(NH_3)_6{}^{2+}[t_{2g}{}^5e_g{}^2](aq) \rightleftharpoons Co(NH_3)_6{}^{2+}[t_{2g}{}^6e_g{}^1](aq) \tag{44}$$

would actually be more favorable than these calculations suggest.

D. Linked Pairs

By *linked pairs* we denote oxidant–reductant pairs that are joined by a chain of atoms serving to fix the mutual distance between certain limits but not interacting electronically with either of the reacting centers, as exemplified in structures **I** to **IV**.

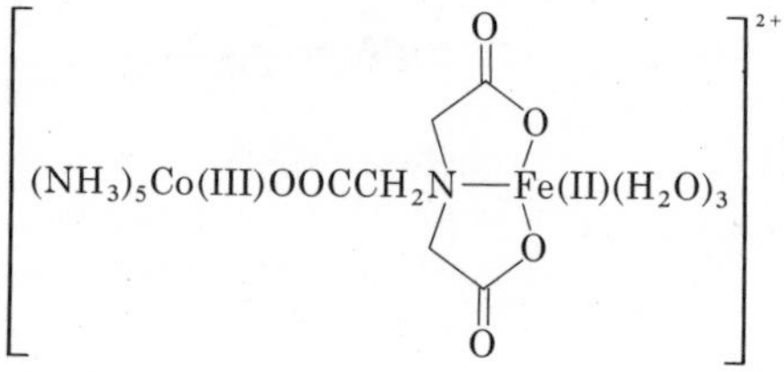

(I)

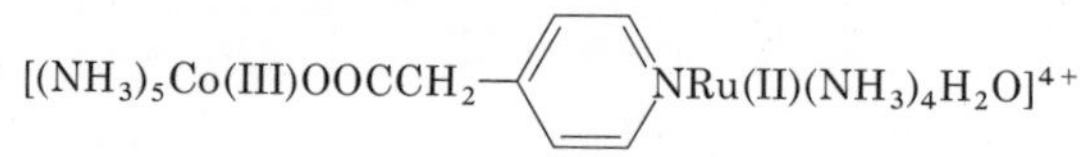

$$[(NH_3)_5Co(III)OOCCH_2 \text{—} \langle \text{ring} \rangle \text{—} NRu(II)(NH_3)_4H_2O]^{4+}$$

(II)

$$[(H_3N)_5Ru(III)N \langle \text{ring} \rangle \text{—} CH_2CH_2 \text{—} \langle \text{ring} \rangle NRu(II)(bip)_2Cl]^{4+}$$

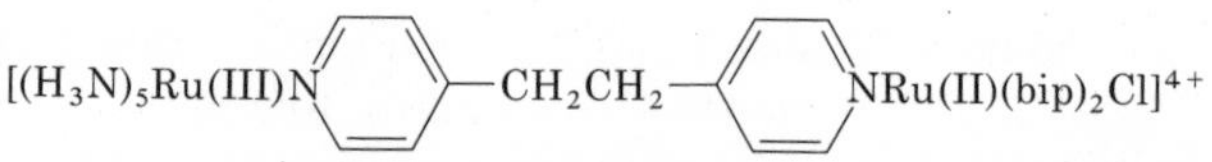

(III)

$$[(C_5H_5)Fe(III)C_5H_4 \cdot CH_2 \cdot C_5H_4Fe(II)(C_5H_5)]^+$$

(IV)

Complexes **I** (20) and **II** (68) are unstable with respect to internal electron transfer, and the rates of the thermal reactions have been measured; complex **III** (2e) is stable with respect to electron transfer. Complex **IV** is symmetrical and presumably subject to rapid internal transfer (103). Optical charge transfer has not been detected in any of these systems. In the case of the two cobalt(III) complexes, comparison with the data of Table III suggests that the bands should be in the visible or near ultraviolet region, and, for complex **III**, at still lower energy. The intensities, however, would be low in view of the weak metal–metal orbital interaction. Complexes of this type are important models for the outer-sphere electron transfer process and further study of analogous systems would be worthwhile. In particular, they could be used to study the dependence of ΔG_{FC} on internuclear distance, by varying the chain length while keeping the same oxidant and reductant ions and the same inner-sphere ligand environments. It seems evident that, as the interionic distance is decreased, ΔG_{FC} will decrease, but so far neither theoretical nor experimental data are available. In the limit of infinite separation, ΔG_{FC} is expected to approach the sum of the two single-ion reorganization energies [Eq. (28)].

E. BRIDGED BINUCLEAR COMPLEXES

In Tables IV–VI are listed spectral data for binuclear complexes in which the metal ions are joined through a single ligand atom or through a bridging molecule that may be expected to modify the electronic structures of oxidant, reductant, or both. These systems may conveniently be classified according to the degree of certainty of the information that may be deduced from them.

TABLE IV

MISCELLANEOUS BINUCLEAR COMPLEXES: CHARGE-TRANSFER SPECTRA

Complex (B·X·A$^+$)	Medium	$\nu_{CT}/10^3$ cm^{-1}	Ref.
U(V)—U(VI)	0.1M HClO$_4$	13.6	(110)
Ti(III)—Ti(IV)	12M HCl	14.9, 20.1	(70)
Ti(III)—Ti(IV)	20% H$_2$SO$_4$	21.2	(52)
(NC)$_6$V(II)OHV(III)(CN)$_5$$^{7-}$	CN$^-$(aq)	17.0	(13)
Fe(II)—Fe(III)	12M HCl	Ca. 12.5–18	(88)
Cu(I)—Cu(II)	10M HCl	Ca. 17	(38, 39, 87, 142)
Cu(I)—Cu(II)	CH$_3$OH, OAc$^-$	11.1	(125)
Sn(II)—Sn(IV)	12M HCl	Ca. 25–30	(146a)
Sb(III)—Sb(V)	HCl	Ca. 20	(146b)^a
Sn(II)—U(IV)	6M HCl	> 24	(101)
Cl$_5$W(III)OW(V)Cl$_5$$^{4-}$	12M HCl	19.5	(79)
Cu(I)—Cu(II)	NH$_3$(aq)	No interaction	(87)
Fe(CN)$_6$$^{4-}$—Fe(CN)$_6$$^{3-}$	aq (?)	No interaction	(87)

^{a} For further references and discussions, see Robin and P. Day (119a), and Allen and Hush (119b).

TABLE V

BINUCLEAR COMPLEXES: ENERGY PARAMETERS FOR THE CHARGE-TRANSFER PROCESS $(A^+ \cdot X \cdot B) \rightarrow (A \cdot X \cdot B^+)$

Complexes							
B·X	A^+	ΔG_{CT}	$\Delta G_E(B^+ \cdot X)$	$-\Delta G_E(A^+)$	ΔG_W	$-\Delta G_W'$	$\Delta G_{FC}{}^a$
$(NC)_5Fe(II) \cdot CN^{4-}$	$(H_2)Fe(III)(CN)_5{}^{2-}$	22.0^b	-8.2^c	$+9.5^d$	—	—	Ca. 20^e
B	$X \cdot A^+$	ΔG_{CT}	$\Delta G_E(B^+)$	$-\Delta G_E(X \cdot A^+)$	ΔG_W	$-\Delta G_W'$	$\Delta G_{FC}{}^a$
$MnO_4{}^-$	Ag^+	71.5^f	-46.0^g	12.8^h	-1.2^i	3.1^j	30.2
$Mo(V)(CN)_6{}^{3-}$	Fe^{2+}	34^k	-17.6^l	$\sim17^m$	-4^n	5.6^k	36
$CrO_4{}^{2-}$	Cr^{3+}	41^o	-50 ± 10^p	7.5 ± 1.5^q	-7°	—	—

[a] See text and Eq. (45).
[b] Glauser et al. (51).
[c] Hanania (59).
[d] Malik and Om (91).
[e] Assuming $\Delta G_W \sim \Delta G_W'$ (see text, Eq. 45).
[f] Symons and Trevalion (133).
[g] Noyes (112).
[h] Carrington and Symons (21).
[i] Assumed to be the same as for $K^+ \cdot TcO_4{}^-$ (124).
[j] Assumed by analogy with $Zn^{2+} \cdot SO_4{}^{2-}$ (94a, b).
[k] Malik and Ali (90).
[l] Zielen and Sullivan (154).
[m] Kolthoff and Tomsicek (78).
[n] Assumed by analogy with various $M^{2+} \cdot Fe(CN)_6{}^{3-}$ complexes (94a, b).
[o] King and Neptune (75).
[p] Haight et al. (56).
[q] Using $E^{\ominus}$ ($HCrO_4{}^- + 2H^+ \rightleftharpoons H_3CrO_4$) = 0.575 ± 0.075 volt (56) and assuming same acid dissociation constants for H_3CrO_4 as for H_3PO_4 (94a, b).

TABLE VI

SYMMETRICAL BRIDGED MIXED-VALENCE COMPLEXES

Complex[a] A·X·A$^+$	Medium	ΔG_{CT} (kcal mole^{-1})	Ref.
$(phen)_2V(II)\cdot OH\cdot V(III)(phen)_2^{4+}$	$SO_4^{2-}(aq)$	—	(106)
$(phen)_2Mn(III)(O)_2Mn(IV)(phen)_2^{3+}$	CH_3CN	—	(115)
$(NC)_4Fe(II)\langle{}^{CN}_{NC}\rangle Fe(III)(CN)_4^{5-}$	aq	22.2	(44)
$(C_5H_5)Fe(II)(C_5H_4{-}C_5H_4)Fe(III)(C_5H_5)^+$	CH_3CN	15.0	(26, 73)
$Fe(II)(C_5H_4{-}C_5H_4)_2Fe(III)^+$	CH_3CN	18.5, 25.1	(105) [cf. (27)]
$(C_5H_5)Fe(II)C_5H_4C{\equiv}CC_5H_4Fe(III)(C_5H_5)^+$	CH_2Cl_2	18.3	(84)
$Fe(II)(C_5H_4{-}CH_2{-}C_5H_4)_2Fe(III)^+$	CH_3CN	38.0	(102)
$(NC)_5Fe(II)(pyz)Fe(III)(CN)_5^{5-}$	$C_6H_5NO_2$	20.3, 24.3	(144)
$L_2Co(I)(Ph_2PCH{:}CHPPh_2)Co(II)L_2^-$	CH_2Cl_2	19.6	(41)[b]
$L_2Co(I)(Ph_2PCH{:}CHPPh_2)Co(II)L_2^-$	CH_2Cl_2	21.0	(41)[c]
$(H_3N)_5Ru(II)NC\cdot CNRu(III)(NH_3)_5^{5+}$	D_2O	20.0	(136)
$(H_3N)_5Ru(II)(pyz)Ru(III)(NH_3)_5^{5+}$	D_2O	19.5	(29a, d)
$py(H_3N)_4Ru(II)(pyz)Ru(III)(NH_3)_4py^{5+}$	D_2O	17.3	(29d)
$L(H_3N)_4Ru(II)(pyz)Ru(III)(NH_3)_4L^{5+}$	D_2O	17.3	(29d)[d]
$(bip)(H_3N)_3Ru(II)(pyz)Ru(III)(NH_3)_3(bip)^{5+}$	D_2O	16.6	(29d)
$(bip)_2ClRu(II)(pyz)Ru(III)Cl(bip)_2^{3+}$	—	22.6	(18)
$(H_3N)_5Ru(II)(py\text{-}py)Ru(III)(NH_3)_5^{5+}$	D_2O	27.5	(137)

[a] Abbreviations: bip = 2:2'-bipyridyl; phen = 9:10-phenanthroline; py = pyridine; py-py = 4:4'-bipyridyl; pyz = pyrazine.

[b] $L^{2-} = S(CF_3)C{:}C(CF_3)S^{2-}$.

[c] $L^{2-} = S(CH_3)C{:}C(CH_3)S^{2-}$.

[d] $L = N{\bigcirc}{-}CONH_2$.

1. Labile Complexes of Unknown Structure

Some of the earliest observations on optical charge transfer were made on solutions containing a metal in two different valency states: interaction colors were observed that could not be attributed to either of the metal ions separately. In most cases, it has been ascertained that only one ion of oxidant and one of reductant is present in the complex, since the intensity of the relevant absorption band is proportional to the concentrations of each. Evidence of the bridging group is indirect but generally convincing; for example, where the medium is shown as hydrochloric acid, it has been found that replacement of chloride ion by some other, noncomplexing, anion discharges the color, and, similarly, with the copper(II)–copper(I) color in methanol, the presence of acetate ion is essential (Table IV). Again, in some cases it has been shown that the addition of other more powerful complexing agents discharges the color, presumably by breaking the binuclear complex down into mononuclear complexes. Thus, the titanium(III)–titanium(IV) system in sulfuric acid is decolorized by the addition of phosphoric acid, and the copper(I)–copper(II) system in methanol is decolorized by acetonitrile.

The actual numbers of anions involved in these complexes have not been determined, and the arrangement of ligands around the two metal ions can only be conjectured. In most cases, it seems likely that the structures are unsymmetrical because the coordination prefer-ences of the metal ions in their two valency states are generally different. Formula $[TiCl^{2+}][T1Cl_6{}^{2-}]$ has been suggested for the titanium(III)–titanium(IV) complex in hydrochloric acid; on the other hand, for the copper(I)–copper(II) acetate complex in methanol, a symmetrical structure has been suggested. (In Table IV, the references for all these investigations are cited.)

2. Complexes with Inert Chromophores

When the constituent metal ions occur in complexes that exchange their ligands only slowly or when their coordination preferences for the medium in question are well established, the structure of the bi-nuclear complex can be written down with some confidence. This appears to be the case for the complex formed by mixing the ions $Fe(II)(CN)_6{}^{4-}$ and $Fe(III)(CN)_5OH_2{}^{2-}$. A strong interaction color is observed, which implies the formation of a bridged species, since no such color is seen on mixing $Fe(CN)_6{}^{4-}$ and $Fe(CN)_6{}^{3-}$; and the rate of formation of the complex is rapid compared with the rates of re-placement exchange of cyanide ligands in either of the two monomers. Hence the most likely structure is one with a single cyanide group connecting the two metals.

Other binuclear complexes with clearly defined structures have been characterized as products of bridged electron transfer reactions, as for example $(NC)_5Fe(II)CNCo(III)(edta)^{4-}$ from the reaction of $Fe(CN)_6{}^{3-}$ and $Co(II)(edta)^{2-}$ (63), and $(H_2O)_5Cr(III)ClIr(III)Cl_5$ from $Cr^{2+}\cdot$ aq and $IrCl_6{}^{2-}$ (132). Charge transfer spectra and photo-electron transfer reactions have recently been reported (141a).

The chromium(III)–chromium(VI) complex listed in Table V almost certainly contains octahedrally coordinated chromium(III) and tet-rahedral chromium(VI), as in the separate ions $Cr(H_2O)_6{}^{3+}$ and $CrO_4{}^{2-}$. It is formed rapidly and reversibly on mixing the solutions of these ions, but the actual rate of formation has not been measured, and since the chromium(VI) ion, but not the chromium(III) ion, is known to undergo substitution rapidly, the experiment does not dis-tinguish between the alternative outer- and inner-sphere structures,

$[Cr(H_2O)_6{}^{3+} \cdot CrO_4{}^{2-}]$ and $[(H_2O)_5Cr(III)OCr(VI)O_3]^+$. Similar ambiguities apply to the complexes $Ag^+ \cdot MnO_4{}^-$ and $Fe^{2+}[Mo(V)(CN)_8{}^{3-}]$ in Table V, in which the cations are labile to substitution whereas the anions are inert.

The energetics of these systems may again be described by the general equation (4), and the unknown ΔG_{PS} may again be estimated by means of a thermodynamic cycle involving separated complexes. There are, however, different versions of this cycle, depending on the treatment of the bridging group X. If X is associated with the metal atom A in both its reduced and oxidized forms, we obtain

$$A^+ \cdot X(\text{env}) + B(\text{env}) \xrightarrow{\Delta Y_w{}'} A^+ \cdot X \cdot B(\text{env}) \xrightarrow{\Delta Y_{CT}} A \cdot X \cdot B^+(\text{env}*)$$

with the vertical steps labeled ΔY_0, ΔY_{PS}, ΔY_{FC}, leading to

$$A \cdot X(\text{env}) + B^+(\text{env}) \xrightarrow{\Delta Y_w{}'} A \cdot X \cdot B^+(\text{env}) \qquad (45)$$

with

$$\Delta Y_0 = \Delta Y_E(A^+ \cdot X) - \Delta Y_E(B^+) \qquad (46)$$

but if X is associated with atom B in both forms, we obtain the analogous cycle with

$$\Delta Y_0 = \Delta Y_E(A^+) - \Delta Y_E(B^+ \cdot X) \qquad (47)$$

The available data are listed in Table V. In the case of complex $(NC)_5Fe(II)CNFe(III)(CN)_5{}^{6-}$, the natural choice of reference complexes is the pair $Fe(CN)_6{}^{3-}$ and $Fe(CN)_5OH_2{}^{2-}$, corresponding to $B^+ \cdot X$ and A of Eq. (47). It happens that the standard potentials of the corresponding redox couples are very nearly the same, i.e., $\Delta G_0 \approx 0$; and, although neither of the two work terms is known, it seems likely that they too will approximately cancel.* If so, the observed charge transfer energy is largely a measure of the Franck–Condon barrier. For the systems $Ag^+ \cdot MnO_4{}^-$, both work terms are unknown and, for $Fe^{2+} \cdot Mo(CN)_8{}^{3-}$, the term ΔG_W is unknown, but these can be estimated by analogy with other complexes, and in both cases the uncertainty in so doing is clearly much smaller than the final estimate of ΔG_{FC}. Data for the Cr(III)—Cr(VI) complex are less satisfactory

---

* In partial support of this suggestion, we note the small value of the equilibrium constant, $\log K \cong -1$, for the reaction $(NC)_5Co(III)NCFe(II)(CN)_5{}^{6-} + Fe(III)(CN)_6{}^{3-} \rightleftharpoons (NC)_5Co(III)NCFe(III)(CN)_5{}^{5-} + Fe(CN)_6{}^{4-}$ (56a).

but more interesting: there is considerable uncertainty in the Cr(IV)/Cr(III) reduction potential, and it is difficult to estimate the work term $\Delta G_W'$ for the equilibrium $Cr(IV) + Cr(V) \rightleftharpoons Cr(IV) \cdot Cr(V)$. We can, however, conclude that $\Delta G_{FC} + \Delta G_W' = -10 \pm 12$ kcal mole^{-1}, so that if ΔG_{FC} is similar to the other values recorded in Table V, $\Delta G_W'$ must be substantially negative, e.g., $\Delta G_W' \leq -20$ kcal mole^{-1}. This argues strongly for an inner-sphere formulation of the successor complex, $[(H_2O)_5Cr(IV)OCr(V)O_3]^+$. It is only because of the stability of this complex that the charge-transfer absorption appears in the visible region at all. No other examples of $Cr(III) \rightarrow Cr(VI)$ charge transfer seem to be known.

3. SYMMETRICAL INERT COMPLEXES

In recent years several complexes have been prepared that contain oxidant and reductant ions of the same element, with the same inner-sphere ligand environment and a symmetrical bridging group (Table VI). In the most-studied cases, analogous compounds with one more and one less electron are known, forming a redox series $A^+ \cdot X \cdot A^+$(env), $A^+ \cdot X \cdot A$(env), $A \cdot X \cdot A$(env), and the electronic spectrum of the mixed-valence species contains a unique band in addition to the bands characteristic of the separate $A^+ \cdot X$ and $A \cdot X$ chromophores.

When the first complex of this type was reported, it was thought that the energy of the lowest observable electronic transition gave a direct measure of the Franck–Condon energy barrier, but it is now recognized that this is not necessarily the case: it depends on the degree of interaction between the zero-order states, as shown in Fig. 1. The three possibilities of weak, moderate, and strong interaction can be neatly distinguished in the series of complexes **V** to **VII**, respectively.

$$[(NH_3)_5RuN \qquad\qquad NRu(NH_3)_5]^{5+}$$

(V)

$$[(NH_3)_5RuN \qquad NRu(NH_3)_5]^{5+}$$

(VI)

$$[(NH_3)_5RuN \equiv C - C \equiv NRu(NH_3)_5]^{5+}$$

(VII)

Complex **V** (*136*) has the characteristic low-energy absorption band, with the predicted half-width, and subject to strong solvent shifts as expected for a long-range charge transfer process; but in the ground state there is evidently very little interaction between the ruthenium atoms. The equilibrium constant K_{con} for the conproportionation reaction among the three members of the redox series,

$$(A \cdot X \cdot A) + (A^+ \cdot X \cdot A^+) \xrightleftharpoons{k_{con}} 2(A \cdot X \cdot A^+) \tag{48}$$

has been shown to be close to the statistical value, $K_{con} = 4$.

Complex **VI** (29) has the low-energy band, but it is relatively narrow and not appreciably solvent-dependent, and there is evidence of appreciable delocalization in the ground state, with $K_{con} \cong 1.3 \times 10^6$. The exact state of this complex is still controversial. There is no doubt that the valencies are subject to a rapid thermal exchange reaction,

$$[\text{Ru(II)} - \text{Ru(III)}]^{5+} \xrightleftharpoons{k_{ex}} [\text{Ru(III)} - \text{Ru(II)}]^{5+} \tag{49}$$

The photoelectron spectrum was initially interpreted (29e) in terms of trapped valencies; but a theoretical study by Hush (67) has shown that a delocalized description is equally consistent with the data available so far. Infrared data are still under discussion, but the Raman spectrum has been interpreted (29f) in terms of localization. It does seem clear that the rate of valence interchange according to Eq. (1) is rapid compared with the NMR time scale (29b); the possibility remains that the rate may be comparable with the frequencies of certain bond vibrations.

Complex **VII** (137) seems quite clearly to be of the average valency type (Fig. 1a, curves 3,3). Again, there is a near infrared electronic absorption band, but it is narrower than that expected for an intervalence band and is not subject to solvent shifts. The conproportionation constant is large ($K_{con} \geq 10^{13}$) and the valences are delocalized at least on the infrared time scale ($k_{ex} > 10^{13} \text{ sec}^{-1}$).

The difficulties of interpretation presented by some of these complexes are well illustrated by the sequence of ferrocene derivatives (**VIII** to **X**). The spectrum of **VIII** shows a near-infrared absorption as expected for a mixed valence complex, but the band has a shoulder, and at low temperatures is resolved into two bands. Complex **IX** has a well-defined band almost certainly ascribable to intervalence transfer, but complex **X** with a similar molecular structure has no such

(**VIII**)   (**IX**)   (**X**)

band. The review paper by Morrison and Hendrickson (*102*) should be consulted for further details and comments on the electronic states of these and related compounds.

Studies have also been made of the effect of nonbridging ligands attached unsymmetrically, i.e., to only one ruthenium atom. Regardless of the nature of the ligand, the charge-transfer absorption is shifted to higher energy (*2e,f,29d*). This is because if ligand L stabilizes ruthenium(III) (e.g., $L = Cl^-$) the ground state of the complex becomes LRu(III)·X·Ru(II)(env), whereas if it stabilizes ruthenium(II) (e.g., L = pyridine), we get Ru(III)·X·Ru(II)L(env). In either case, the charge-transfer energy now includes a contribution due to the difference in stability, ΔG_{PS}, between precursor and successor states, e.g.,

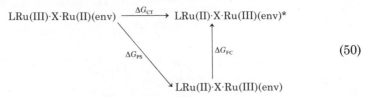

$$\text{(50)}$$

Values of ΔG_{PS} have not been obtained experimentally, but free-energy data for analogous systems such as the redox couple $LRu(III)X + e^-$ $\rightleftharpoons LRu(II)X$ are known and a plot of ΔG_{CT} against $\Delta G_E[LRu(III)X]$ has been found to be linear with unit slope, as expected (*29d*).

F. DIRECTLY BONDED COMPLEXES

Systems in which the donor and acceptor atoms are directly linked are extremely common and were among the first electron transfer systems to be spectroscopically characterized (*48, 69*). The transition may be between localized states as in the aqueous ion $FeCl^{2+}$ or delocalized states as in $OsCl_6$. In the former case, the ground state can be written $(H_2O)_5Fe^{3+}·Cl^-$, with excitation to $(H_2O)_5Fe^{2+}Cl$; in the latter case, whether one writes an ionic or a covalent formula, the transferring electron may be said to leave a "hole" which is symmetrically distributed over the 6 chlorine atoms (*69a*). No case is known in which donor and acceptor atoms are of the same element. There are several binuclear complexes with metal–metal bonds that could be formulated as mixed valence species, e.g., $Re(III)Re(IV)Cl_8{}^-$, $Ru(II)$ $Ru(III)(OAc)_4Cl(25)$, $Rh(II)Rh(III)(OAc)_4{}^+(148)$, but in each case there is evidence that the valence states are equivalent on the infrared time scale. With suitable distortions of bond lengths, some localization would occur, for example, by compressing the bonds at one end of the molecules, and lengthening the bonds at the other end, and these would represent modes of vibration. Such modes would be expected, therefore, to show electronic coupling and this might be revealed by a study of the infrared spectrum, but no such studies have been reported as yet.

1. *Octahedral Metal Complexes: Ligand-to-Metal Charge Transfer*

We shall discuss the energetics of these processes by referring first to a specific example, the complex $Ru(NH_3)_5I^{2+}$. This has an intense absorption in the visible region and a recent MCD study confirms its assignment as iodide → ruthenium charge transfer (*47*). From known equilibrium data, we may obtain the free energy of the state $Ru(NH_3)_5$ $OH_2{}^{2+}(aq) + I(aq)$, and the question is whether or not we can link this to the excited charge-transfer complex in a meaningful way. It is suggested that we may do so as follows. In the convention already established, we write the excited complex as $[Rh(II)NH_3)_5 \cdot I(env^*)]^{2+}$, indicating a species with the electronic configuration as shown but with all atoms and solvent molecules in the positions appropriate to the ground state $Ru(NH_3)_5I^{2+}$. Let this be dissociated into $I(env)^*$ and $Ru(NH_3)_5{}^{2+}(env^*)$, where the former is the same species as previously defined, i.e., an iodine atom with a solvation environment characteristic of the iodide ion, and the latter is a ruthenium(II) ion with one vacant position in the inner sphere, with five inner-sphere ammonia ligands attached at positions characteristic of ruthenium (III), and with an outer-sphere solvent shell likewise characteristic of ruthenium(III). On relaxing the solvent and ammonia molecules, but not filling the sixth ligand position, we arrive at $Ru(NH_3)_5{}^{2+}(env)$, a five-coordinate ruthenium(II) complex with outer sphere solvation. This, in turn, closely parallels the definition of the activated complex involved in ligand exchange reactions by the dissociative or SN_1 mechanism. For example, in the aquation of $Co(III)(NH_3)_5X$ complexes, there is evidence that in the transition state the CoN_5 framework remains intact, the Co—X group is fully broken, and that incoming groups have little specific interaction with the Co atom, while at the same time there is an appreciable stabilization of the five-coordinate system by solvation (*9*). We may, therefore, estimate the energy of this species by using the free energy of activation for the water-exchange process:

$$Ru(NH_3)_5OH_2{}^{2+} + H_2O^{18} \xrightarrow{\ k\ } Ru(NH_3)_5O^{18}H_2{}^{2+} + H_2O \qquad (51)$$

$$Ru(NH_3)_5OH_2{}^{2+} \xrightleftharpoons{\ \Delta G^{\ddagger}\ } [Ru(NH_3)_5{}^{2+}]^{\ddagger} + H_2O \qquad (52)$$

where $\Delta G^{\ddagger}$ is related to the second-order rate constant k by

$$k = Z \exp(-\Delta G^{\ddagger}/RT) \qquad (53)$$

and Z is the specific rate for the diffusion-controlled reaction (*92a*). The completed cycle (Scheme 1) contains three unknowns that may

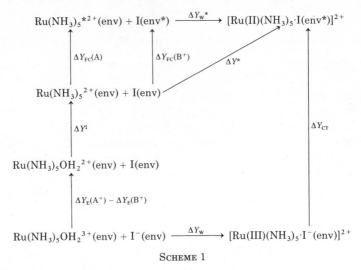

SCHEME 1

be grouped together as

$$\Delta Y^* = \Delta Y_{FC}(A) + \Delta Y_{FC}(B^+) + \Delta Y_W^* \qquad (54)$$

giving

$$\Delta Y^* = \Delta Y_{CT} + \Delta Y_W - \Delta Y^{\ddagger} - \Delta Y_E(A^+) + \Delta Y_E(B^+) \qquad (55)$$

where, in this case, $A = Ru(NH_3)_5OH_2{}^{2+}$ and $B = I^-$.

Calculations of ΔG^* based on this cycle are shown in Table VII. We can only comment qualitatively on the trends observed in these results, and it must be remembered that ΔG_W^*, and, therefore, also ΔG^*, may contain substantial attrative or repulsive contributions from the bond term [see Eq. (13)].

It does, however, seem significant that the highest ΔG^* are associated with the most ionic complexes, $Fe(H_2O)_5X^{2+}$, and that there is little difference between the complexes with the two halogens, $X = Cl$, Br (as with the outer-sphere complexes shown in Table III). If, in these cases, ΔG_W^* were close to zero, the values $\Delta G^* \sim 43 \, kcal \, mole^{-1}$ would represent the sum of the two reorganization energies, and again, the values are not very different from those observed in the outer-sphere case. The very much smaller values of ΔG^* for $Ru(NH_3)_5X^{2+}$ complexes ($X = Cl$, Br, I) might suggest that the reorganization energy of $Ru(NH_3)_5{}^{2+}$ is correspondingly smaller, but the appreciable difference

DIRECTLY BONDED COMPLEXES: ENERGY PARAMETERS
FOR THE CHARGE-TRANSFER PROCESS $(A^+—B) \rightarrow (A—B^+)$

Complex $A^+—B$	ΔG_{CT}	$-\Delta G_E(A^+)$	$\Delta G_E(B^+)^a$	ΔG_W	$-\Delta G^{\ddagger\,b}$	ΔG^*
$(NH_3)_5Ru(III)I^{2+}$	52.6^c	$+3.7^d$	-31.5	-2.3^d	-15.2^e	7.7
$(NH_3)_5Ru(III)Br^{2+}$	72.1^c	$+3.7^d$	-45.0	-1.9^d	-15.2^e	13.7
$(NH_3)_5Ru(III)Cl^{2+}$	87.2^c	$+3.7^d$	-57.4	-3.0^d	-15.2^e	15.7
$(H_2O)_5Ru(III)Cl^{2+}$	91.0^f	$+5.0^g$	-57.4	-2.0^h	—	—
$(H_2O)_5Fe(III)Br^{2+}$	75.3^i	$+17.6^j$	-45.0	$+0.4^k$	-8.5^l	39.8
$(H_2O)_5Fe(III)Cl^{2+}$	89.5^i	$+17.6^j$	-57.4	-0.8^k	-8.5^l	40.4
$(NC)_5Co(III)I^{3-}$	86.7^m	-27.1^n	-31.5	-2.2^o	0.0^p	25.9
$(NC)_5Co(III)Br^{3-}$	97.0^m	-27.1^n	-45.0	0.0^o	0.0^p	24.9
$(NC)_5Co(III)Cl^{3-}$	108.0^m	-27.1^n	-57.4	$+1.5^q$	0.0^p	~ 25
$(H_3N)_5Co(III)I^{2+}$	74.7^m	-8.6^r	-31.5	$+0.8^s$	—	—
$(H_3N)_5Co(III)Br^{2+}$	91.0^m	-8.6^r	-45.0	$+0.6^s$	—	—
$(H_3N)_5Co(III)Cl^{2+}$	103.1^m	-8.6^r	-57.4	-0.1^s	—	—
$I_2 \cdot I^-$	81.3^t	3.0^u	-31.5	-4.0^v	0.0^w	49.8
$I_2 \cdot Br^-$	106.0^t	3.0^u	-45.0	-1.6^x	0.0^w	62.4
$I_2 \cdot Cl^-$	115.4^t	3.0^u	-57.4	-1.0^t	0.0^w	60.4
$Br_2 \cdot Br^-$	108.0^t	—	-45.0	-1.9^y	—	—
$Br_2 \cdot Cl^-$	117.7^t	—	-57.4	-0.3^z	—	—
$Cl_2 \cdot Cl^-$	122.8^t	—	-57.4	$+1.9^{aa}$	—	—

[a] Berdinikov and Bazhin (14).

[b] Calculated using Eq. (53), with $k = k'[H_2O]^{-1}$, where k' is the first-order specific rate of water exchange. We have used $2.303\,RT = 1.36$ kcal mole^{-1}, $[H_2O] = 55.5\,M$, $Z = 10^{10.5}M^{-1}$ sec^{-1}.

[c] Hartmann and Buschbeck (60).

[d] Endicott and Taube (46).

[e] J. A. Stritar (129).

[f] Cady and Connick (17).

[g] Mercer and Buckley (97).

[h] Estimated from $Ru(NH_3)_5Cl^{2+}$, by analogy with $Rh(H_2O)_5Cl^{2+}$ and $Rh(NH_3)_5Cl^{2+}$ (94).

[i] Rabinowitch (117).

[j] Zielen and Sullivan (154).

[k] Rabinowitch and Stockmayer (118).

[l] Swift and Connick (131).

[m] Miskowski and Gray (100).

[n] Taking the polarographic half-wave potential ($E_{1/2} = -1.45$ volts, versus standard caloriel electrode hence $E^{\ominus}_{1/2} = -1.18$ volts) in $1.0M$ CN$^-$ as the measure of the reversible electrode potential (64).

[o] At 40°C (57).

[p] Assumed. The $Co(CN)_5{}^{3-}$ ion has a tetragonal pyramidal structure equivalent to a regular octahedral structure with one ligand removed. There is, however, a change in color in going from the solid state to the aqueous solution, which suggests weak hydration to give $[Co(CN)_5{}^{3-}\cdot H_2O](aq)$ (15a).

[q] Assumed.

[r] Yalman (150).

[s] Langford (80).

[t] Meyerstein and Treinin (99).

[u] Baxendale and Bevan (9a).

[v] Katzin and Gebert (72).

[w] Assumed. The bimolecular decay constant k for $I_2{}^- + I_2{}^- \rightarrow I_2 + 2I^-$ is close to the diffusion-controlled limit, viz. $k = 10^9\,M^{-1}$ sec^{-1} (9a).

[x] Daniele (31).

[y] Daniele (32).

[z] Daniele (33).

[aa] Zimmerman and Strong (155).

between the iodo and bromo complexes, and the much higher value $\Delta G_{FC} = 42 \pm 2$ kcal mole^{-1} already obtained for the ion pair Ru $(NH_3)_6^{2+} \cdot I$, suggest that the metal–halogen bond is responsible for the difference. Possibly the Ru^{2+} ion and I atom should be considered as "soft" acid and base, respectively, so that ΔG_W^* would become more negative along the sequence X = Cl, Br, I. A similar comment could apply to complexes $Co(CN)_5X^{3-}$, but here there is the difference that the pentacoordinate cobalt(II) fragment $Co(CN)_5^{3-}$ is a known chemical species (15a), weakly hydrated in the sixth-ligand position, but showing very little tendency to add another ligand (15b). Thus, ΔG_W^* is more likely to be small or positive than to be substantially negative, and we may conclude that, for a given halogen, the differ-ence in ΔG^* between $Co(CN)_5X^{3-}$ and $Fe(H_2O)_5X^{2+}$ is largely a difference in the reorganization energy. The fact that ΔG_{FC} is lower for $Co(CN)_5^{3-}$ than for $Fe(H_2O)_5^{2+}$, in spite of higher ionic charge, suggests that the inner-sphere bond adjustment contributes less in the pentacyano than in the pentaaquo cases. It would be interesting to have data for $Fe(III)(CN)_5X^{3-}$ complexes for comparison with the cobalt systems and with $Fe(CN)_6^{3-}$.

2. *Interhalogen Complexes*

A cycle analogous to Scheme 1 is shown in Scheme 2 for systems with iodine as oxidant and halide ion X^- as reductant. Formulas $(I_2 \cdots X^-)(env)$, $I_2^-(env)$, and $I_2(env)$ denote the species I_2X^-, I_2^-, and I_2 with atomic configurations (i.e., bond distances and solvation) at equilibrium, whereas $(I_2^- \cdots X)(env^*)$, $I_2^-(env^*)$, and $I(env^*)$ denote I_2X^-, I_2^-, and I with the atomic configurations of $(I_2 \cdots X^-)$, I_2, and I^-, respectively.

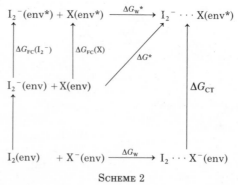

SCHEME 2

Calculations of ΔG^* are shown in Table VII. Because ΔG_W^* is prob-ably similar to ΔG_W, the high value of ΔG^*, as compared with those

for octahedral complexes, is to be attributed to inner-sphere reorganization of I_2^- i.e., to differences between the equilibrium bond
distances in I_2^- and I_2. The magnitude of this effect can be roughly
estimated as follows. First, from thermodynamic data on the reaction
$I + I^- \rightleftharpoons I_2^-$ in aqueous solution, and with the aid of an assumed
radius for the ion I_2^-, Baxendale (9a) has deduced the gas-phase
electron affinity of the I_2 molecule, namely, $\Delta H \cong -56$ kcal mole^{-1} for

$$I_2(g) + e^-(g) \rightarrow I_2^-(g) \tag{56}$$

where the molecule ion $I_2^-(g)$ has the same bond length as I_2^- at
equilibrium in solution. Second, from a correlation of charge-transfer
spectra of complexes of benzene with various electron acceptors,
Jortner and Sokolov (71) have estimated the *vertical* electron affinity
of the iodine molecule, namely $\Delta H = -36$ kcal mole^{-1} for the reaction

$$I_2(g) + e^-(g) \rightarrow I_2^-(g)^* \tag{57}$$

where $I_2^-(g)^*$ has the same bond distance as I_2. Strictly speaking, the
latter value refers to solvated species in benzene, not in the gas phase,
but, if this difference can be neglected, we obtain the enthalpy of
compression of the bond in I_2^- in the gas phase, namely $\Delta H \cong 20$ kcal
mole^{-1} for

$$I_2^-(g) \rightarrow I_2^-(g)^* \tag{58}$$

If we accept this as an estimate of the inner-sphere contribution to
the term $\Delta G_{FC}(I_2^-)$ in Scheme 2, it suggests that not more than half
of the observed Franck–Condon barrier can be attributed to solvent
reorganization effects.

G. Further Single-Ion Processes

1. *Electron Transfer to Solvent*

Various strongly reducing ions such as I^-, $Fe(CN)_6^{4-}$, and Cr^{2+}
exhibit absorption bands in solution whose energies correlate with
the oxidation potentials, and it is well-established that these arise
from a process in which one electron is lost. Earlier workers (120)
described the process in different ways, postulating, for example, the
generation of a "free electron," or of a reduced water molecule, H_2O^-;
but Platzman and Franck (116) argued convincingly that the excited
state is one in which the electron, although effectively detached from
the central ion, remains localized in the potential field of the orientated
solvent molecules. This description, which has been retained in all
subsequent analyses (49, 127), is close to that of the solvated electron;

and, indeed, the hydrated electron was first introduced (49) as a hypo-thetical entity in the discussion of these spectra, shortly before it was discovered experimentally.* In the cycle,

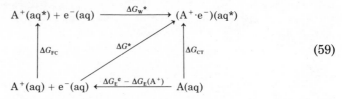

$$A^+(aq^*) + e^-(aq) \xrightarrow{\Delta G_w^*} (A^+ \cdot e^-)(aq^*)$$

$$(59)$$

the term $(A^+ \cdot e^-)(aq^*)$ denotes the ion-to-solvent charge-transfer ex-cited state, $e^-(aq)$ is the solvated electron, and $A^+(aq^*)$ is the hypo-thetical state of the A^+ ion with solvent molecules reorganized (or, rather, disorganized) into the positions characteristic of the reduced ion $A(aq)$. Term ΔG_w^* is analogous to the work terms introduced in previous cycles in that it represents the work done in bringing the solvated electron from infinity and superimposing it on the configura-tion of $A^+ (aq^*)$. Platzman and Franck (116) took $\Delta G_w^* = 0$, assuming that in the excited state the electron is so far from the nucleus that there is no coulombic interaction. Later studies of the spectra have confirmed that the excited electron has a large effective radius. The process labeled ΔG_w^* could, in turn, be subdivided into two stages—first desolvating the electron and placing it in the continuum level and, then, relocalizing the electron at the position of $A^+(aq^*)$. On the con-trary, however, we shall consider the Franck–Condon and work terms together, introducing the quantity ΔG^*, defined by

$$\Delta G^* = \Delta G_w^* + \Delta G_{FC} \qquad (60)$$

which is directly related to the experimentally measurable quantities

$$\Delta G^* = \Delta G_{CT} - \Delta G_E^e + \Delta G_E(A^+) \qquad (61)$$

where ΔG_E^e is the standard free energy of the process

$$\tfrac{1}{2}H_2(g) \rightarrow e^-(aq) + H^+(aq) \qquad (62)$$

and $\Delta G_E(A^+)$ is defined by Eq. (18) (Section II,A).

The available data are listed in Table VIII. The best-studied systems are the halide ions, and there is a considerable literature (127) on the

---

* Friedman wrote "$e^-(aq)$ is defined as the electron in equilibrium with respect to solvation by the water, it being assumed that such an equilibrium is possible because the reaction $e^- + H_2O = H + OH^-$ is slow compared with the rate of orientation of water molecules about an ion (49)."

TABLE VIII

SINGLE-ION CHARGE-TRANSFER PROCESSES

Complex (reduced form) A	Radii $(10^{-10}$ m$)^a$		Free-energy terms (kcal mole^{-1})				
			Ion-to-solvent^b			Solvent-to-ion^b	
	Complex A	Complex A$^+$	$\Delta G_E(A^+)$	ΔG_{CT}	$\Delta G^\star$	$\Delta G_{CT}'$	$\Delta G^{\star\prime}$
Cl$^-$	1.81	0.99	-57.4^h	158.2^s	34.6	91.1^y	215
Br$^-$	1.95	1.14	-45.0^h	143.2^s	30.2	103.8^y	215
I$^-$	2.16	1.33	-31.5^h	126.7^s	29.0	111.8^y	210
OH$^-$	1.40^c	0.74^g	-42.8^h	153.2^s	44.1	124.5^y	234
CN$^-$	1.82^c	—	-42.8^h	$>150^s$	>41	—	—
NO$_2^-$	1.55^c	1.55^e	-22.5^h	135.7^s	47.0	—	—
ClO$_2^-$	1.95^d	1.95^e	-21.6^i	$>150^s$	>62	—	—
NCS$^-$	1.95^c	1.95^e	-39.6^{aa}	132.6^{bb}	26.7	—	—
Fe(CN)$_6^{4-}$	4.3^e	4.3^e	-8.2^j	106^t	31.7	—	—
Ru(CN)$_6^{4-}$	4.5^e	4.5^e	-19.7^k	115 ± 10^u	~ 30	—	—
Cr^{3+}	0.69	0.56	-50 ± 10^l	$\sim 150^v$	~ 35	—	—
Ce^{3+}	1.11	1.01	-40^m	—	—	92^z	198
Mn^{2+}	0.80	0.66	-34^n	$\sim 155^v$	~ 60	—	—
Fe^{2+}	0.76	0.64	-18°	$\sim 150^v$	~ 70	122^z	206
Ru(NH$_3$)$_6^{2+}$	0.74^f	0.7^e	-2^p	$104^{p,w}$	36	—	—
Ti^{3+}	0.76	0.68	$+1^q$	$\sim 135^v$	~ 60	—	—
Cr^{2+}	0.84	0.69	$+9^r$	106^x	49	—	—

^{a} Pauling radii. Unless otherwise stated, these are taken from Phillips and Williams(114, pp. 109, 152).

^{b} The $\Delta G^\star$ and $\Delta G^{\star\prime}$ are calculated using Eqs. (59) and (63) of the text, and taking $\Delta G_E(e) = 66.3$ kcal mole^{-1} (70a).

^{c} "Thermochemical radii" (151).

^{d} Estimated from radius of NO$_2^-$ by comparison of Cl—O distance in NH$_4$ClO$_2$ with N—O distance in NaNO$_2$ (130).

^{e} Estimated.

^{f} Assumed same as for Co^{2+}.

^{g} Taken as half the O—O bond distance in H$_2$O$_2$.

^{h} Berdnikov and Bazhin (14).

^{i} Troitskaya et al. (140).

^{j} Hanania et al. (59).

^{k} DeFord and Davidson (34).

^{l} Haight (56).

^{m} Conley (24).

^{n} Vetter and Manecke (141).

^{o} Zielen and Sullivan (154).

^{p} Meyer and Taube (98a).

^{q} Calculated using $E^\ominus$ (TiOH^{3+} + H$^+$ + e$^- \rightleftharpoons$ Ti^{3+}) = -0.055 volt (139) and assuming K(Ti$^{4+} \rightleftharpoons$ TiOH^{3+} + H$^+$) = $1M$.

^{r} Latimer (82).

^{s} Friedman (49).

^{t} Shirom and Stein (123).

^{u} Guttel and Shirom (55).

^{v} Schläfer (121). (Values of ν_{CT} estimated from a rather small-scale diagram printed in the text.)

^{w} Hintze and Ford (62).

^{x} Dainton and James (30c).

^{y} Treinin and Hayon (138).

^{z} Rabinowitch (117).

^{aa} Wilmarth et al. (147).

^{bb} Gusarsky and Treinin (54).

nature of the excited state, and the dependence of ΔG_{CT} on temperature (*127b*) and solvent composition (*127e, f, g*). The more highly reducing complex ions, such as $Fe(CN)_6^{4-}$ and $Ru(CN)_6^{4-}$, also have charge transfer to solvent states as evidenced by their photochemical behavior (*1, 30d*), but in some cases, such as $IrCl_6^{3-}$, it is difficult to distinguish these states from the alternative possibility of metal-to-ligand charge transfer (*1e*).

There has also been some discussion of the spectra of cations. As early as 1951 it was shown that for a series of transition metal ions, the low-energy threshold of the ultraviolet absorption correlates with the thermodynamic redox potential in the sequence $V^{2+} < Cr^{2+} < Fe^{2+} < Co^{2+} < Mn^{2+} < Ni^{2+}$ (*30*), and photoelectron detachment processes have been detected on irradiation in the near ultraviolet region (*30d, 23*); but with the exception of $Cr^{2+}\cdot aq$, which in perchlorate media has a well-defined shoulder at ca. 37.0 kK, the actual charge-transfer bands are not easy to locate. Prior to this, the spectra of B-group ions had been discussed. For example, $Tl^+\cdot aq$ ion shows a well-defined peak at 46.7 kK, which was originally assumed to be due to charge transfer (*48b*); but it is now assigned to an internal $6s^2 \rightarrow 6s6p$ transition (*69b*). Similar bands are seen in spectra of the other s^2 ions, Sn^{2+}, Pb^{2+}, Bi^{3+}, etc. (*69b*). Ion Ag^+ has peaks in the range 44.7, 47.9, and 51.9 kK (*69c*), but these seem too low in energy to be due to metal-to-solvent charge transfer and should, perhaps, be assigned to the reverse transition, that is, solvent-to-metal. Several ruthenium(II) complexes (*98*), however, show absorption peaks at around 37.0 kK which have been described as "formally ligand field in assignment but having significant charge-transfer character" (*98b*). The case of the $Ru(bip)_3^{2+}$ complex is more difficult to assess: the luminescent excited state was assigned as metal-to-ligand charge transfer because the absorption band shows no solvent shifts (*86*); but there is also no change in frequency when the ligand is deuterated, and, on this basis, some charge transfer to solvent character is proposed (*98c*). However, the lifetime of the excited state is such that the solvent molecules will be equilibrated in the excited state, so these data are not directly relevant to the present study.

2. *Electron Transfer from Solvent*

The reverse process, photochemical transfer of an electron from the solvent molecule to the central metal ion, has been postulated in the case of several of the more strongly oxidizing metal ions, and, more recently, on the basis of an extensive study of the spectra of free radicals such as halogen atoms and OH. By analogy with Eq.

(59), we may construct the cycle

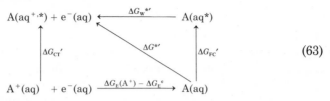

$$\text{(63)}$$

Here $A(aq^{+}\cdot *)$ denotes the solvent-to-ion charge-transfer excited state, and $A(aq*)$ is the hypothetical state of the A ion with the solvent molecules reorganized in the configuration appropriate for $A^{+}\cdot(aq)$. Term $\Delta G^{*\prime}$ is a work term analogous to ΔG^{*}. Values of $\Delta G^{*\prime}$ calculated from the cycle are shown in Table VIII.

In the case where both processes can be studied for the same atom or ion (e.g., transfer from iodide ion to solvent and from solvent to iodine atom), the two cycles may be combined to give

$$(\Delta G_{\text{CT}} + \Delta G_{\text{CT}}') = (\Delta G_{\text{FC}} + \Delta G_{\text{FC}}') + (\Delta G_{\text{W}}^{*} + \Delta G_{\text{W}}^{*\prime}) \qquad (64)$$

and

$$(\Delta G^{*} - \Delta G^{*\prime}) = (\Delta G_{\text{FC}} - \Delta G_{\text{FC}}') + (\Delta G_{\text{W}}^{*} - \Delta G_{\text{W}}^{*\prime}) \qquad (65)$$

From the discussion that follows in Section III, it may be expected that the terms ΔG_{FC} and $\Delta G_{\text{FC}}'$ in Eq. (65) will roughly cancel. It is apparent, however, that terms ΔG_{W}^{*} and $\Delta G_{\text{W}}^{*\prime}$ do not cancel even approximately. The difference $(\Delta G^{*} - \Delta G^{*\prime})$ remains roughly constant for the four negative ions A listed in Table VIII (i.e., -180 ± 5 kcal mole^{-1} for A = halide or hydroxide ion) and changes substantially on going to the cation (-135 kcal mole^{-1} for A = $Fe^{2+}\cdot aq$).

III. Theory

A. Ionic Solvation

In Section II, A we defined the processes of electron ionization in the gas phase and of electron transfer to the continuum in the solution phase:

$$A(g) \rightarrow A^{+}(g) + e^{-}(g) \qquad (15)$$

$$A(aq) \rightarrow A^{+}(aq) + \{e^{-}\}_{aq} \qquad (16a)$$

In the electrostatic theory (16) of solvation, these processes are compared with the classical process of charging. Starting with a body A$'$ having the same boundary as the desired ion A^{+}, electric charge is removed by infinitesimal stages to yield the charge distribution characteristic of A^+. This may be written as

$$A'(g) \to A^+(g) + \{-\}_0 \tag{66}$$

where the state symbols (g) have the same meaning as before, and $\{-\}_0$ denotes 1 mole of "electricity," infinitely dispersed *in vacuo* and at zero potential (*53*). For the analogous process in solution, the solvent is considered as a uniform dielectric medium, the body A' is replaced by a cavity of the same shape, and an electric field is slowly applied to give the field characteristic of $A^+(aq)$. This may be written as

$$A'(aq) \to A^+(aq) + \{-\}_{aq} \tag{67}$$

where A' and A^+ denote cavities as described, (aq) denotes the dielectric medium with the equilibrium polarization where appropriate; and $\{-\}_{aq}$ denotes 1 mole of electricity infinitely dispersed in the medium.

It will be noted that, if the charged species in equations (15) and (66) are deemed to be at zero potential, then it may be supposed that those in equations (16a) and (67) are at some other potential, namely, the "inner potential" or the average potential over the bulk solvent phase. However, there is no way of unambiguously defining this potential, and it does not enter into the calculations (*53*).

All of the preceding can be restated, with appropriate verbal changes, for the formation of a negative species $A^-(aq)$.

The work done in process (66) is given by

$$W^0 = (\epsilon_0/2)\int_V \mathbf{E}_C \cdot \mathbf{E}_C \, dV \tag{68}$$

where the vector $\mathbf{E}_C$ is the field set up in the vacuum by the ion A^+, and ϵ_0 is the permittivity of free space. The integration is carried over all space.* Similarly (*85*) for process (67) the work, subject to certain conditions,† is

$$W = (\epsilon_0/2D_s)\int_V \mathbf{E}_C \cdot \mathbf{E}_C \, dV \tag{69}$$

---

* This statement needs qualification when the charge distribution of A^+ involves point charges, but the conclusions are not affected (see Ref. *98b*).

† The principal assumptions involved in the derivation of Eq. (69) are (i) that dielectric image effects are negligible; (ii) that the charges, dipoles, etc., are the same in the solvated ion $A^+(env)$ as in the gaseous ion $A^+(g)$, i.e., the ion A^+ is nonpolarizable; and (iii) that the dielectric is isotropic and unsaturated. Condition (i) is always fulfilled if the surface of the ion is an equipotential surface. In some more detailed treatments, image effects are accounted for by assigning a dielectric constant D_i to the interior of the ion. If however $D_i \ll D_s$, the more detailed equation reduces again to Eq. (69); e.g., see Ref. *76*.

where D_s is the static dielectric constant. It is then assumed that the difference in work between processes (15) and (16a) is the same as the difference in work between processes (66) and (67) so that for the chemical process

$$A^+(g) + e^-(g) \rightarrow A^+(aq) + \{e^-\}_{aq} \qquad (70)$$

we have[‡]

$$\Delta G_{env} = \frac{L\epsilon_0}{2}\left[1 - \frac{1}{D_s}\right]\int \mathbf{E}_C \cdot \mathbf{E}_C \, dV \qquad (71)$$

where L is Avogadro's constant.

For a spherical ion of radius a with a centrally distributed charge ze, Eq. (71) yields the well-known expression (16a):

$$\Delta G_{env} = \frac{L}{4\pi\epsilon_0}\left[1 - \frac{1}{D_s}\right]\frac{z^2 e^2}{2a} \qquad (72)$$

B. Continuum Theory for Electron Transfer Processes

Marcus (92) elaborated the continuum theory by separating the polarization of the dielectric into two superimposed polarizations: the "nonequilibrium polarization," and the "equilibrium polarization." In the precursor complex p there is a characteristic charge distribution giving a field $\mathbf{E}_C{}^p$, and both polarizations are at equilibrium. When the photon is absorbed the charge distribution rapidly changes to a new value with an associated field $\mathbf{E}_C{}^s$. The nonequilibrium polarization remains at its old value, but the equilibrium polarization changes to a new value, jointly determined by the charges and the nonequilibrium polarization.

The reorganization energy for the successor state ΔG_{FC}, is given by the difference in environmental energy between the excited state s^* and the ground state s [Eq. (3)]. Thus (65), subject to the same conditions as govern Eq. (69),

$$\Delta G_{FC} = \frac{L\epsilon_0}{2}\left[\frac{1}{D_0} - \frac{1}{D_s}\right]\oint [\mathbf{E}_C{}^p - \mathbf{E}_C{}^s]^2 \, dV \qquad (73)$$

[‡] The energy change for process (70) is identified as a free-energy change on the grounds that both processes (66) and (67) are reversible. The question remains as to whether it is the Gibbs or the Helmholtz free energy. For rigid dielectrics, there is no differenc. For liquids at constant pressure, it is commonly assumed (see, e.g., Refs. 47 and 114) to be the Gibbs free energy, but this is not strictly correct since it fails to take account of the compression of the dielectric under the influence of the applied field. For a rigorous discussion, see Ref. 47a.

Vector $(\mathbf{E}_C{}^p - \mathbf{E}_C{}^s)$ is itself a field and it defines a charge distribution, namely the distribution which *in vacuo* would exert that field. Expression (73) may, therefore, be read as the free energy of transfer of a body bearing the appropriate charge distribution, from a medium of dielectric constant D_s to another medium of dielectric constant D_0. For the case of charge transfer to or from the continuum, $(\mathbf{E}_C{}^s - \mathbf{E}_C{}^p)$ is centrally distributed with a net charge of 1 unit; for transfer to or from the solvent, it is centrally distributed but uncharged, with a "core" of charge of one sign, surrounded by a region of opposite sign. For transfer between two localized centers, it approximates to a finite dipole of charges $\pm e$. It is important to note that, although the shapes and sizes of the molecules affect the result, the overall ionic charges do not. It also follows directly from Eq. (73) that, in the spherically symmetrical cases, or in dipolar cases provided only that the boundary is symmetrical about the mid-point of the line joining the two charges and regardless of the actual shape, the reorganization energy is the same for transfer in either direction:

$$\Delta G_{FC} = \Delta G_{FC}' \tag{74}$$

This would apply, for example, to two metal ions of equal radius, either separate from each other or joined by a symmetrical bridging group, provided that the oxidized and reduced forms of the ions had the same radii.

We conclude this section by mentioning some important integrated forms of Eq. (73).

Case a. Isolated Spherical Ion

In the photoemission process (Section II,A), the "electron acceptor" is the supposed completely delocalized state of the electron. For a spherical ion of radius a, Eq. (73) then reduces to

$$\Delta G_{FC} = \frac{Le^2}{4\pi\epsilon_0}\left(\frac{1}{D_0} - \frac{1}{D_s}\right) \cdot \frac{1}{2a} \tag{75}$$

Case b. Spherical Ions at Long Distance

The approximation used by Marcus (*92a*) and Hush (*66*) in the discussion of outer-sphere electron transfer between spherical ions assumes that the ions are far enough apart so that the fields around each one can be considered spherically symmetrical, but not so far apart that the coulombic force between them can be neglected. This leads to

$$\Delta G_{FC} = \frac{Le^2}{4\pi\epsilon_0}\left(\frac{1}{D_0} - \frac{1}{D_s}\right)\left(\frac{1}{2a_1} + \frac{1}{2a_2} - \frac{1}{R}\right) \tag{76}$$

where a_1 and a_2 are the ionic radii, and R is the interionic distance.

Case c. Spherical Ions at Short Distance

When the distance R is not large compared with the radius sum $(a_1 + a_2)$, the problem of evaluating ΔG_{FC} becomes very complicated; there is, in fact, no possible solution in closed form (95). Kharkats (74a) has recently proposed the equation

$$\Delta G_{FC} = \frac{Le^2}{4\pi\epsilon_0}\left(\frac{1}{D_0} - \frac{1}{D_s}\right)\left(\frac{1}{2a_1} + \frac{1}{2a_2} - \frac{1}{R} - f(R,a_1) - f(R,a_2)\right) \tag{77}$$

where functions $f(R,a_1)$, $f(R,a_2)$ are defined by

$$f(R,r) = \frac{1}{4}\cdot\frac{R}{R^2 - r^2}\left[\frac{r}{R} - \frac{1}{2}\left(1 - \frac{r^2}{R^2}\right)\ln\frac{(R + r)}{(R - r)}\right] \tag{78}$$

For the case of a "contact ion pair" $(R = a_1 + a_2)$, the additional terms account for ca. 6% of ΔG_{FC} when $a_1 = a_2$, or 11% when $a_1 = a_2/3$. Nevertheless, Eq. (77) is only an approximation; it is obtained by allowing for the volumes of the ions, but it still depends on taking the field around each ion as spherically symmetrical and neglecting the effect of the other ion; that is to say, the boundary conditions are not obeyed.

In a more sophisticated treatment (74b), Kharkats has attempted to allow for the mutual influence of the fields of the two ions by assuming the ions to be completely polarizable (i.e., analogous to metallic spheres) and by calculating as a first approximation the dipole induced on each ion by the point charge of the other. The resulting equation is extremely complex, but from the authors' graphs it can be seen that the modification has a significant effect. For the contact case, $R = a_1 + a_2$, the additional terms contribute a 27% increase in ΔG_{FC} when $a_1 = a_2$, and a larger increase when $a_1 \neq a_2$. But this is still only a first approximation—a complete calculation would consider not only the dipole, quadrupole, etc., induced on sphere 1 by the charge of sphere 2; but also the new dipole, etc., induced on sphere 1 by the induced dipole, etc., of sphere 2; and so on *ad infinitum*. And for a completely polarizable ion the net effect would be that as the ions approached toward contact, the effective centers of the two point charges that constitute the dipole would move to the point of contact, and the expression for ΔG_{FC} would diverge to infinity (19).

Case d. Ellipsoidal Model

A more realistic model for complexes of the bridged or directly bonded types (see Sections II, E and F) is suggested in Fig. 2. The boundary surface is a prolate ellipsoid of revolution, and the field $(E_C^p - E_C^s)$ is taken to be that of a finite dipole due to equal and opposite charges placed at the foci. If the distance between oxidizing and reducing centers is R, the dipole moment change (assuming complete localization of valencies) is $p = eR$. The problem of integrating equations such as Eq. (73) for an ellipsoidal boundary was thoroughly treated by Westheimer and Kirkwood (145) in connection with their studies of the activity coefficients of dipolar solutes.

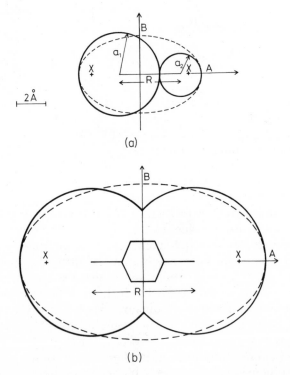

(a)

(b)

FIG. 2. Scale drawings of (a) the outer-sphere successor complex $Ru(NH_3)_6{}^{2+}\cdot I$ and (b) the bridged mixed-valence complex $Cl(bip)_2Ru(pyz)Ru(bip)_2Cl^{3+}$, with assumed dimensions as shown in Tables IX and X. The dotted lines are boundaries of ellipsoids (raddi A, B) having the same volumes as the complexes. Points X, X are foci of the ellipsoids.

Adapting their result to the present case, the present writer (19a) obtained:

$$\Delta G_{FC} = \frac{Lp^2}{8\pi\epsilon_0 A^2 B} \left(\frac{1}{D_0} - \frac{1}{D_s} \right) S(\lambda_0) \tag{79}$$

$$S(\lambda_0) = \sum_{n=0}^{\infty} \tfrac{1}{2} (1 - (-1)^n)(2n + 1)\lambda_0(\lambda_0^2 - 1)Q_n(\lambda_0)/P_n(\lambda_0) \tag{80}$$

where A and B are the semimajor and semiminor axes of the ellipsoid, $\lambda_0^2 = A^2/(A^2 - B^2)$, and $P_n(\lambda_0)$ and $Q_n(\lambda_0)$ are Legendre polynomials of the first and second kinds, of degree n. Expression $S(\lambda_0)$ depends only on the eccentricity of the ellipsoid: it varies from $S \cong 0.65$ when $\lambda_0 = 1$ to $S = 1$ in the limit $\lambda_0 \to \infty$.

Case e. Spherical Ion, Point Dipole

Proceeding to the limit $\lambda_0 \to \infty$, as the radii A and B become equal, the foci of the ellipsoid converge at the center. The result is a spherical ion, with a point dipole located at the center. Equation (73) then reduces to

$$\Delta G_{FC} = \frac{Lp^2}{8\pi\epsilon_0 a^3} \left(\frac{1}{D_0} - \frac{1}{D_s} \right) \tag{81}$$

where $a = A = B$. The same expression can also be deduced from an earlier calculation by Kirkwood (76a).

IV. Conclusions

A. SINGLE-ION PROCESSES

The processes discussed in Sections II, A and G are formally similar in that the electron donor has approximately spherical symmetry. In the photoemission process, the electron acceptor is the supposedly completely delocalized state of the electron, whereas in the charge-transfer to solvent process, properly so called, the acceptor is the spherically symmetrical polarization field set up by the donor ion in the solvent.

For the former process, Eq. (75) may be applied, and, on comparing this with the Born equation (72), we see that the Franck–Condon energy is expected to be proportional to the solvation energy

$$\Delta G_{FC} = \left(\frac{1}{D_0} - \frac{1}{D_s} \right) \left(1 - \frac{1}{D_s} \right)^{-1} \Delta G_{env} \tag{82}$$

TABLE IX

COMPARISON OF CALCULATED AND OBSERVED REORGANIZATION ENERGIES, USING THE SEPARATE-SPHERES MODEL

Successor complex[a] $(A \cdots B^+)$	Assumed ionic and atomic radii[b,c] a_1	a_2	Assumed internuclear distance[c] R	ΔG_{FC} (kcal mole^{-1}) calc.[d]	obs.	χ^e
$Ru(NH_3)_6{}^{2+} \cdot Cl$	0.74[f]	0.99	3.55[j]	164	41[o]	0.14
$Ru(NH_3)_6{}^{2+} \cdot Br$	0.74[f]	1.14	3.70[j]	154	45[o]	0.16
$Ru(NH_3)_6{}^{2+} \cdot I$	0.74[f]	1.33	3.89[j]	145	41[o]	0.15
$Ru(en)_3{}^{2+} \cdot I$	0.74[f]	1.33	4.23[k]	149	36[o]	0.13
$MnO_4{}^{2-} \cdot Ag^{2+}$	0.52[g]	0.97[h]	3.63[l]	219	40[p]	0.10
$Cl(bip)_2Ru(II)(pyz)Ru(III)(bip)_2Cl^{3+}$	0.74[f]	0.63[i]	6.99[m]	243	23[q]	0.06
$(NH_3)_5Ru(II)(py-py)Ru(III)(NH_3)_5{}^{5+}$	0.74[f]	0.63[i]	11.3[m]	330	27.5[q]	0.03
$(C_5H_5)Fe(II)(C_5H_4-C_5H_4)Fe(III)(C_5H_5)^+$	0.76	0.64	6.4[n]	233	18.5[q]	0.03

[a] Abbreviations: bip = 2:2′-bipyridyl; pyz = 1:4-pyrazine; py—py = 4:4′-bipyridyl.

[b] Pauling radii (114, pp. 109, 152).

[c] Units are 10^{-10} m.

[d] Equation (76).

[e] Equation (84).

[f] Assumed same as for Co^{2+}.

[g] Assumed same as for Cr^{6+}.

[h] Assumed same as for Ag^{2+}.

[i] Assumed same as for Co^{3+}.

[j] Using outer-sphere radius of $Ru(NH_3)_6{}^{2+}$, assumed same as that of $Co(NH_3)_6{}^{2+}$, i.e., 2.56 Å, estimated from the Co—I distance in solid $Co(NH_3)_6I_2$ (4.72 Å) by subtracting the radius of I^- (149).

[k] Using outer-sphere radius of $Ru(en)_3{}^{2+}$, assumed same as that of $Co(en)_3{}^{3+}$, i.e., 2.90 Å, estimated from the mean Co—Br distance in solid $Co(en)_3Br_3 \cdot H_2O$ (4.85 Å), by subtracting the radius of Br^- (149).

[l] Sum of Mn—O distance in $KMnO_4$ (1.55 Å) and Ag—O distance in AgO (2.08 Å) (130).

[m] Bond lengths in pyrazine and bipyridyl estimated from suitable molecules; Ru—N distance assumed same as R—NO in $Ru(NH_3)_4(NO)(OH)Cl_2$ (130).

[n] Calculated as $R = (x^2 + y^2)^{1/2}$, where x is Fe—Fe distance and y is inter-ring distance in $Fe(C_5H_4—C_5H_4)_2Fe$ (89).

[o] Table III.

[p] Table V.

[q] Table VI.

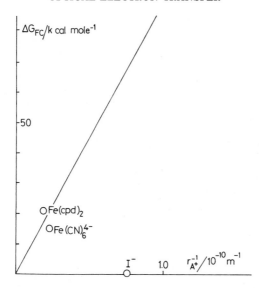

FIG. 3. Single-ion reorganization energies ΔG_{FC} (Table II) compared with the predictions of Eq. (75). Radii $r(A^+)$ are those of the oxidized forms of the complexes, i.e., of I, $Fe(CN)_6^{3-}$ and $Fe(cpd)_2^+$.

The values of ΔG_{FC} calculated above are listed in Table IX and plotted against $(1/r_{A+})$ in Fig. 3. The straight line is drawn with slope calculated from Eq. (75), using $D_s = 78.5$, $D_0 = 1.78$, for water at 25°C. The points for iodide ion is clearly unacceptable, as already discussed, but it seems unlikely that any revision will bring $\Delta G_{FC}(I)$ up from zero to the predicted value of ca. 70 kcal mole^{-1}. Clearly more data are needed before the continuum theory can be subjected to even an approximate quantitative test, but already we may forecast that the dependence of ΔG_{FC} on ionic radius will be less sensitive than Eq. (75) implies.

A similar conclusion follows from the more extensive data on charge transfer to solvent and vice versa. Quantities ΔG^* and $\Delta G^{*\prime}$ calculated for these systems in Section II,G are not the Franck–Condon energies because they include the unknown interaction energies ΔG_w^* and $\Delta G_w^{*\prime}$ which may vary from system to system; nevertheless, the main sources of variation are likely to be ΔG_{FC} and $\Delta G_{FC}'$.

The data for charge transfer to solvent are shown in Fig. 4 as a plot of ΔG^* against $1/a$. It appears that there is a general correlation, with the following notable exceptions. First, the bent polyatomic ions NO_2^- and ClO_2^-, but not the linear NCS^-, have reorganization energies significantly higher than predicted. Friedman showed (49) that this could be accounted for by bond strain in the excited state,

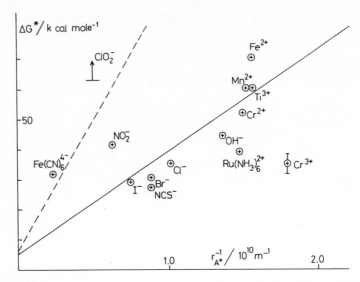

FIG. 4. Single-ion charge-transfer to solvent data. Correlation of ΔG^* (Table VIII) with $1/r(A^+)$, where $r(A^+)$ is the crystal radius of the oxidized form of the ion. The dashed line is drawn with slope predicted from Eq. (75).

and this is supported by later structural studies. The bond angles in the gaseous molecules NO_2, ClO_2 are, respectively, 18° and 7° smaller than those of the corresponding anions in ionic crystals, and there are measurable differences in the bond lengths (130). Second, for $Fe(CN)_6^{4-}$ (and for $Ru(CN)_6^{4-}$ not shown in Fig. 4), ΔG^* is again more positive than might have been expected in view of the very large ionic radius; and this may reflect a positive work term ΔG_W^* due to the high overall negative charge. Lastly, the Cr^{3+} ion is significantly out of line even when full allowance is made for the uncertainty in the Cr^{4+}/Cr^{3+} reduction potential.

The remaining data can be fitted to a straight line, and, for comparison with other data and with theory, we may write

$$\Delta G^* \cong \frac{Le^2}{4\pi\epsilon_0} \cdot \chi \cdot \frac{1}{2a} + \text{constant} \tag{83}$$

with $\chi = 0.21$ to be compared with $(D_0^{-1} - D_s^{-1}) = 0.55$. The uncertainty as to the size and variability of ΔG_W^* inhibits further discussion. What is required here is a more extended study including donor ions of diverse charge types so that the ΔG_W^* and ΔG_{FC} can be estimated separately.

The less extensive data for charge transfer in the reverse direction are similarly plotted in Fig. 5. Within the series of four equally charged

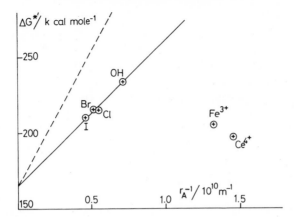

FIG. 5. Single-ion charge-transfer from solvent data. Correlation of $\Delta G^{*\prime}$ (Table VIII) with $1/r(A)$, where $r(A)$ is the crystal radius of the reduced form of the ion. The dashed line is drawn with the slope predicted from Eq. (75).

species there is a good correlation of $\Delta G^{*\prime}$ with $(1/a)$; but more data on cationic complexes are needed before any further comments can be made.

B. DONOR–ACCEPTOR PROCESSES

Here we consider together all systems in which the donor and acceptor sites are both separate and localized, i.e., those discussed in Sections II, B to F. A representative selection is shown in Table IX. The complexes are written as successor complexes $(A \cdots B^+)(env)$. The internuclear distances have been estimated from bond distances in analogous known complexes, there being no crystal structure data for any of the actual species listed in Table IX.

To compare the experimental values of ΔG_{CF} with the predictions of the Marcus–Hush theory, there would seem to be two worthwile approaches.

1. Separate Sphere Model

We consider the central atoms of the complexes as spheres, using the Pauling radii, and treat the solvent, the inner-sphere ligands, and the bridging ligands where present, as the medium within which electron transfer takes place. This is a very gross approximation, since, apart from the lack of homogeneity of the medium as thus defined, there are important electronic interactions. Such interactions will be largely of two sorts: the charges on the oxidant and reductant ions will be partially delocalized onto the neighboring ligands; and the wave functions of the two centers will sometimes be mixed, either directly or via bridging ligand orbitals, so that the effective charge transferred is less than 1 electron. We subsume all these effects under the general headings of polarity and polarizability of the ligands.

The equation appropriate to this approximation is Eq. (76), and for each system we have, therefore, calculated an empirical coefficient χ, defined by

$$\Delta G_{FC} = \frac{Le^2}{4\pi\epsilon_0} \cdot \chi \cdot \left[\frac{1}{2a_1} + \frac{1}{2a_2} - \frac{1}{R} \right] \qquad (84)$$

Values of χ are listed in Table IX. As with the single-ion values, they are all less than 0.55, and this is evidently a measure, although a rather crude measure, of the electronic effect just mentioned. The inner-sphere ligands can be arranged in order of decreasing χ, in the series $NH_3 \approx en > O > $ pyrazine $\approx$ biscyclopentadienyl $\approx$ 4,4′-bipyridyl and this coincides roughly with increasing polarizability or covalent bonding. A more refined treatment might seek to relate χ to the polarizabilities of the ligands as measured from the refractive indexes of suitable compounds.

2. Ellipsoid Model

In this case we consider the whole complex, including inner-sphere ligands and bridging ligands if any, as an ellipsoidal body (Fig. 2) and use Eqs. (79) and (80). In deriving these equations, it has been assumed that the average dielectric constant over the interior of the molecule is small compared with that of the solvent [see footnote in Section III, A regarding derivation of Eq. (69)]. This is a considerable oversimplification, because the polarizabilities of inner-sphere ligands such as I^- and pyrazine are considerably greater than that of water, as estimated from the refractive indices of suitable compounds; but, on the other hand, if the inner-sphere ligands are dielectrically saturated the error in our assumption is less.

The radii used, and the resulting values of ΔG_{FC}^{out}, are shown in Table X. The semimajor axes A have been estimated from crystallographic data, and the semiminor axes B have been calculated so that the volume of the ellipsoid is equal to the estimated volume of the complex. The dipole moments have been taken as $p = eR$.

For the four outer-sphere complexes listed in Table X the calculated ΔG_F^{out} are similar to, although slightly larger than, the experimental total ΔG_{FC}. This can be considered satisfactory agreement, and it

TABLE X

COMPARISON OF CALCULATED AND OBSERVED REORGANIZATION ENERGIES,
USING THE ELLIPSOID MODEL

Successor complex[a] A $\cdots$ B$^+$	Assumed radii of ellipsoids[b,c]			Reorganization energies[e]			
				Calcd.[f]	Observed		
	A	B	$S(\lambda_0)$[d]	ΔG_{FC}^{out}	ΔG_{FC}	ΔG_{FC}^{in}	ΔG_{FC}^{out}
Ru(NH$_3$)$_6$$^{2+}$·Cl	3.55[g]	2.24[k]	0.77	50	41[m]	—	—
Ru(NH$_3$)$_6$$^{2+}$·Br	3.70[g]	2.22[k]	0.75	51	45[m]	—	—
Ru(NH$_3$)$_6$$^{2+}$·I	3.89[g]	2.22[k]	0.73	53	41[m]	—	—
Ru(en)$_3$$^{2+}$·I	4.23[g]	2.51[k]	0.75	46	36[m]	—	—
Cl(bip)$_2$Ru(pyz)Ru(bip)$_2$Cl^{3+}	8.4[h]	5.1[l]	0.76	15	22.6[n]	15.6[o]	7.0[o]
(NH$_3$)$_5$Ru(py—py)Ru(NH$_3$)$_5$$^{5+}$	8.2[i]	2.7[l]	0.70	137	27.5[n]	10.0[p]	17.5[p]
(C$_5$H$_5$)Fe(C$_5$H$_4$—C$_5$H$_4$)Fe(C$_5$H$_5$)$^+$	6.5[j]	3.4[l]	0.70	35	18.5[n]	—	—

[a] Abbreviations: bip = 2:2′-bipyridyl; pyz = 1:4 pyrazine; py—py = 4:4′-bipyridyl.

[b] Units are 10^{-10} m.

[c] Equation (79).

[d] Equation (80).

[e] Aqueous medium 25°C. Units are kcal mole^{-1}.

[f] Calculated from Eqs. (79) and (80), using $p = eR$, where R is the internuclear distance as shown in Table IX.

[g] $A = R$.

[h] $A = \frac{1}{2}R + r$, where r is the radius of the Ru(bip)$_2$Cl moieties, assumed to be same as that of V(bip)$_3$, i.e., half the nearest-neighbor V—V distance in V(bip)$_3$ (3).

[i] $A = \frac{1}{2}R + r$, where r is the outer-sphere radius of Ru(NH$_3$)$_6$$^{2+}$ (see Table IX, footnote j).

[j] $A = \frac{1}{2}R + r$, where r is the radius of the Fe(C$_5$H$_5$) moiety, assumed to be same as that of Fe(C$_5$H$_5$)$_2$, i.e., half the mean nearest-neighbor Fe—Fe distance in solid Fe(C$_5$H$_5$)$_2$ (42).

[k] $B = (a_1{}^2 - a_1 a_2 + a_2{}^2)^{1/2}$, where a_1 and a_2 are outer-sphere radii of component ions.

[l] Calculated so that $(4/3)\pi AB^2$ is the volume of the complex.

[m] Table III.

[n] Table VI.

[o] Callahan and Meyer (18).

[p] Tom et al. (137).

confirms that in these cases at least the model is a reasonable one and that the outer-sphere part of the reorganization energy is a substantial contribution to the whole. The lack of dependence of ΔG_{FC} on the nature of the halide ion is particularly noteworthy.

The three bridged complexes show rather diverse behavior. For Cl(bip)$_2$Ru(pyz)Ru(bip)$_2$Cl^{3+}, the calculated ΔG_F^{out} is quite small and is less than the experimental ΔG_{FC}; for (NH$_3$)$_5$Ru(py—py)Ru(NH$_3$)$_5$$^{5+}$, the calculated ΔG_F^{out} is much greater, and this is due about equally to the greater internuclear distance and to the smaller thickness of the inner coordination sphere surrounding each ruthenium atom. Such a large discrepancy between theory and experiment does not seem explicable wholly by deficiencies in the model, and it may be suggested that here there is some delocalization of valencies in the ground state. It will be noted that ΔG_{FC}^{out} depends on the square of the effective dipole moment change; hence a reduction of p by a factor of about 2 would be sufficient to bring this complex into line with the pyrazine-bridged complex. For the diiron complex $(C_5H_5)Fe(C_5H_4 \cdot C_5H_4)Fe(C_5H_5)^+$, a reduction of p by about 1.7 is implied.

Several authors have pointed out that it is possible to distinguish the inner- and outer-sphere parts of ΔG_{FC} by studying solvent effects. It is assumed that ΔG_{FC}^{in} [Eq. (14)] is independent of the nature of the solvent, whereas ΔG_{FC}^{out} is proportional to $(D_0^{-1} - D_s^{-1})$ as shown in Section III,B. Thus, when ΔG_{FC} is plotted against $(D_0^{-1} - D_s^{-1})$, a straight-line dependence is expected, and extrapolation to $(D_0^{-1} - D_s^{-1}) = 0$ gives ΔG_{FC}^{in}. In this way it was shown that ΔG_{FC}^{out} is negligible for complexes $(NH_3)_5Ru(pyz)Ru(NH_3)^{5+}$ (29d) and $(NH_3)_5RuNC \cdot CN Ru(NH_3)_5^{5+}$ (136) consistent with complete delocalization of valencies. Values of ΔG_{FC}^{out} for two other ruthenium(II)–ruthenium(III) complexes are shown in Table X. The closest agreement between theory and experiment occurs with complex $Cl(bip)_2Ru(pyz)Ru(bip)_2Cl^{3+}$.

Variations in transition energy $h\nu_{CT}$ with solvent are, of course, well established in charge-transfer spectroscopy of organic systems, and have been used many times to establish scales of solvent polarity (10). However, all this work has been based on unsymmetrical charge-transfer systems: typical examples being the ion pair $CH_3NC_5H_5^+ \cdot I^-$ yielding the radical pair $CH_3NC_5H_5 \cdot I$ on excitation. The charge-transfer energy is then the sum of two contributions, as shown in Eq. (3), and it seems likely that the main contribution to the solvent shift lies in the thermal free-energy difference ΔG_{PS} rather than in the Franck–Condon energy ΔG_{FC}. Callaghan et al. have synthesized the unsymmetrical bridged complex $(H_3N)_5Ru(III)(pyrazine)Ru(II)(bip)_2Cl^{4+}$ and found an appreciable dependence of $h\nu_{CT}$ on solvent (2e); and the same comment would apply.

In summary, it appears from this discussion that Franck–Condon energies can now be calculated for a diverse group of inorganic charge-transfer systems and that, although the accuracy of individual values is uncertain, it is possible qualitatively to rationalize the differences between analogous systems. Absolute predictions are much less satisfactory at the present time, and the electrostatic theory based on a dielectric continuum has only very limited applicability to the systems that have so far been studied. When inner-sphere reorganization effects are dominant, as is frequently the case, a treatment based on the distortion of individual bonds, using force constants from vibrational spectroscopy, would seem to be more promising. In the calculation of equilibrium solvation energies ΔG_{env}, significant improvements have been achieved by considering ion–dipole interactions in the inner-sphere region, and using continuum theory only for the outer region [9, 114 (Chapter 5)]. The application of this method to thermal activation energies of electron transfer has been considered in some detail, although not yet exhaustively, by Hush (65). An alternative approach would be to retain the continuum model but to introduce the concept of dielectric saturation. Sophisticated treatments of this kind, again for the equilibrium situation, have been reviewed by Padova (113).

The outstanding need, however, is for more experimental data, especially on systems simple enough to be treated theoretically. Photoemission data are of particular interest, but their interpretation will call for more thermodynamic data on oxidation–reduction systems in nonaqueous solvents. Theoretical treatments of single-ion solvation energies for solvents other than water, and at temperatures other than room temperature, are also necessary if the arguments outlined in the foregoing are to be developed. More data are needed on mixed-valence binuclear metal complexes with known structures, and with negligible or small interaction between oxidizing and reducing centers in the ground state. A point of particular interest here is the effect of changing internuclear distance: long-range electron transfer occurs in the gas phase (104), and as semiconduction in the solid state (28), but it has never been unambiguously demonstrated for metal complex ions in solution. Yet, long-range transfer is frequently discussed in connection with biological systems; and the decisive experiments may yet come from the study of appropriately synthesized inorganic models.

ACKNOWLEDGMENTS

I am grateful to Professor J. Endicott, Dr. D. R. Rosseinsky and Dr. F. Wilkinson who read the first draft and made many useful comments. I have not accepted all the opinions offered and take full responsibility for any remaining errors.

REFERENCES

1. (a) Adamson, A. W., Waltz, W. L., Zinato, E., Watts, D. W., Fleischauer, P. D., and Lindholm, R. D., *Chem. Rev.* **68**, 541 (1968); (b) Airey, P. L., and Dainton, F. S., *Proc. Roy. Soc., Ser. A* **291**, 478 (1966); (c) Ohno, S., *Bull. Chem. Soc. Jpn.* **40**, 1770, 1776, 1779 (1967); (d) Waltz, W. L., Adamson, A. W., and Fleischauer, P. D., *J. Am.*

Chem. Soc. **89**, 3923 (1967); (e) Waltz W. L., and Adamson, A. W., *J. Phys. Chem.* **73**, 4250 (1969); (f) Guttel, C., and Shirom, M., *J. Photochem.* **1**, 197 (1973).

2. (a) Adeyimi, S. A., Braddock, J. N., Brown, G. M., Ferguson, J. A., Miller, F. J., and Meyer, T. J., *J. Am. Chem. Soc.* **94**, 300 (1972); (b) Adeyimi, S. A., Johnson, E. C., Miller, F. J., and Meyer, T. J., *Inorg. Chem.* **12**, 2371 (1973); (c) Callahan, R. W., Brown, G. M., and Meyer, T. J., *J. Am. Chem. Soc.* **96**, 7829 (1974); (d) Brown, G. M., Meyer, T. J., Cowan, D. O., LeVanda, C., Kaufman, F., Roling, P. V., and Rausch, M. D., *Inorg. Chem.* **14**, 506 (1975); (e) Callahan, R. W., Brown, G. M., and Meyer, T. J., *ibid.* **14**, 1443 (1975); (f) Powers, M. J., Callahan, R. W., Salmon, D. J., and Meyer, T. J., *ibid.* **15**, 894 (1976); (g) Callahan, R. W., and Meyer, T. J., *Chem. Phys. Lett.* **39**, 82 (1976).

3. Albrecht, G., *Z. Chem.* **3**, 182 (1963).

3a. Armor, J. N., *J. Inorg. Nucl. Chem.* **35**, 2067 (1973).

4. (a) Ballard, R. E., and Griffiths, G., *Chem. Commun.* p. 1472 (1971); (b) *J. Chem. Soc. A* p. 1960 (1971).

5. Ballhausen, C. J., and Jørgensen, C. K., *Acta Chem. Scand.* **9**, 397 (1955).

6. (a) Balzani, V., and Carassiti, V., "Photochemistry of Coordination Compounds." Academic Press, New York, 1970; (b) Endicott, J. F., *in* "Concepts of Inorganic Photochemistry" (Adamson, A. W., and Fleischauer, P. D., eds.), p. 81. Wiley, New York, 1975.

7. (a) Baron, B., Chartier, P., Delahay, P., and Lugo, R., *J. Chem. Phys.* **51**, 2562 (1969); (b) Delahay, P., Chartier, P., and Nemec, L., *ibid.* **53**, 3126 (1970); (c) Nemec, L., Baron, B., and Delahay, P., *Chem. Phys. Lett.* **16**, 278 (1972); (d) Bomchil, G., Delahay, P., and Levin, I., *J. Chem. Phys.* **56**, 5194 (1972).

8. (a) Baron, B., Delahay, P., and Lugo, R., *J. Chem. Phys.* **53**, 1399 (1970); (b) **55**, 4180 (1971); (c) Aulich, H., Baron, B., Delahay, P., and Lugo, R., *ibid.* **58**, 4439 (1973); (d) Aulich, H., Nemec, L., and Delahay, P., *ibid.* **61**, 4235 (1974).

9. Basolo, F., and Pearson, R. G., "Mechanisms of Inorganic Reactions," 2nd ed., p. 129 ff. Wiley, New York, 1967.

9a. Baxendale, J. H., and Bevan, P. L. T., *J. Chem. Soc. A* p. 2240 (1969).

10. (a) Bayliss, N. S., *J. Chem. Phys.* **18**, 292 (1950); (b) McRae, E. G., *J. Phys. Chem.* **61**, 562 (1957); (c) Kosower, E. M., *J. Am. Chem. Soc.* **80**, 3253 (1958); (d) *J. Chim. Phys.* **61**, 230 (1964); (e) Reichardt, C., *Angew. Chem. Int. Ed. Engl.* **4**, 29 (1965).

11. Beattie, J. K., Hush, N. S., and Taylor, P. R., *Inorg. Chem.* **15**, 992 (1976).

12. Beck, M. T., *Coord. Chem. Rev.* **3**, 91 (1968).

13. Bennett, B. G., and Nicholls, D., *J. Chem. Soc. A* p. 1204 (1971).

14. Berdnikov, V. M., and Bazhin, N. M., *Russ. J. Phys. Chem.* **44**, 395 (1970).

15. (a) Brown, L. D., and Raymond, K. N., *Inorg. Chem.* **14**, 2590 (1975); (b) Adamson, A. W., *J. Am. Chem. Soc.* **73**, 5710 (1951).

16. (a) Born, M., *Z. Physik*, **1**, 45 (1920); (b) Rosseinsky, D. R., *Electrochim. Acta* **16**, 19 (1971).

17. Cady, H. H., and Connick, R. E., *J. Am. Chem. Soc.* **80**, 2646 (1958).

18. Callahan, R. W., and Meyer, T. J., *Chem. Phys. Lett.* **39**, 82 (1976).

19. Cannon, R. D., unpublished calculations.

19a. Cannon, R. D., *Chem. Phys. Letters*, **49**, 299 (1977).

20. Cannon, R. D., and Gardiner, J., *Inorg. Chem.* **13**, 390 (1974).

21. Carrington, A., and Symons, M. C. R., *J. Chem. Soc.* p. 3373 (1956).

22. (a) Clark, R. J. H., and Trumble, W. R., *J. Chem. Soc., Chem. Commun.* p. 318 (1975); (b) Strekas, T. C., and Spiro, T. G., *Inorg. Chem.* **15**, 974 (1976).

23. Collinson, E., Dainton, F. S., and Malati, M. A., *Trans. Faraday Soc.* **55**, 2096 (1959).

24. Conley, H. L., Thesis, UCRL-9332.

25. Cotton, F. A., and Pedersen, E., *Inorg. Chem.* **14**, 388 (1975).

26. Cowan, D. O., and Kaufman, F., *J. Am. Chem. Soc.* **92**, 219 (1970).

27. Cowan, D. O., and LeVanda, C., *J. Am. Chem. Soc.* **94**, 9271 (1972).

28. (a) Cowan, D. O., Park, J., Pittman, C. U., Sasaki, Y., Mukherjee, T. K., and Diamond, N. A., *J. Am. Chem. Soc.* **94**, 5110 (1972); (b) Pittman, C. U., and Surynarayanan, B., *ibid.* **96**, 7916 (1974).

29. (a) Creutz, C., and Taube, H., *J. Am. Chem. Soc.* **91**, 3988 (1969); (b) Elias, J. H., and Drago, R. S., *Inorg. Chem.* **11**, 415 (1972); (c) Creutz, C., Good, M. L., and Chandra, S., *Inorg. Nucl. Chem. Lett.* **9**, 171 (1973); (d) Creutz, C., and Taube, H., *J. Am. Chem. Soc.* **95**, 1086 (1973); (e) Citrin, P. H., *ibid.* **95**, 6472 (1973); (f) Strekas, T. C., and Spiro, T. G., *Inorg. Chem.* **15**, 974 (1976); (g) Beattie, J. K., Hush, N. S., and Taylor, P. R., *ibid.* **15**, 992 (1976).

30. (a) Dainton, F. S., and James, D. G. L., *J. Chim. Phys.* **48**, C17 (1951); (b) Dainton, F. S., *J. Chem. Soc.* p. 1533 (1952); (c) Dainton, F. S., and James, D. G. L., *Trans. Faraday Soc.* **54**, 649 (1958); (d) Airey, P. L., and Dainton, F. S., *Proc. Roy. Soc., Ser. A* **291**, 340, 478 (1966).

31. Daniele, G., *Gazzetta* **90**, 1082 (1960).

32. Daniele, G., *Gazzetta* **90**, 1585 (1960).

33. Daniele, G., *Gazzetta* **90**, 1597 (1960).

34. DeFord, D. D., and Davidson, A. W., *J. Am. Chem. Soc.* **73**, 1469 (1951).

35. (a) Delahay, P., *J. Chem. Phys.* **55**, 4188 (1971); (b) Nemec, L., *ibid.* **59**, 6092 (1973).

36. (a) Delahay, P., *in* "Electrons in Fluids" (Jortner, J., and Kestner, N. R., *eds.*), p. 131. Springer, New York, 1973. (b) Dainton, F. S., *Chem. Soc. Rev.* **4**, 323 (1975).

37. Delahay, P., private communication.

38. Diehl, H., Carlson, P. A., Christian, D., Dewel, E. H., Emerson, M. R., Heumann, F. K., and Standage, H. W., *Proc. Iowa Acad. Sci.* **55**, 241 (1948).

39. Doehlemann, E., and Fromherz, H., *Z. phys. Chem. A* **171**, 353 (1934).

40. Dogonadze, R. R., Krishtalik, L. I., and Pleskov, Yu. V., *Sov. Electrochem.* **10** (4), 489 (1974).

41. Eaton, G. R., and Holm, R. H., *Inorg. Chem.* **10**, 805 (1971).

42. Eiland, P. F., and Pepinsky, R., *J. Am. Chem. Soc.* **74**, 4971 (1952).

43. Elsbernd, H., and Beattie, J. K., *Inorg. Chem.* **7**, 2468 (1968).

44. Emschwiller, G., and Jørgensen, C. K., *Chem. Phys. Lett.* **5**, 561 (1970).

45. Endicott, J. F., and Ferraudi, G. J., *Inorg. Chem.* **14**, 3133 (1975).

46. Endicott, J. F., and Taube, H., *Inorg. Chem.* **4**, 437 (1965).

47. Evans, R. S., and Schreiner, A. F., *Inorg. Chem.* **14**, 1705 (1975).

47a. Frank, H. S., *J. Chem. Phys.* **23**, 2023 (1955).

48. (a) Franck, J., and Scheibe, G., *Z. phys. Chem. A* **139**, 22 (1929); (b) Rabinowitch, E., *Rev. Mod. Phys.* **14**, 112 (1942); (c) Katzin, L. I., *J. Chem. Phys.* **23**, 2055 (1955).

49. Friedman, H. L., *J. Chem. Phys.* **21**, 319 (1953).

50. Gaswick, D., and Haim, A., *J. Am. Chem. Soc.* **93**, 7347 (1971).

51. Glauser, R., Hauser, U., Herren, F., Ludi, A., Roder, P., Schmidt, E., Siegenthaler, H., and Wenk, F., *J. Am. Chem. Soc.* **95**, 8457 (1973).

52. Goroshchenko, Ya. G., and Godneva, M. M., *Russ. J. Inorg. Chem.* **6**, 744 (1961).

53. Guggenheim, E. A., "Thermodynamics: An Advanced Treatment for Chemists and Physicists," 5th ed., p. 300. North-Holland Publ., Amsterdam, 1967.

54. Gusarsky, E., and Treinin, A., *J. Phys. Chem.* **69**, 3176 (1965).

55. Guttel, C., and Shirom, M., *J. Photochem.* **1**, 197 (1973).

56. Haight, G. P., Huang, T. J., and Platt, H., *J. Am. Chem. Soc.* **96**, 3137 (1974).

56a. Haim, A., and Wilmarth, W. K., *J. Am. Chem. Soc.* **83**, 509 (1961).

57. Haim, A., Grassie, R. J., and Wilmarth, W. K., *Adv. Chem. Ser.* **49**, 31 (1965).
58. Halliwell, H. F., and Nyburg, S. C., *Trans. Faraday Soc.* **59**, 1126 (1963).
59. Hanania, G. I. H., Irvine, D. H., Eaton, W. A., and George, P., *J. Phys. Chem.* **71**, 2022 (1967).
60. Hartmann, H., and Buschbeck, C., *Z. phys. Chem. (Frankfurt)* **11**, 120 (1957).
61. Hepler, L. G., Sweet, J. R., and Jesser, R. A., *J. Am. Chem. Soc.* **82**, 304 (1960).
62. Hintze, R. E., and Ford, P. C., *J. Am. Chem. Soc.* **97**, 2664 (1975).
63. Huchital, D. H., and Wilkins, R. G., *Inorg. Chem.* **6**, 1022 (1967).
64. Hume, D. N., and Kolthoff, I. M., *J. Am. Chem. Soc.* **71**, 867 (1949).
65. Hush, N. S., *Progr. Inorg. Chem.* **8**, 391 (1967).
66. Hush, N. S., *Trans. Faraday Soc.* **57**, 557 (1961).
67. Hush, N. S., *Chem. Phys.* **10**, 361 (1975).
68. Isied, S. S., and Taube, H., *J. Am. Chem. Soc.* **95**, 8198 (1973).
69. (a) Jørgensen, C. K., "Absorption Spectra and Chemical Bonding in Complexes," Chapter 9. Pergamon Press, 1962; (b) *ibid.*, p. 185; (c) *ibid.*, Table 27.
70. Jørgensen, C. K., *Acta Chem. Scand.* **11**, 73 (1967).
70a. Jortner, J., and Noyes, R. M., *J. Phys. Chem.* **70**, 770 (1966).
71. Jortner, J., and Sokolov, U., *Nature (London)* **190**, 1003 (1961).
72. Katzin, L. I., and Gebert, E., *J. Am. Chem. Soc.* **77**, 5814 (1955).
73. Kaufman, F., and Cowan, D. O., *J. Am. Chem. Soc.* **92**, 6198 (1970).
74. (a) Kharkats, Yu. I., *Sov. Electrochem.* **9**, 845 (1973); (b) **10**, 588 (1974).
75. King, E. L., and Neptune, J. A., *J. Am. Chem. Soc.* **77**, 3186 (1955).
76. (a) Kirkwood, J. G., *J. Chem. Phys.* **2**, 351 (1934); (b) Kirkwood, J. G., and Westheimer, F. H., *ibid.* **6**, 506 (1938).
77. Kolthoff, I. M., and Thomas, F. G., *J. Phys. Chem.* **69**, 3049 (1965).
78. Kolthoff, I. M., and Tomsicek, W. J., *J. Phys. Chem.* **40**, 247 (1936).
79. König, E., *Inorg. Chem.* **2**, 1238 (1963); **8**, 1278 (1969).
80. Langford, C. H., *Inorg. Chem.* **4**, 265 (1965).
81. Larsson, R., *Acta Chem. Scand.* **21**, 257 (1967).
82. Latimer, W. M., "Oxidation Potentials," 2nd ed. Prentice-Hall, New York, 1952.
83. Latimer, W. M., Pitzer, K. S., and Slansky, C. M., *J. Chem. Phys.* **7**, 108 (1939).
84. LeVanda, C., Cowan, D. O., Leitch, C., and Bechgaard, K., *J. Am. Chem. Soc.* **96**, 6788 (1974).
85. (a) Lorrain, P., and Corson, D. R., "Electromagnetic Fields and Waves," 2nd ed., pp. 77, 124. Freeman, San Francisco, 1970; (b) Padova, J., *Electrochim. Acta* **12**, 1227 (1967).
86. Lytle, F. E., and Hercules, D. M., *J. Am. Chem. Soc.* **91**, 253 (1969).
87. McConnell, H., and Davidson, N., *J. Am. Chem. Soc.* **72**, 3168 (1950).
88. McConnell, H., and Davidson, N., *J. Am. Chem. Soc.* **72**, 5557 (1950).
89. MacDonald, A. C., and Trotter, J., *Acta Crystallogr.* **17**, 872 (1964).
90. Malik, W. U., and Ali, S. I., *Talanta* **8**, 737 (1961).
91. Malik, W. V., and Om. H., *Indian J. Chem.* **4**, 106 (1966).
92. (a) Marcus, R. A., *J. Chem. Phys.* **24**, 966 (1956); (b) *ibid.*, 979.
93. Marcus, R. A., *Ann. Rev. Phys. Chem.* **15**, 155 (1964).
94. (a) Martell, A. E., and Sillén, L. G., "Stability Constants of Metal-Ion Complexes," Chem. Soc. Spec. Publ. No. 17, London, 1964; (b) Suppl. No. 1, Spec. Publ. No. 25, London, 1971.
95. Maxwell, J. C., "A Treatise on Electricity and Magnetism," 3rd ed., Vol. I, p. 272. Oxford Univ. Press, London and New York, 1904.
96. Mayoh, B., and Day, P., *J. Am. Chem. Soc.* **94**, 2885 (1972).
97. Mercer, E. E., and Buckley, R. R., *Inorg. Chem.* **4**, 1692 (1965).

98. (a) Meyer, T. J., and Taube, H., *Inorg. Chem.* **7**, 2369 (1968); (b) Hintze, R. E., and Ford, P. C., *J. Am. Chem. Soc.* **97**, 2664 (1975); (c) Van Houten, J., and Watts, R. J., *ibid.* **97**, 3843 (1975); (d) Hintze, R. E., and Ford, P. C., *Inorg. Chem.* **14**, 1211 (1975).

99. Meyerstein, D., and Treinin, A., *Trans. Faraday Soc.* **59**, 1114 (1963).

100. Miskowski, V. M., and Gray, H. B., *Inorg. Chem.* **14**, 401 (1975).

101. Moore, R. L., *J. Am. Chem. Soc.* **77**, 1504 (1955).

102. Morrison, W. H., and Hendrickson, D. N., *Inorg. Chem.* **14**, 2331 (1975).

103. Morrison, W. H., Krogsrud, S., and Hendrickson, D. N., *Inorg. Chem.* **12**, 1998 (1973).

104. See, e.g., Mott, N. F., and Massey, H. S. W., "The Theory of Atomic Collisions," 3rd ed., p. 655. Oxford Univ. Press, London and New York, 1965.

105. Mueller-Westerhof, U. T., and Eilbracht, P., *J. Am. Chem. Soc.* **94**, 9272 (1972).

106. Murray, K. S., and Sheahan, R. M., *J. Chem. Soc., Dalton Trans.* p. 1182 (1973).

107. Mussini, T., and Faita, C., *Ric. Sci.* **36**, 175 (1966).

108. National Bureau of Standards, "Selected Values of Thermodynamic Properties," NBS Circ. 500. Washington, D.C., 1961.

109. (a) Nemec, L., and Delahay, P., *J. Chem. Phys.* **57**, 2135 (1972); (b) Aulich, H., Delahay, P., and Nemec, L., *ibid.* **59**, 2354 (1973); (c) Aulich, H., Baron, B., and Delahay, P., *ibid.* **58**, 603 (1973); (d) Chia, L., Nemec, L., and Delahay, P., *Chem. Phys. Lett.* **32**, 90 (1975); (e) Delahay, P., *J. Chem. Phys.*, to be published (communicated privately).

110. Newton, T. W., and Baker, F. B., *Inorg. Chem.* **4**, 1166 (1965).

111. Noyes, R. M., *J. Am. Chem. Soc.* **86**, 971 (1964).

112. Noyes, A. A., De Vault, D., Coryell, C. D., and Deahl, T. J., *J. Am. Chem. Soc.* **59**, 1326 (1937).

113. Padova, J. I., *in* "Modern Aspects of Electrochemistry" (B. E. Conway and J. O' M. Bockris, eds.), No. 7. Butterworths, London, 1971.

114. Phillips, C. S. G., and Williams, R. J. P., "Inorganic Chemistry," Vol. 1, Oxford Univ. Press, London and New York, 1965).

115. Placsin, R. M., Stoufer, R. C., Mathew, M., and Palenik, G. J., *J. Am. Chem. Soc.* **94**, 2121 (1967).

116. (a) Platzman, R. L., and Franck, J., *in* "L. Farkas Memorial Volume" (A. Farkas and E. P. Wigner, eds.), Res. Council Isr. Spec. Publ. No. 1, pp. 21-36. Jerusalem, 1952; (b) *Z. phys. Chem.* **138**, 411 (1954).

117. Rabinowitch, E., *Rev. Mod. Phys.* **14**, 112 (1942).

118. Rabinowitch, E., and Stockmayer, W. H., *J. Am. Chem. Soc.* **64**, 335 (1942).

119. (a) Robin, M. B., and Day, P., this series **10**, 247 (1967); (b) Allen, G. C., and Hush, N. S., *Progr. Inorg. Chem.* **8**, 357 (1967).

120. (a) Scheibe, G., *Sitzber. phys. med. Sozietät Erlangen* **58/9**, 342 (1927); (b) Franck J., and Scheibe, G., *Z. phys. Chem. A* **139**, 22 (1929); (c) Pauling, L., *Phys. Rev.* **34**, 954 (1929); (d) Franck, J., and Haber, F., *Sitzber. Preuss. Akad. Wiss. Phys. Math. Kl.* p. 250 (1931); (e) Farkas, A., and Farkas, L., *Trans. Faraday Soc.* **34**, 1113 (1938).

121. Schläfer, H. L., *Z. phys. Chem. (Frankfurt)* **3**, 263 (1955).

122. Schmidtke, H. H., *Z. phys. Chem. (Frankfurt)* **38**, 170 (1963).

123. Shirom, M., and Stein, G., *Isr. J. Chem.* **7**, 405 (1969).

124. Shvedov, V. P., and Kotegov, K. V., *Radiokhimiya* **5**, 374 (1963).

125. Sigwart, C., Hemmerich, P., and Spence, J. T., *Inorg. Chem.* **7**, 2545 (1968).

126. Sillén, L. G., and Martell, A. E., "Stability Constants," Chem. Soc. Spec. Publ. No. 17, London, 1967.

127. (a) Smith, M., and Symons, M. C. R., *Trans. Faraday Soc.* **54**, 338, 346 (1958); (b) Stein, G., and Treinin, A., *ibid.* **55**, 1086, 1091 (1959); (c) **56**, 1393 (1960); (d)

Jortner, J., and Treinin, A., *ibid.* **58**, 1503 (1962); (e) Burak, I., and Treinin, A., *ibid.* **59**, 1490 (1963); (f) Treinin, A., *J. Phys. Chem.* **68**, 893 (1964); (g) Barker, B. E., Fox, M. F., Walton, A., and Hayon, E., *J. Chem. Soc., Faraday Trans. A* **72**, 344 (1976).

128. Stranks, D. R., *Disc. Faraday Soc.* **29**, 73 (1960).
129. Stritar, J. A., Ph.D. Thesis, Stanford University, 1967 [quoted in Taube (135)].
130. Sutton, L. E., (ed.), "Interatomic Distances," Chem. Soc. Spec. Publ. No. 11, London, 1958; Suppl. Chem. Soc. Spec. Publ. No. 18, London, 1965.
131. Swift, T. J., and Connick, R. E., *J. Chem. Phys.* **37**, 307 (1962).
132. Sykes, A. G., and Thornely, R. N. F., *J. Chem. Soc. A* p. 232 (1970).
133. Symons, M. C. R., and Trevalion, P. A., *J. Chem. Soc.* p. 3503 (1962).
134. Sytilin, M. S., *Russ. J. Phys. Chem.* **42**, 595 (1968).
135. Taube, H., "Electron Transfer Reactions of Complex Ions in Solution," p. 4. Academic Press, New York, 1970.
136. Tom, G. M., and Taube, H., *J. Am. Chem. Soc.* **97**, 5310 (1975).
137. Tom, G. M., Creutz, C., and Taube, H., *J. Am. Chem. Soc.* **96**, 7827 (1974).
138. Treinin, A., and Hayon, E., *J. Am. Chem. Soc.* **97**, 1716 (1975).
139. Tribalat, S., and Delafosse, D., *Anal. Chim. Acta* **19**, 74 (1958).
140. Troitskaya, N. V., Mishchenko, K. P., and Fleis, I. E., *Russ. J. Phys. Chem.* **33**, 77 (1959).
141. Vetter, K. J., and Manecke, G., *Z. phys. Chem. (Leipzig)* **195**, 270 (1950).
141a. Vogler, A., and Kunkely, H., *Ber. Bunsengesellschaft phys. Chem.*, **79** 83, 301 (1975).
142. Von Stackelberg, M., *Z. Elektrochem.* **46**, 125 (1940) [cited in McConnell and Davidson (*87*)].
143. Waysbort, D., Evenor, M., and Navon, C., *Inorg. Chem.* **14**, 514 (1975).
143a. Welber, B., *J. Chem. Phys.* **42**, 4262 (1965).
144. Wenk, F., unpublished; quoted in Glauser *et al.*, (*51*).
145. Westheimer, F. H., and Kirkwood, J. G., *J. Chem. Phys.* **6**, 513 (1938).
146. (a) Whitney, J. E., and Davidson, N., *J. Am. Chem. Soc.* **69**, 2076 (1947); (b) **71**, 3809 (1949).
147. Wilmarth, W. K., Stanbury, D., Khalaf, S., and Po, H., 172nd ACS Meeting, San Francisco, 1976.
148. (a) Wilson, C. R., and Taube, H., *Inorg. Chem.* **14**, 2276 (1975); (b) Cannon, R. D., Powell, D. B., Sarawek, K., and Stillman, J. S., *J. Chem. Soc., Chem. Commun.* p. 31 (1976).
149. Wyckoff, R. W. G., "Crystal Structures," 2nd ed., Vols. 1–5. Interscience, New York 1963-1966.
150. Yalman, R. G., *Inorg. Chem.* **1**, 16 (1962).
151. Yatsimirskii, *Bull. Acad. Sci. URSS, Cl. Sci. Chim.* p. 453 (1947) [*CA* **42**, 2168d (1948)].
152. Yokoyama, H., and Yamatera, H., *Bull. Chem. Soc. Jpn.* **44**, 1725 (1971).
153. Yoneda, H., *Bull. Chem. Soc. Jpn.* **28**, 125 (1955).
154. Zielen, A. J., and Sullivan, J. C., *J. Phys. Chem.* **66**, 1065 (1962).
155. Zimmerman, G., and Strong, F. C., *J. Am. Chem. Soc.* **79**, 2063 (1957).

VIBRATIONAL SPECTRA OF THE BINARY FLUORIDES OF THE MAIN GROUP ELEMENTS

N. R. SMYRL* and GLEB MAMANTOV

Department of Chemistry, University of Tennessee, Knoxville, Tennessee

I. Introduction   231
II. Inorganic Binary Fluorides   232
III. Matrix Isolation Studies of Transient, Inorganic, Binary Fluoride Species .   246
    References   250

I. Introduction

Vibrational spectroscopy is an important tool for the characterization of various chemical species. Valuable information regarding molecular structures as well as intra- and intermolecular forces can be extracted from vibrational spectral data. Recent advances, such as the introduction of laser sources to Raman spectroscopy, the commercial availability of Fourier transform infrared spectrometers, and the continuing development and application of the matrix-isolation technique to a variety of chemical systems, have greatly enhanced the utility of vibrational spectroscopy to chemists.

Inorganic fluorine compounds are of considerable interest and importance not only because fluorides of nearly all known elements, including some of the rare gases, have been synthesized, but because a number of these compounds are quite different from other halides. Consider, for example, compounds SF_4 and SF_6 and their chloro analogs. The SF_6 is chemically a very inert and a very stable compound, whereas SF_4, also quite stable, is quite reactive. The corresponding tetrachloride molecule is stable only at low temperatures, whereas the hexachloride molecule is nonexistent.

Many fluorides, particularly the interhalogens, are very reactive and require special handling techniques. Matrix isolation has proven to be a very useful technique in the study of inorganic fluorides; it is particularly helpful in the study of very reactive species, such as

* Union Carbide Corporation, Nuclear Division, P.O. Box Y, Oak Ridge, Tennessee 37830.

the free radicals and high-temperature vapor species. Such molecules can be stabilized in a low-temperature matrix environment that effectively prevents reaction with cell and window materials. Because the molecules are effectively isolated from each other, problems involving molecular association are minimized. There is generally a good correlation between matrix and gas phase spectral data except in cases where strong interaction between the matrix material and the isolated molecule exists. Narrow bandwidths are a characteristic feature of matrix spectra, making it frequently possible to obtain structural information from bands involving different isotopes. Matrix-isolation data will be utilized, wherever possible, throughout this paper.

In this review we summarize and attempt to correlate vibrational spectral data amassed from the literature for the main group, inorganic binary fluorides. In addition, a brief review of the matrix-isolation studies of both reactive intermediates and high-temperature fluoride vapor species is included, placing particular emphasis on the inter-halogen molecules.

II. Inorganic Binary Fluorides

In this section available vibrational data for the main group binary fluorides will be reviewed in an attempt to establish trends in both structures and bond-stretching force constants. The review is restricted primarily to molecular entities that were observed either in the gas phase or in inert matrices. The emphasis on the matrix data in this review is in contrast with the more general approach taken by Reynolds in an earlier review of the vibrational spectra of inorganic fluorides (1).

Vibrational spectra, besides being utilized to establish molecular structures, can be used in the evaluation of molecular force constants. Considerable effort has been expended to relate bond-stretching force constants to bond order. Although no specific theoretical model has been found that will cover all types of chemical bonds, there exists a qualitative relationship between bond-stretching force constants and bond orders, permitting a comparison of bond strengths for molecules such as the binary fluorides. In making such a comparison, a consistent force field model must be adhered to. The general valence force field (GVFF) or an approximation of this field generally referred to as the modified valence force field (MVFF) are the potential models that have become increasingly popular with chemists. Valence force constants normally exhibit good transferability between molecules possessing similar types of bonds. Valence stretching force constants further bear a close relationship to the chemical bonding forces. Force constants utilizing the valence force field were, therefore, chosen for comparison in this paper for the reasons just mentioned and because of the wealth of data available in the literature. Although a unique set of force constants rarely exists for a given polyatomic molecule, one preferred set can normally be chosen on a physical basis. Relating stretching force constants to bond energies should be treated with caution, particularly for molecules exhibiting a large percentage of ionic bonding, since the magnitude of the force constant is primarily a reflection of only the covalent character of a particular bond.

Table I lists the symmetry point groups and X—F-stretching force constants for those main group binary fluorides for which the oxidation state of the element combined with fluorine corresponds to the group number. In Tables II–VI vibrational spectral data (using matrix-isolation results if available), including the X—F-stretching force constants, are presented for the main group binary fluorides having formulas ranging from XF to XF_5, respectively. Two force constant trends are apparent. First, the force constants decrease in magnitude from top to bottom for a given group in each of these tables with certain exceptions, as noted in the following. For molecules that possess both equatorial and axial X—F bonds, the axial stretching force constant exhibits the top-to-bottom trend opposite to that previously mentioned. Going from left to right, the force constants in the periodic table first increase, reaching a maximum in a given period somewhere in the vicinity of the transition point from metal to nonmetal; this is followed by a decrease in the force constant. These force constant trends are much more apparent for Tables I–III where a more complete set of data is available.

From the data presented in Tables I–VI, it appears that a particular structure is maintained for molecules of the same formula and within a given group with the exception of BeF_2 and MgF_2, which are linear whereas the remaining members of that group have bent (C_{2v}) structures. The vibrational data for the diatomic fluorides presented in Table II are taken from gas-phase electronic and matrix-isolation spectral studies; both sets of data are available for a number of the diatomic molecules in this table. The gas-phase frequencies have all been corrected for anharmonicity except in the case of fluorine; this is apparently the reason for its force constant being lower than that for ClF. If the matrix data for the diatomic molecules of Group VIIA are examined, however, the expected top-to-bottom trend is observed. The apparent reversal in the magnitudes of the gas phase and matrix frequencies for MgF is a rather unusual occurrence that is also worth noting.

TABLE I

Stretching Force Constants and Symmetry for Main Group Binary Fluorides[a]

Parameter[b]	IA	IIA	IIIA	IVA	VA	VIA	VIIA	VIIIA
	LiF* (23, 24)	BeF$_2$* (26)	BF$_3$* (29, 30)	CF$_4$* (33, 34)	—	—	—	—
Sym.	$C_{\infty v}$	$D_{\infty h}$	D_{3h}	T_d	—	—	—	—
$K_{\text{F-X}}$	2.12	5.15	7.29	6.70	—	—	—	—
	NaF* (9)	MgF$_2$* (8, 27, 28)	AlF$_3$* (31, 32)	SiF$_4$* (35, 36)	PF$_5$ (39)	SF$_6$* (40, 41)	—	—
Sym.	$C_{\infty v}$	$D_{\infty h}$	D_{3h}	T_d	D_{3h}	O_h	—	—
$K_{\text{F-X}}$	1.51	3.23	4.91	6.11	6.32,[c] 4.47[d]	5.26	—	—
	KF* (25)	CaF$_2$* (27)	—	GeF$_4$ (37, 38)	AsF$_5$* (4, 39)	SeF$_6$ (42)	—	—
Sym.	$C_{\infty v}$	C_{2v}	—	T_d	D_{3h}	O_h	—	—
$K_{\text{F-X}}$	1.18	2.21	—	5.45	5.47,[c] 4.51[d]	5.01	—	—
	RbF* (9)	SrF$_2$* (27)	—	—	SbF$_5$* (3, 4, 39, 152)	TeF$_6$ (43)	IF$_7$ (44–46)	—
Sym.	$C_{\infty v}$	C_{2v}	—	—	D_{3h}	O_h	D_{5h}	—
$K_{\text{F-X}}$	1.01	1.97	—	—	5.38,[c] 4.75[d]	5.07	3.01,[c] 4.10[d]	—
	CsF* (9)	BaF$_2$* (27)	—	—	—	—	—	—
Sym.	$C_{\infty v}$	C_{2v}	—	—	—	—	—	—
$K_{\text{F-X}}$	0.95	1.59	—	—	—	—	—	—

[a] Asterisk (*) indicates that matrix data are available; references are given in parentheses.

[b] Units for force constants are in millidynes per angström.

[c] Equatorial fluorine atoms.

[d] Axial fluorine atoms.

TABLE II
VIBRATIONAL DATA FOR MAIN GROUP BINARY FLUORIDES WITH FORMULA XF[a]

Parameter[b]	IA	IIA	IIIA	IVA	VA	VIA	VIIA
	LiF (23, 24, 47)	BeF (47)	BF (47)	CF (53)	NF (55, 56)	OF (5, 25, 59)	F$_2$ (6, 47)
K_{F-X}	2.48 2.12*	5.77	8.05	7.09*	6.18 5.90*	5.42*	4.45^c 4.43*
ω_e	906 840* [Ar]	1266	1400	1279* [Ar]	1141 1115* [Ar]	1029* [Ar]	892^c 890* [N$_2$]
	NaF (9, 47)	MgF (47, 48)	AlF (31, 47)	SiF (54)	PF (57)		ClF (6, 47)
K_{F-X}	1.76 1.51*	3.22 3.41*	4.36 4.05*	4.91	4.98	—	4.56 4.24*
ω_e	536 496* [Ar]	718 738* [Kr]	815 785* [Ne]	858	847	—	793 764* [N$_2$]
	KF (25, 47)	CaF (47)	GaF (49, 50)	GeF (47)	AsF (58)		BrF (47, 60)
K_{F-X}	1.37 1.18*	2.62	3.25*	3.93	4.20	—	4.07 3.93*
ω_e	426 396* [Ar]	587	609* [Ne]	665	686	—	671 660* [Ar]
	RbF (9, 47)	SrF (47)	InF (51)	SnF (47)	SbF (47)		IF (61)
K_{F-X}	1.30 1.01*	2.30	2.75	3.28	3.65	—	3.62
ω_e	376 332* [Ar]	500	535	583	614	—	610
	CsF (9, 47)	BaF (47)	TlF (47, 52)	PbF (47)	BiF (47)		
K_{F-X}	1.22 0.95*	2.16	2.31 1.99*	2.64	2.68	—	—
ω_e	353 312* [Ar]	469	475 441* [Ar]	507	511	—	—

[a] Asterisk (*) indicates that matrix data are available; references are given in parentheses; matrix material is indicated within brackets.
[b] Force constant units are in millidynes per angstrom, and frequency is expressed per centimeter.
[c] Uncorrected for anharmonicity.

235

TABLE III

Vibrational Data for Main Group Binary Fluorides with Formula XF_2[a]

Parameter[b]	IA	IIA	IVA	VA	VIA	VIIA	VIIIA
	$^6LiF_2$* (9)	BeF_2* (26)	CF_2* (33)	NF_2* (33, 65, 66)	OF_2* (5, 25)	—	—
Sym.	C_{2v}	$D_{\infty h}$	C_{2v}	C_{2v}	C_{2v}	—	—
K_{F-x}	—	5.15	6.00	4.83	4.10	—	—
Obs. freq.	v_1, 452	v_2, 330	v_1, 1222	v_1, 1070	v_1, 925	—	—
	v_2, 708 [Ar]	v_3, 1542 [Ne]	v_2, 668	v_2, 573	v_2, 446	—	—
			v_3, 1102 [Ar]	v_3, 931 [N$_2$]	v_3, 826 [Ar]	—	—
	NaF_2* (9)	MgF_2* (8, 27, 28)	SiF_2* (37, 62)	PF_2* (36)	SF_2* (67)	ClF_2* (6, 20)	—
Sym.	C_{2v}	$D_{\infty h}$	C_{2v}	C_{2v}	C_{2v}	C_{2v}	—
K_{F-x}	—	3.23	4.90	4.93		2.48	—
Obs. freq.	v_1, 475	v_1, 550	v_1, 843	v_1, 831	v_3, 847 [Ar]	v_1, 500	—
	v_2, 454 [Ar]	v_2, 249	v_2, 343	v_3, 852 [Ar]		v_2, 242	—
		v_3, 842 [Ar]	v_3, 855 [Ar]			v_3, 578 [N$_2$]	—
	KF_2* (9)	CaF_2* (27)	GeF_2* (7, 37)	—	—	BrF_2* (21)	KrF_2* (5, 68, 69)
Sym.	C_{2v}	C_{2v}	C_{2v}	—	—	C_{2v}	$D_{\infty h}$
K_{F-x}	—	2.21	4.16	—	—	—	2.46
Obs. freq.	v_1, 464	v_1, 485	v_1, 685	—	—	v_3, 569 [Ar]	v_1, 452
	v_2, 342 [Ar]	v_2, 163	v_2, 263^d	—	—		v_2, 236
		v_3, 554 [Kr]	v_3, 653 [Ne, N$_2$]	—	—		v_3, 580 [Ar, Xe]

	RbF$_2$* (9)	SrF$_2$* (27)	SnF$_2$* (63, 64)	XeF$_2$* (5, 68, 69)
Sym.	C_{2v}	C_{2v}	C_{2v}	$D_{\infty h}$
K_{F-x}	—	1.97	3.41	2.84
Obs. freq.	v_1, 462 v_2, 266^c [Ar]	v_1, 441 v_2, 82 v_3, 442 [Kr]	v_1, 605 v_2, 201 v_3, 584 [Ne]	v_1, 512 v_2, 213^e v_3, 547 [Ar, Xe]

	CsF$_2$* (9)	BaF$_2$* (27)	PbF$_2$* (63, 64)	
Sym.	C_{2v}	C_{2v}	C_{2v}	
K_{F-x}	—	1.59	2.92	
Obs. freq.	v_1, 459 v_2, 248^c [Ar]	v_1, 390 v_3, 413 [Kr]	v_1, 546 v_2, 170 v_3, 523 [Ne]	

^{a} There are no data for Group IIIA XF$_2$ compounds. Asterisk (*) indicates that matrix data are available; references are given in parentheses; matrix material is indicated within brackets.

^{b} Stretching force constants units are in millidynes per angström, and frequency is expressed per centimeter.

^{c} These assignments are subject to uncertainty.

^{d} Estimated from electronic spectra.

^{e} Gas-phase frequency.

TABLE IV

VIBRATIONAL DATA FOR MAIN GROUP BINARY FLUORIDES WITH FORMULA XF_3[a]

Parameter[b]	IIIA	IVA	VA	VIA	VIIA
	BF_3 * (29, 30, 70)	CF_3 * (33)	NF_3 * (60, 72, 73)	—	—
Sym.	D_{3h}	C_{3v}	C_{3v}	—	—
K_{F-X}	7.29	5.48	4.16	—	—
Obs. freq.	ν_1, 888[c] ν_3, 1439 [Kr] ν_2, 680 ν_4, 480	ν_1, 1087 ν_3, 1251 [Ar] ν_2, 703 ν_4, 512	ν_1, 1029 ν_3, 898 [Ar] ν_2, 649 ν_4, 495		
	AlF_3 * (31, 32)	SiF_3 * (71)	PF_3 * (36, 74)	SF_3 * (67)	ClF_3 * (77)
Sym.	D_{3h}	C_{3v}	C_{3v}	C_s	C_{2v}
K_{F-X}	4.91	5.49	5.34		4.19,[e] 2.70
Obs. freq.	ν_2, 280 ν_4, 252 [Ne] ν_3, 949	ν_1, 832 ν_3, 954 [Ar] ν_2, 406 ν_4, 290	ν_1, 887 ν_3, 850 [Ar] ν_2, 484 ν_4, 349	ν_5, 682 [Ar]	ν_1, 751 ν_4, 677 [Ar] ν_2, 523 ν_5, 431 ν_3, 328 ν_6, 332
	GaF_3 * (49)		AsF_3 (75)		BrF_3 * (77, 78)
Sym.	D_{3h}	—	C_{3v}	—	C_{2v}
K_{F-X}	—	—	4.64	—	4.08[c] 3.01
Obs. freq.	ν_2, 190 ν_4, 190 [Ne] ν_3, 759		ν_1, 740[d] ν_3, 702[d] ν_2, 336[d] ν_4, 262[d]		ν_1, 674 ν_4, 597 [Ar] ν_2, 547 ν_5, 347 ν_3, 238 ν_6, 253
			SbF_3 * (76)		IF_3 (79)
Sym.	—	—	C_{3v}	—	C_{2v}
K_{F-X}	—	—	3.89	—	
Obs. freq.	—	—	ν_1, 654 ν_3, 624 [N₂] ν_2, 259		ν_1, $\left\{\begin{array}{l}640^f\\628\end{array}\right.$ ν_4, 480[f] ν_2, 550[f] ν_5, 331[f] ν_3, 228[f] ν_6, $\left\{\begin{array}{l}245^f\\240\end{array}\right.$

[a] Asterisk (*) indicates that matrix data are available; references are given in parentheses; matrix material is indicated within brackets.
[b] Stretching force constant units are in millidynes per angström, and frequency is expressed per centimeter.
[c] Calculated from combination bands in the gas-phase IR spectrum.
[d] Gas-phase frequencies. [e] Unique fluorine atom. [f] Solid-state frequencies.

TABLE V

Vibrational Data for Main Group Binary Fluorides with Formula XF_4[a]

Parameter[b]	IVA	VIA	VIIIA
	CF_4* (33, 34, 80)	—	—
Sym.	T_d	—	—
K_{F-X}	6.70	—	—
Obs. freq.	ν_1, 909[c] ν_3, 1267 [Ar] ν_2, 435[c] ν_4, 631	—	—
	SiF_4* (35, 36, 81)	SF_4* (77, 83, 84)	—
Sym.	T_d	C_{2v}	—
K_{F-X}	6.11	5.51,[d] 3.30[e]	—
Obs. freq.	ν_1, 800[c] ν_3, 1022 [Ar] ν_2, 268[c] ν_4, 384	ν_1, 884 ν_4, 228[c] ν_7, 529 [Ar, N$_2$] ν_2, 552 ν_5, 474[c] ν_8, 858 ν_3, 360 ν_6, 707 ν_9, 357	—
	GeF_4* (37, 38, 82)	SeF_4* (85)	—
Sym.	T_d	C_{2v}	—
K_{F-X}	5.45	5.04,[d] 3.38[e]	—
Obs. freq.	ν_1, 740[c] ν_3, 799 [Ne] ν_2, 200[c] ν_4, 260[c]	ν_1, 743 ν_4, ~160[c] ν_8, 723 [N$_2$] ν_2, 589 ν_5, 596 ν_9, 254 ν_3, 406 ν_7, 364	—
	SnF_4* (64)	TeF_4* (85)	XeF_4* (69, 86, 87)
Sym.	T_d	C_{2v}	D_{4h}
K_{F-X}	—	4.62,[d] 3.53[e]	3.00
Obs. freq.	ν_3, 687 [Ar]	ν_1, 695 ν_6, 587 [N$_2$] ν_2, 572 ν_7, 273 ν_3, 333 ν_8, 682	ν_1, 554[c] ν_4, 216[c,f] ν_7, 161[c,f] ν_2, 290 ν_5, 524[c] [Ar] ν_3, 218[c] ν_6, 568
	PbF_4* (64)	—	—
Sym.	T_d	—	—
K_{F-X}	—	—	—
Obs. freq.	ν_3, 663 [Ar]	—	—

[a] There are no data for Groups VA and VIIA XF_4 compounds. Asterisk (*) indicates that matrix data are available; references are given in parentheses; matrix material is indicated within brackets.

[b] Stretching force constants units are in millidynes per angström, and frequency is expressed per centimeter.

[c] Gas-phase frequencies.

[d] Equatorial fluorine atoms.

[e] Axial fluorine atoms.

[f] Evaluated from overtone observed in the Raman spectrum.

In Table III there are several points to be considered. For both Groups VA and VIIIA, there appear to be deviations from the apparent normal top-to-bottom trend with PF_2 and XeF_2 having slightly larger force constants than NF_2 and KrF_2, respectively. All molecules in Table III have bent (C_{2v}) structures with the exception of KrF_2 and

TABLE VI

VIBRATIONAL DATA FOR THE MAIN GROUP BINARY FLUORIDES WITH FORMULA XF_5[a]

Parameter[b]	VA[c]	VIA	VIIA
	PF_5 (39, 88)	SF_5* (89)	ClF_5* (90, 91)
Sym.	D_{3h}	C_{4v}	C_{4v}
K_{F-X}	6.32,[d] 4.47[e]	—	2.75, 3.66[g]
Obs. freq.	ν_1, 817[f] ν_4, 576[f] ν_7, 175[f]	ν_7, 812 ν_8, 552 [Ar]	ν_1, 722 ν_4, 480[f] ν_8, 482
	ν_2, 640[f] ν_5, 1026[f] ν_8, 514[f]	—	ν_2, 539 ν_6, 375[f] ν_9, 299 [N$_2$]
	ν_3, 945[f] ν_6, 533[f]	—	ν_3, 493 ν_7, 726
	AsF_5* (4, 39, 88)	—	BrF_5* (77, 78, 91)
Sym.	D_{3h}	—	C_{4v}
K_{F-X}	5.47,[d] 4.51[e]	—	3.26, 4.01[g]
Obs. freq.	ν_1, 734[f] ν_4, 397 ν_7, 130[f]	—	ν_1, 681 ν_4, 535[f] ν_8, 415
	ν_2, 644[f] ν_5, 808 ν_8, 366 [Ar]	—	ν_2, 582 ν_6, 312[f] ν_9, 240 [Ar]
	ν_3, 782 ν_6, 372[f]	—	ν_3, 366 ν_7, 636
	SbF_5* (3, 4, 39, 152)	—	IF_5 (91)
Sym.	D_{3h}	—	C_{4v}
K_{F-X}	5.38,[d] 4.75[e]	—	3.64, 4.68[g]
Obs. freq.	ν_1, 683[f] ν_4, 285 ν_8, 272[f]	—	ν_1, 698[f] ν_4, 575[f] ν_8, 374[f]
	ν_2, 636[f] ν_5, 726 [Ar]	—	ν_2, 593[f] ν_6, 273[f] ν_9, 189[f]
	ν_2, 726 ν_6, 263	—	ν_3, 315[f] ν_7, 640[f]

[a] There are no data for Group VIIIA XF_5 compounds. The asterisk (*) indicates that matrix data are available; references are given in parentheses; matrix material is indicated in brackets.

[b] Stretching force constant units are in millidynes per angström, and frequency is expressed per centimeter.

[c] See text for a discussion of SbF_5.

[d] Equatorial fluorine atoms.

[e] Axial fluorine atoms.

[f] Gas-phase frequencies.

[g] Unique fluorine atom.

XeF_2, which are linear ($D_{\infty h}$) as are BeF_2 and MgF_2, as previously noted, and the MF_2 molecules of Group IA, which are apparently triangular (C_{2v}). For the series of molecules, CF_2, NF_2, and OF_2, the ν_1 mode is observed to have a higher frequency than the ν_3 mode; this is an apparent deviation from normal trends. This behavior may be a result of dynamic effects due to the similarity in the masses of the various atoms making up this series of triatomic molecules (2). The assignment of the ν_1 mode to a higher frequency value than ν_3 has been made for a number of other molecules in Table III. In some cases, however, these assignments are subject to some uncertainty. Raman data, which could be utilized to clarify such situations, are largely unavailable for molecules in Table III because the majority of the data come from infrared matrix-isolation studies.

In Table IV, NF_3 appears to be an exception as its stretching force constant is lower than that for PF_3. It should be noted also that the force constants for CF_3 and SiF_3 are very nearly the same. The XF_3 compounds of Group VIIA possess two equivalent axial X—F bonds and one equatorial or unique bond. The equatorial-type bonds are stronger than the axial bonds for all molecules in Tables IV–VI that exhibit bonding of this type with the exception of the XF_5 molecules of Group VIIA. The range of point-group symmetry for the molecules in Table III is D_{3h} for Group IIIA, C_{3v} for Groups IVA and VA, C_s for Group VIA, and C_{2v} for Group VIIA. The frequency data for Group VA indicate that v_1 is greater than v_3 for each member of that series.

Vibrational data for the main group binary fluorides with the formula XF_4 are restricted to compounds of Groups IVA, VIA, and VIIIA as exemplified in Table V. Molecules of Group VIA possess two axial and two equatorial X—F bonds. The range of point-group symmetry for the molecules in Table V is T_d for Group IVA, C_{2v} for Group VIA, and D_{4h} for Group VIIIA.

The XF_5 molecules of Group VIIA possess C_{4v} symmetry and have four equivalent equatorial X—F bonds and one unique axial bond, as indicated in Table VI. The equatorial stretching force constant exhibits the reverse of the normal top-to-bottom trend previously noted for bonds of this type. The XF_5 molecules of Group VA have D_{3h} symmetry. It appears appropriate at this point to present a brief discussion of SbF_5, since there has been considerable uncertainty in the interpretation of the vibrational spectra and the assignment of a structure to the monomer. A recent report involving a temperature-dependent study (3) of the Raman gas-phase spectrum of SbF_5 indicates that previous vibrational assignments were apparently in error.* This study shows that SbF_5 is strongly associated in the gas phase even at temperatures as high as 250°C. The association apparently occurs through bonds involving cis-fluorine bridging. An assignment of a C_{4v} structure to the monomer for SbF_5 from a previous matrix-isolation study (4) appears to be negated, because the experimental conditions that were utilized could not preclude the predominance of the associated species. The force constants presented in Table VI for SbF_5 are based on an incorrect vibrational assignment. It would appear, however, that a corrected assignment might produce very little

---

* Additional matrix studies of monomeric SbF_5 have been recently reported (152).

TABLE VII

Vibrational Data for Various Isoelectronic Binary Fluoride Series with Formula XF_4 [a]

Parameter [b]	IIIA	IVA [d]	VA	VIA [d]	VIIA
	BF_4^- (34, 92, 93)	CF_4 (33, 34, 80)	NF_4^+ (95, 96)	—	—
Sym.	T_d	T_d	T_d	—	—
K_{F-X}	4.57	6.70	6.06	—	—
Obs. freq.	v_1, 769 v_3, 1084 [c] [aqueous sol.] v_2, 353 v_4, 524		v_1, 849 v_3, 1162 [solid $NF_4^+BF_4^-$] v_2, 450 v_4, 611	—	—
	AlF_4^- (94)	SiF_4 (35, 36, 81)	—	SF_4 (77, 83, 84)	ClF_4^+ (84, 97)
Sym.	T_d	T_d	—	C_{2v}	C_{2v}
K_{F-X}	4.33	6.11	—	5.51,[e] 3.30[f]	4.78,[e] 3.73[f]
Obs. freq.	v_1, 622 v_3, 760 [MF–AlF₃ melts] v_2, 210 v_4, 322	—	—	—	v_1, 800 v_4, 250 v_7, 515 [ClF₅·AsF₅ ClF₅·SbF₅ adducts] v_2, 571 v_5, 475 v_8, 829 v_3, 385 v_6, 795 v_9, 385
				SeF_4 (85)	BrF_4^+ (97)
Sym.	—			C_{2v}	C_{2v}
K_{F-X}	—			5.04,[e] 3.38[f]	
Obs. freq.	—			—	v_1, 733 v_4, 216 v_8, 736 [BrF₅·2SbF₅ adduct] v_2, 606 v_6, 704 v_9, 369 v_3, 385 v_7, 419
			SbF_4^- (76)	TeF_4 (85)	IF_4^+ (97)
Sym.			C_{2v}	C_{2v}	C_{2v}
K_{F-X}			3.3,[e] 2.0[f]	4.62,[e] 3.53[f]	—
Obs. freq.			v_1, 596 v_4, 163 v_7, 257 [MeCN sol.] v_2, 449 v_5, ~220 v_8, 566 v_3, 285 v_6, 431 v_9, 180	—	v_1, 704 v_4, 151 v_8, 720 [IF₅·SbF₅ adduct] v_2, 609 v_5, 655 v_9, 316 v_3, 341 v_7, 385

[a] References are given in parentheses; the state for which the frequencies were observed is denoted in brackets.

[b] Stretching force constant units are in millidynes per ångström, and frequency is expressed per centimeter.

[c] KBr matrix value.

[d] Observed frequencies are given in Table V.

[e] Equatorial fluorine atoms.

[f] Axial fluorine atoms.

effect on magnitudes of these reported stretching force constants, and so they are included for comparison.

Vibrational data for various isoelectronic binary fluorides with the formula XF_4 are listed in Table VII. For the first series in the upper left of this table, the force constant increases in going from BF_4^- to CF_4 and then decreases slightly for NF_4^+. It would appear that the next series beginning with AlF_4^- would follow a similar trend except that data for PF_4^+ are unavailable. The fluorides of these two series all have T_d symmetry. Very little can be said with regard to the three series in the lower right section of Table VII since the data are largely incomplete. The symmetry, however, for the various species in these series is C_{2v}.

Vibrational data are presented in Table VIII for various isoelectronic series of the formula XF_6. All the species in Table VIII possess O_h symmetry. The left-to-right, horizontal, force constant trends are similar for each of these series, as the magnitudes are observed to increase to Group VIA and then decrease slightly for Group VIIA. The exception to this particular trend occurs in the lower series that increases continuously from left to right. There is a general decrease in the force constants in going from top to bottom for all groups in this table with the exception of VIIA that exhibits the reverse trend.

In the compilation of force constant data for the inorganic binary fluorides, it would be of interest to examine how bonding of additional atoms such as oxygen might affect the bond strengths of selected X—F bonds. Vibrational data have been compiled in Table IX for NF, ClF, NF_3, and ClF_3 along with various oxygenated forms of these compounds to illustrate this effect. The N—F and Cl—F stretching force constants are observed to be reduced to approximately one-half of their original values in going from NF to FNO and from ClF to FClO, respectively. Further addition of oxygen atoms in both series appears to have little effect on the corresponding X—F stretching force constants, except for an eventual slight increase over the value for the mono-oxygenated species. The N—F stretching force constant is essentially unaffected in going from NF_3 to NF_3O, whereas the Cl—F force constants are mildly reduced from ClF_3 to ClF_3O. The force constant for the unique bond of ClF_3O exhibits a much larger reduction than that for the axial bonds. On addition of a second oxygen atom to yield the molecule ClF_3O_2, both unique and axial force constants increase over the values for ClF_3O, with latter force constant having a value equal to that of ClF_3.

Of interest is also the effect that additional electrons have on the structures and stretching force constants for a given fluoride series

TABLE VIII

Vibrational Data for Various Isoelectric Binary Fluoride Series of Formula XF_6 [a]

Parameter [b]	IIIA	IVA	VA	VIA	VIIA
	AlF_6^{3-} (98)	SiF_6^{2-} (99)	PF_6^{-} (43)	SF_6 (40,100,101)	ClF_6^{+} (40)
Sym.	O_h	O_h	O_h	O_h	O_h
K_{F-x}	—	3.01	4.39	5.26	4.68
Obs. freq.	ν_1, 555 ν_5, 345 [MF–AlF₃ melts] ν_2, 390	ν_1, 663 ν_4, 483 [solid Na₂SiF₆] ν_2, 477 ν_5, 408 ν_3, 741	ν_1, 746 ν_4, 557 [PF₆⁻ salt] ν_2, 561 ν_5, 475 ν_3, 817	ν_1, 774 ν_4, 614 [gas] ν_2, 642 ν_5, 525 ν_3, 939	ν_1, 679 ν_4, 582 [solid ClF₆⁺PtF₆⁻] ν_2, 630 ν_5, 513 ν_3, 890
		GeF_6^{2-} (99)	AsF_6^{-} (43)	SeF_6 (42,100,101)	BrF_6^{+} (43,102)
Sym.		O_h	O_h	O_h	O_h
K_{F-x}		2.71	3.98	5.01	4.90
Obs. freq.		ν_1, 624 ν_4, 359, 339 ν_2, 471 ν_5, 335 [solid K₂GeF₆] ν_3, 603	ν_1, 682 ν_4, 385 [solid BrF₆⁺AsF₆⁻] ν_2, 568 ν_5, 369 ν_3, 696	ν_1, 707 ν_4, 437 [gas] ν_2, 659 ν_5, 405 ν_3, 780	ν_1, 658 ν_4, 430 [solid BrF₆⁺AsF₆⁻] ν_2, 668 ν_5, 405 ν_3, 775
		SnF_6^{2-} (99)	SbF_6^{-} (43)	TeF_6 (43,100,101)	IF_6^{+} (43,103)
Sym.		O_h	O_h	O_h	O_h
K_{F-x}		2.77	3.90	5.07	5.42
Obs. freq.		ν_1, 592 ν_4, 300 [solid Na₂SnF₆] ν_2, 477 ν_5, 252 ν_3, 559	ν_1, 653 ν_4, 280 [solid BrF₆⁺SbF₆⁻] ν_2, 561 ν_5, 273 ν_3, 667	ν_1, 697 ν_4, 325 [gas] ν_2, 670 ν_5, 314 ν_3, 752	ν_1, 708 ν_4, 343 [solid IF₆⁺AsF₆⁻] ν_2, 732 ν_5, 340 ν_3, 794

[a] References are given in parentheses; the state for which the frequencies were observed is denoted in brackets.

[b] Stretching force constant units are in millidynes per angström, and frequency is expressed per centimeter.

TABLE IX

VIBRATIONAL DATA FOR SELECTED NITROGEN AND CHLORINE FLUORIDES
CONTAINING OXYGEN[a]

Compound	Symmetry	Observed frequencies (cm^{-1})	F—X stretching force constant $(mdyn/Å)$
1. NF			
NF* $(55, 56)^b$	$C_{\infty v}$	—	5.90
FNO (104)	C_s	ν_1, 1877	2.26
		ν_2, 775 [gas]	
		ν_3, 521	
FNO$_2$ (105)	C_{2v}	ν_1, 1310 ν_4, 1792	2.66
		ν_2, 822 ν_5, 560 [gas]	
		ν_3, 569 ν_6, 742	
NF$_3$* $(72, 73)^c$	C_{3v}	—	4.16
NF$_3$O $(106, 107)$	C_{3v}	ν_1, 1689 ν_4, 880	4.25
		ν_2, 740 ν_5, 522 [gas]	
		ν_3, 534 ν_6, 398	
2. ClF			
ClF* $(6, 47)^b$	$C_{\infty v}$	—	4.24
FClO* (108)	C_s	ν_1, 1038	2.59
		ν_2, 594 [Ar matrix]	
		ν_3, 315	
FClO$_2$ (109)	C_s	ν_1, 1106 ν_4, 402	2.53
		ν_2, 630 ν_5, 1271 [gas]	
		ν_3, 547 ν_6, 351	
FClO$_3$ $(110, 111)$	C_{3v}	ν_1, 1061 ν_4, 1315	2.79
		ν_2, 715 ν_5, 589 [gas]	
		ν_3, 549 ν_6, 405	
ClF$_3$* $(77)^c$	C_{2v}	—	$4.19,^e$ 2.70
ClF$_3$O* (112)	C_s	ν_1, 1223 ν_4, 478 ν_7, 652	$3.16,^e$ 2.34
		ν_2, 686 ν_5, 323 ν_8, 499 $\left[N_2\ matrix\right]$	
		ν_3, 484 ν_6, 224^d ν_9,414	
ClF$_3$O$_2$* (113)	C_{2v}	ν_1, 1093 ν_5, 287 ν_9, 372	$3.35,^e$ 2.70
		ν_2, 683 ν_6, 417 ν_{10}, 1331 $\begin{bmatrix}N_2\\ matrix\end{bmatrix}$	
		ν_3, 519 ν_7, 686 ν_{11}, 531	
		ν_4, 487 ν_8, 591 ν_{12}, 222^d	

[a] Asterisk (*) indicates that matrix data are available; references are given in parentheses; the state for which the frequencies were observed is denoted in brackets.

[b] Observed frequencies are given in Table II.

[c] Observed frequencies are given in Table IV.

[d] Gas-phase frequency.

[e] Unique fluorine atom.

TABLE X

VIBRATIONAL AND STRUCTURAL DATA FOR ClF_2^n ($n = +1, 0, -1$) SERIES[a]

ClF_2 species	Symmetry	Observed frequencies (cm^{-1})	Bond angle	F—Cl stretching force constant (mdyn/Å)
ClF_2^+ (6, 114, 115)	C_{2v}	ν_1, 807 $\begin{bmatrix} solid \\ ClF_2^+SbF_6^- \end{bmatrix}$ ν_2, 387 ν_3, 830	100°	4.74
ClF_2 (6, 20)[b]	C_{2v}	—	~150°	2.48
ClF_2^- (116)	$D_{\infty h}$	ν_3, 476 $\begin{bmatrix} solid \\ Na^+ClF_2^- \\ Rb^+ClF_2^- \end{bmatrix}$ ν_3, 635	180°	2.35

[a] References are given in parentheses; the state for which the frequencies were observed is denoted in brackets.
[b] Observed frequencies given in Table III.

having the same molecular formula. This effect is illustrated in Table X for the ClF_2^+, ClF_2, and ClF_2^- series. The structures range from bent (C_{2v}) for ClF_2^+ and ClF_2 with bond angles of 100° and ~150°, respectively, to linear ($D_{\infty h}$) for ClF_2^-. The stretching force constants for ClF_2 and ClF_2^- are 2.48 and 2.35 mdyn/Å, respectively, which is approximately half the value for that of ClF_2^+ and indicates that the additional electrons for ClF_2 and ClF_2^- are largely antibonding. A more complete discussion of ClF_2 will be made in the next section.

In concluding this section, a list of other main group binary fluorides for which there are either complete or partial vibrational data available is presented in Table XI.

III. Matrix Isolation Studies of Transient, Inorganic, Binary Fluoride Species

The area in which matrix isolation is perhaps of greatest value is the stabilization of transient species such as free radicals and high-temperature vapors. Until quite recently, infrared spectroscopy was utilized almost exclusively for the vibrational studies of matrix-isolated species. With the introduction of laser sources and the development of more sensitive, electronic, light detection systems, Raman matrix-isolation studies are now feasible and have recently been applied to a limited number of unstable inorganic fluoride species including the molecules OF (5) and ClF_2 (6). Both of these species were formed for Raman study by a novel technique that utilizes the

TABLE XI

List of Other Binary Fluorides for Which Vibrational Data are Available[a]

IA	IIA	IIIA	IVA	VA	VIA	VIIA	VIIIA
$(LiF)_2$* (24, 117)	BeF_4^{2-} (120)	B_2F_4* (121)	C_2F_4* (33, 123)	N_2F^+ (126)	O_2F_2* (134, 135)	F_3^-* (10)	—
$(LiF)_3$* (118)		$(BF_3)_2$* (29)	C_2F_6 (33, 124)	N_2F_2* (56, 127, 128)	O_4F_2* (136)		—
				$N_2F_3^+$ (129)			—
				N_2F_4* (65, 130, 131)			—
				N_3F*			
$(NaF)_2$* (9, 119)	$(MgF_2)_2$* (8, 28)	$(AlF)_2$* (31, 49)	Si_2F_6 (125)	P_2F_4* (36, 132, 133)	S_2F_{10}* (11, 137, 138)	ClF_4^- (142)	KrF* (147)
		$(AlF_3)_2$* (31, 122)		$PF_3{=}PF$* (36)	S_2F_2 (139)	Cl_2F^+ (143, 144)	KrF^+ (148, 149)
					$(SF_4)_2$* (77)	Cl_2F* (21, 145)	$Kr_2F_3^+$ (149)
					S_2F_4 (140)	Cl_2F_2* (21, 145)	
					SF_5^- (141)	Cl_2F_3* (21, 145)	
						$(ClF_3)_2$* (145)	
$(KF)_2$* (9)	—	$(GaF)_2$* (49)	$(GeF_2)_2$* (7, 37)		$(SeF_4)_2$* (85)	BrF_2^+ (115)	—
		$(GaF_3)_2$* (49)			SeF_5^- (141)	BrF_4^- (142)	
						Br_2F* (21)	
						Br_2F_2* (21)	
						$(BrF_3)_2$* (77)	
$(RbF)_2$* (9)	—		—	SbF_5^{2-} (76)	TeF_5^- (76, 141)	IF_4^- (142)	XeF* (147)
				$Sb_2F_{11}^-$ (102)	Te_2F_{10} (137)	IF_6^- (146)	XeF^+ (150)
					$(TeF_4)_2$* (85)	IF_6^{3-} (142)	$Xe_2F_3^+$ (150)
						I_2F^+ (21)	XeF_6* (151)
						I_2F_2* (21)	
$(CsF)_2$* (9)	—	$(TlF)_2$* (52)	—				—

[a] Asterisk (*) indicates that matrix data are available; reference numbers are given in parentheses.

4880-Å output from an Ar ion laser as both the photolysis source and the Raman excitation source. The examples of ClF_2 (6), GeF_2 (7), and MgF_2 (8) illustrate the complementary data that Raman matrix-isolation spectroscopy can provide. Other binary fluoride molecules for which Raman matrix data are available include OF_2 (5), MF_2 and MF_3 (M = alkali metals) (9, 10), S_2F_{10} (11), KrF_2 and XeF_2 (5), with the latter two molecules having been formed by the same laser photolysis method as previously mentioned. Raman matrix-isolation spectroscopy has recently been reviewed by Ozin (12), and, although merely in the early stages of development, it is anticipated that this area will continue to grow and become a valuable counterpart to the infrared technique.

This section represents a brief review of matrix-isolation studies of the main group, binary fluoride free radicals and high-temperature vapor species. A discussion of the halogen fluoride molecules will be stressed since this is an area of considerable interest to our laboratory. A list of main group binary fluorides that have been characterized by matrix isolation appears in Table XII. It is apparent from Table XII that there are large gaps in the existing data for species such as the halogen fluoride radicals, which must be filled in order to gain a better understanding of the structural properties and bonding characteristics of these systems. Presently, ClF_2 and Cl_2F are the only halogen fluoride radicals for which vibrational data have been published. The ClF_4 (13, 14), ClF_6 (14, 15), BrF_6 (15, 16), and IF_6 (15) radicals have recently been prepared by γ-radiolysis and characterized by ESR, although no vibrational spectroscopic data are available. Other interhalogen radicals, Cl_3 and Br_3, which were reportedly formed by microwave discharge (17, 18) may in fact be the ionic $X_3{}^-$ species (19).

The ClF_2 radical is of sufficient chemical significance to merit a brief discussion. This particular radical is an interesting species because it possesses 21 valence electrons and appears to have an intermediate bond angle of $\sim 150°$. Such a bond angle is apparently unique for interhalogen species since the bond angles for most of these molecules are very near 90° or 180°. The ClF_2 radical was first prepared in our laboratory through the matrix reaction $ClF + F \rightarrow ClF_2$ (20). This reaction was initiated by production of F atoms through in situ UV photolysis of molecular fluorine. From the infrared data, the three vibrational fundamentals were assigned. On this basis and on the basis of ^{35}Cl, ^{37}Cl isotopic shift data for v_3, compound ClF_2 was assigned a bent (C_{2v}) structure with a bond angle of $\sim 135°$. Very recently, however, Andrews' group (6) has repeated the infrared work and, in addition, has successfully prepared ClF_2 by the laser photolysis technique for Raman study. Based on the new Raman data, the v_1 mode

TABLE XII

Inorganic Binary Fluoride Free Radicals, Intermediates, and High-Temperature Vapor Species Characterized by Matrix Isolation[a]

IA	IIA	IIIA	IVA	VA	VIA	VIIA
LiF (23, 24)	BeF_2 (26)	—	CF (53); CF_2 (33); CF_3 (33)	NF (56); NF_2 (33, 65)	OF (5, 25, 59); O_2F (134, 136)	—; —
NaF (9, 119)	MgF (48); MgF_2 (8, 27, 28)	AlF (31); AlF_3 (31, 32)	SiF_2 (37, 62); SiF_3 (71)	PF_2 (36)	SF_2 (67); SF_3 (67)	ClF_2 (6, 20); Cl_2F (21, 145)
KF (9)	CaF_2 (27)	GaF (49, 50); GaF_3 (49)	GeF_2 (7, 37)	—	SF_5 (89)	—; —
—	SrF_2 (27)	—	SnF_2 (63, 64); SnF_4 (64)	SbF_3 (76)	—	BrF_2 (21); Br_2F (21)
—	BaF_2 (27)	TlF (52)	PbF_2 (63, 64); PbF_4 (64)	—	—	I_2F (21); —

[a] There are no listings for Group VIIIA. References are given in parentheses.

was reassigned from a value of 536 cm^{-1} to that of 500 cm^{-1}. It appears that the principle of mutual exclusion holds for the ClF_2 radical, which is indicative of a linear triatomic species, but, as Andrews points out, mutual exclusion is valid only within the limits of detectability, which may be a problem for weakly active modes in dilute matrices. In a reassessment of the data, Andrews favors a slightly bent structure with a bond angle of $\sim 150°$.

Recently we have completed (in collaboration with E. S. Prochaska and L. Andrews of the University of Virginia) a matrix-isolation study of the halogen–fluorine systems (21). Evidence for the formation of BrF_2, Br_2F, Br_2F_2, I_2F and I_2F_2 as well as more definitive data for Cl_2F and Cl_2F_2 were obtained. These experiments involved UV photolysis of matrix mixtures and microwave discharge experiments. The microwave experiments were stimulated by a recent mass spectrometric kinetic study (22) that appeared to present a method producing BrF in a much purer form. The data for BrF_2 are of considerable interest for the purposes of comparison with ClF_2. A doublet near 569 cm^{-1} with a 2.2 cm^{-1} isotopic splitting was attributed to BrF_2, and a calculation of $152° \pm 8°$ for the lower limit of the F—Br—F valence angle strongly indicates that, like ClF_2, compound BrF_2 is also obtusely bent. Bands observed at 555 and 507 cm^{-1} were attributed to Br_2F_2 and Br_2F, respectively, based on photolysis behavior similar to their chlorine analogs.

Vibrational data for the majority of the molecules in Table XII were presented in Section I, and these species will not be discussed further.

ACKNOWLEDGMENTS

We would like to acknowledge the support of the Army Research Office and of The Air Force Office of Scientific Research.

REFERENCES

1. Reynolds, D. J., *Adv. Fluorine Chem.* **7**, 1 (1973).
2. Andrews, L. *in* "Vibrational Spectroscopy of Trapped Species" (H. E. Hallam, ed.), p. 197. Wiley, New York, 1973.
3. Alexander, L. E., and Beattie, I. R., *J. Chem. Phys.* **56**, 5829 (1972).
4. Aljibury, A. L. K., and Redington, R. L., *J. Chem. Phys.* **52**, 453 (1970).
5. Andrews, L., *Appl. Spectrosc. Rev.* **11**, 125 (1976).
6. Prochaska, E. S., and Andrews, L., *Inorg. Chem.* **16**, 339 (1977).
7. Huber, H., Künig, E. P., Ozin, G. A., and Vander Voet, A., *Can. J. Chem.* **52**, 95 (1974).
8. Lesiecki, M. L., and Nibler, J. W., *J. Chem. Phys.* **64**, 871 (1976).
9. Howard, W. F., Jr., and Andrews, L., *Inorg. Chem.* **14**, 409 (1975).
10. Ault, B. S., and Andrews, L., *J. Am. Chem. Soc.* **95**, 1591 (1976).
11. Smardzewski, R. R., Noftle, R. E., and Fox, W. B., *J. Mol. Spectrosc.* **62**, 449 (1976).

12. Ozin, G. A., *in* "Vibrational Spectroscopy of Trapped Species" (H. E. Hallam, ed.), p. 373. Wiley, 1973.
13. Morton, J. R., and Preston, K. F., *J. Chem. Phys.* **58**, 3112 (1973).
14. Nishikida, K., Williams, F., Mamantov, G., and Smyrl, N., *J. Am. Chem. Soc.* **97**, 3526 (1975).
15. Boate, A. R., Morton, J. R., and Preston, K. F., *Inorg. Chem.* **12**, 3127 (1975).
16. Nishikida, K., Williams, F., Mamantov, G., and Smyrl, N., *J. Chem. Phys.* **63**, 1693 (1975).
17. Nelson, L., and Pimentel, G. C., *J. Chem. Phys.* **47**, 3671 (1967).
18. Boal, D. H., ar.d Ozin, G. A., *J. Chem. Phys.* **55**, 3598 (1971).
19. Wight, C. A., Ault, B. S., and Andrews, L., *J. Chem. Phys.* **65**, 1244 (1976).
20. Mamantov, G., Vasini, E. J., Moulton, M. C., Vickroy, D. G., and Maekawa, T., *J. Chem. Phys.* **54**, 3419 (1971).
21. Prochaska, E. S., Andrews, L., Smyrl, N. R., and Mamantov, G., *Inorg. Chem.* (in press).
22. Appleman, E. H., and Clyne, M. A. A., *Trans. Faraday Soc.* **71**, 2072 (1975).
23. Schlick, S., and Schnepp, O., *J. Chem. Phys.* **41**, 463 (1964).
24. Linevsky, M. J., *J. Chem. Phys.* **38**, 658 (1963).
25. Andrews, L., and Raymond, J. I., *J. Chem. Phys.* **55**, 3078 (1971).
26. Snelson, A., *J. Phys. Chem.* **70**, 3208 (1966).
27. Calder, V., Mann, D. E., Seshadri, K. S., Allavena, M., and White, D., *J. Chem. Phys.* **51**, 2093 (1969).
28. Hauge, R. H., Margrave, J. L., and Kana'an, A. S., *Trans. Faraday Soc.* **71**, 1082 (1975).
29. Bassler, J. M., Timms, P. L., and Margrave, J. L., *J. Chem. Phys.* **45**, 2704 (1966).
30. Duncan, J. L., *J. Mol. Spectrosc.* **13**, 338 (1964).
31. Snelson, A., *J. Phys. Chem.* **71**, 3302 (1967).
32. Yang, Y. S., and Shirk, J. S., *J. Mol. Spectrosc.* **54**, 39 (1975).
33. Milligan, D. E., and Jacox, M. E., *J. Chem. Phys.* **48**, 2265 (1968).
34. Krebs, B., Müller, A., and Fadini, A., *J. Mol. Spectrosc.* **24**, 198 (1967).
35. Beattie, I. R., Livingston, K. M. S., and Reynolds, D. J., *J. Chem. Phys.* **51**, 4269 (1969).
36. Burdett, J. K., Hodges, L., Dunning, V., and Current, J. H., *J. Phys. Chem.* **74**, 4053 (1970).
37. Hastie, J. W., Hauge, R., and Margrave, J. L., *J. Phys. Chem.* **72**, 4492 (1968).
38. Müller, A., and Krebs, B., *J. Mol. Spectrosc.* **24**, 180 (1967).
39. Wendling, E. J. L., Mahmoudi, S., and MacCordick, H. J., *J. Chem. Soc. A* p. 1747 1971.
40. Christe, K. O., *Inorg. Chem.* **12**, 1580 (1973).
41. Shirk, J. S., and Claassen, H. H., *J. Chem. Phys.* **54**, 3237 (1971).
42. Abramowitz, S., and Levin, I. W., *Inorg. Chem.* **6**, 538 (1967).
43. Christe, K. O., and Wilson, R. D., *Inorg. Chem.* **14**, 694 (1975).
44. Khanna, R. K., *J. Mol. Spectrosc.* **8**, 134 (1972).
45. Lord, R. C., Lynch, M. A., Schumb, W. C., and Slowinski, E. J., *J. Am. Chem. Soc.* **72**, 522 (1950).
46. Eyseland, H. H., and Seppelt, K., *J. Chem. Phys.* **56**, 5081 (1972).
47. Jones, L. H., "Inorganic Vibrational Spectroscopy," pp. 183–187. Dekker, New York, 1971.
48. Mann, D. E., Calder, G. V., Seshadri, K. S., White, D., and Linevsky, M. J., *J. Chem. Phys.* **46**, 1138 (1967).
49. Hastie, J. W., Hauge, R. H., and Margrave, J. L., *J. Fluorine Chem.* **3**, 285 (1973).

50. Hastie, J. W., Hauge, R. H., and Margrave, J. L. *in* "Spectroscopy in Inorganic Chemistry" (Rao, C. N. R., and Ferraro, J. R. eds.), Vol. 1, p. 69. Academic Press, New York, 1970.

51. Barrow, R. F., Glaser, D. V., and Zeeman, P. B., *Proc. Phys. Soc., London, Sect. A* **68**, 962 (1955).

52. Brom J. M., and Franzen, H. F., *J. Chem. Phys.* **54**, 2874 (1971).

53. Jacox, M. E., and Milligan, D. E., *J. Chem. Phys.* **50**, 3252 (1969).

54. Appelbald, O., Barrow, R. F., and Verma, R. D., *J. Phys. B* **1**, 274 (1968).

55. Jones, A. E., and Jones, W. E., *Can. J. Phys.* **44**, 2251 (1966).

56. Milligan, D. E., and Jacox, M. E., *J. Chem. Phys.* **40**, 2461 (1964).

57. Douglas, A. E., and Frackowiak, M., *Can. J. Phys.* **40**, 832 (1962).

58. Liu, D. S., and Jones, W. E., *Can. J. Phys.* **50**, 1230 (1972).

59. Arkell, A., Reinhard, R. R., and Larson, L. P., *J. Am. Chem. Soc.* **87**, 1016 (1965).

60. Mamantov, G., and Smyrl, N., unpublished results.

61. Stein, L., *in* "Halogen Chemistry" (V. Gutman, ed.), Vol. 1, p.176. Academic Press, New York, 1967.

62. Milligan, D. E., and Jacox, M. E., *J. Chem. Phys.* **49**, 4269 (1968).

63. Margrave, J. L., Hastie, J. W., and Hauge, R. H., *Am. Chem. Soc., Div. Petroleum Chem., Preprint.* **14**, E11–E13, (1969).

64. Hauge, R. H., Hastie, J. W., and Margrave, J. L., *J. Mol. Spectrosc.* **45**, 420 (1973).

65. Harmony M. D., and Myers, R. J., *J. Chem. Phys.* **37**, 636 (1962).

66. Selig, H., and Holloway, J. H., *J. Inorg. Nucl. Chem.* **33**, 3169 (1971).

67. Smardzewski, R. R., and Fox, W. B., *J. Fluorine Chem.* **7**, 353 (1976).

68. Claassen, H. H., Goodman, G. L., Malm, J. G., and Schreiner, F., *J. Chem. Phys.* **42**, 1229 (1965).

69. Hallam, H. E. ed., *in* "Vibrational Spectroscopy of Trapped Species," p. 107. Wiley, New York, 1973.

70. Brown, C. W., and Overend, J., *Spectrochim. Acta*, Part A **25**, 1535 (1969).

71. Milligan, D. E., and Jacox, M. E., *J. Chem. Phys.* **49**, 5330 (1968).

72. Jacox, M. E., and Milligan, D. E., *J. Mol. Spectrosc.* **52**, 322 (1974).

73. Allan, A., Duncan, J. L., Holloway, J. H., and McKean D. C., *J. Mol. Spectrosc.* **31**, 368 (1969).

74. Mirri, A. M., Scappini, F., and Favero, P. G., *Spectrochim. Acta* **21**, 965 (1965).

75. Brieux De Mandirola, O., *J. Mol. Struct.* **1**, 203 (1967).

76. Adams, C. J., and Downs, A. J., *J. Chem. Soc. A* p. 1534 (1971).

77. Frey, R. A., Redington, R. L., and Aljibury, A. L. K., *J. Chem. Phys.* **54**, 344 (1971).

78. Christe, K. O., Curtis, E. C., and Pilipovich, D., *Spectrochim. Acta, Part A* **27**, 931 (1971).

79. Schmeisser, M., Naumann, D., and Lehmann, E., *J. Fluorine Chem.* **3**, 441 (1973).

80. Monostori, B., and Weber, A., *J. Chem. Phys.* **33**, 1867 (1960).

81. Jones, E. A., Kirby-Smith, J. S., Woltz, P. J. H., and Nielsen, A. H., *J. Chem. Phys.* **19**, 242 (1951).

82. Woltz, P. J. H., and Nielsen, A. H., *J. Chem. Phys.* **20**, 307 (1952).

83. Christe, K. O., and Sawodny, W., *J. Chem. Phys.* **52**, 6320 (1970).

84. Christe, K. O., Curtis, E. C., Schack, C. J., Cyvin, S. J., Brunvoll, J., and Sawodny, W., *Spectrochim. Acta, Part A* **32**, 1141 (1976).

85. Adams, C. J., and Downs, A. J., *Spectrochim. Acta*, **28A**, 1841 (1972).

86. Claassen, H. H., Chernick, C. L., and Malm, J. G., *J. Am. Chem. Soc.* **85**, 1927 (1963).

87. Tsao, P., Cobb, C. C., and Claassen, H. H., *J. Chem. Phys.* **54**, 5247 (1971).

88. Selig, H., Holloway, J. H., Tyson, J., and Claassen, H. H., *J. Chem. Phys.* **53**, 2559 (1970).

89. Smardzewski, R. R., and Fox, W. B., *J. Fluorine Chem.* **7**, 456 (1976).
90. Christe, K. O., *Spectrochim. Acta, Part A* **27**, 631 (1971).
91. Begun, G. M., Fletcher, W. H., and Smith, D. F., *J. Chem. Phys.* **42**, 2236 (1965).
92. Goubeau, J., and Bues, W., *Z. Anorg. Allg. Chem.* **268**, 221 (1952).
93. Bonadeo, H., and Silberman, E., *J. Mol. Spectrosc.* **32**, 214 (1969).
94. Gilbert, B., Mamantov, G., and Begun, G. M., *Inorg. Nucl. Chem. Lett.* **10**, 1123 (1974).
95. Goetschel, C. T., Campanile, V. A., Curtis, R. M., Loos, K. R., Wagner, C. D., and Wilson, J. N., *Inorg. Chem.* **7**, 1696 (1972).
96. Christe, K. O., Guertin, J. P., Pavlath, A. E., and Sawodny, W., *Inorg. Chem.* **6**, 533 (1967).
97. Christe, K. O., and Sawodny, W., *Inorg. Chem.* **12**, 2879 (1973).
98. Gilbert, B., Mamantov, G., and Begun, G. M., *J. Chem. Phys.* **62**, 950 (1975).
99. Begun, G. M., and Rutenberg, A. C., *Inorg. Chem.* **6**, 2212 (1967).
100. Claassen, H. H., Goodman, G. L., Holloway, J. H., and Selig, H., *J. Chem. Phys.* **53**, 341 (1970).
101. Weinstock, B., and Goodman, G., *Adv. Chem. Phys.* **9**, 169 (1965).
102. Gillespie, R. J., and Schroblgen, G. J., *Inorg. Chem.* **13**, 1230 (1974).
103. Christe, K. O., and Sawodny, W., *Inorg. Chem.* **6**, 1783 (1967).
104. Ryan, R. R., and Jones, L. H., *J. Chem. Phys.* **50**, 1492 (1969).
105. Bernitt, D. L., Miller, R. H., and Hisatsune, I. C., *Spectrochim. Acta, Part A* **23**, 237 (1967).
106. Hirschmann, R. P., Harnish, D. F., Holmes, J. R., MacKenzie, J. S., and Fox, W. B., *Appl. Spectrosc.* **23**, 333 (1969).
107. Abramowitz, S., and Levin, I. W., *J. Chem. Phys.* **51**, 463 (1969).
108. Andrews, L., Chi, F. K., and Arkell, A., *J. Am. Chem. Soc.* **96**, 1997 (1974).
109. Smith, D. F., Begun, G. M., and Fletcher, W. H., *Spectrochim. Acta,* **20**, 1763 (1964).
110. Lide, D. R., and Mann, D. E., *J. Chem. Phys.* **25**, 1128 (1956).
111. Sawodny, W., Fadini, A., and Ballein, K., *Spectrochim. Acta* **21**, 995 (1965).
112. Christe, K. O., and Curtis, E. C., *Inorg. Chem.* **11**, 2196 (1972).
113. Christe, K. O., and Curtis, E. C., *Inorg. Chem.* **12**, 2245 (1973).
114. Gillespie, R. J., and Morton, M. J., *Inorg. Chem.* **9**, 616 (1970).
115. Christe, K. O., and Schack, C. J., *Inorg. Chem.* **9**, 2296 (1970).
116. Christe, K. O., Sawodny, W., and Guertin, J. P., *Inorg. Chem.* **6**, 1159 (1967).
117. Abramowitz, S., Acquista, N., and Levin, I. W., *J. Res. Nat. Bur. Stand.* **72A** 487 (1968).
118. Snelson, A., *J. Chem. Phys.* **46**, 3652 (1967).
119. Cyvin, S. J., Cyvin, B. N., and Snelson, A., *J. Phys. Chem.* **74**, 4338 (1970).
120. Quist, A. S., Bates, J. B., and Boyd, G. E., *J. Phys. Chem.* **76**, 78 (1972).
121. Nimon, L. A., Sheshadri, K. S., Taylor, R. C., and White, D., *J. Chem. Phys.* **53**, 2416 (1970).
122. Cyvin, S. J., *Spectrosc. Lett.* **7**, 255 (1974).
123. Nielsen, J. R., Claassen, H. H., and Smith, D. C., *J. Chem. Phys.* **18**, 812 (1950).
124. Nielsen, J. R., Richards, C. M., and McMurry, H. L., *J. Chem. Phys.* **16**, 67 (1948).
125. Timms, P. L., Kent, R. A., Ehlert, T. C., and Margrave, J. L., *J. Am. Chem. Soc.* **87**, 2824 (1965).
126. Moy, D., and Young, A. R., II, *J. Am. Chem. Soc.* **87**, 1889 (1965).
127. King, S., and Overend, J., *Spectrochim. Acta* **22**, 689 (1966).
128. King, S., and Overend, J., *Spectrochim. Acta, Part A* **23**, 61 (1967).
129. Young, A. R., II, and Moy, D., *Inorg. Chem.* **6**, 178 (1967).
130. Durig, J. R., and Clark, J. W., *J. Chem. Phys.* **48**, 3216 (1968).

131. Koster, D. F., and Miller, F. A., *Spectrochim. Acta, Part A* **24**, 1487 (1968).

132. Rudolph, R. W., Taylor, R. C., and Parry, R. W., *J. Am. Chem. Soc.* **88**, 3728 (1966).

133. Rhee, K. H., Snider, A. M., and Miller, F. A., *Spectrochim. Acta, Part A* **29**, 1029 (1973).

134. Spratley, R. D., Turner, J. J., and Pimentel, G. C., *J. Chem. Phys.* **44**, 2063 (1966).

135. Loos, K. R., Goetschel, G. T., and Campanile, V. A., *J. Chem. Phys.* **52**, 4418 (1970).

136. Arkell, A., *J. Am. Chem. Soc.* **87**, 4057 (1965).

137. Dodd, R. E., Woodward, L. A., and Roberts, L. H., *Trans. Faraday Soc.* **53**, 1545 (1957).

138. Wilmhurst, J. K., and Bernstein, H. J., *Can. J. Chem.* **35**, 191 (1957).

139. Brown, R. D., and Pez, G. P., *Spectrochim. Acta, Part A* **26**, 1375 (1970).

140. Seel, F., and Budenz, R., *J. Fluorine Chem.* **1**, 117 (1971).

141. Christe, K. O., Curtis, E. C., Schack, C. J., and Pilipovich, D., *Inorg. Chem.* **11**, 1679 (1972).

142. Christe, K. O., and Naumann, D., *Inorg. Chem.* **12**, 59 (1973).

143. Gillespie, R. J., and Morton, M. J., *Inorg. Chem.* **9**, 811 (1970).

144. Christe, K. O., and Sawodny, W., *Inorg. Chem.* **8**, 212 (1969).

145. Clarke, M. R., Fletcher, W. H., Mamantov, G., Vasini, E. J., and Vickroy, D. G., *Inorg. Nucl. Chem. Lett.* **8**, 611 (1972).

146. Christe, K. O., *Inorg. Chem.* **11**, 1215 (1972).

147. Ault, B. S., and Andrews, L., *J. Chem. Phys.* **64**, 3075 (1976).

148. Selig, H., and Peacock, R. D., *J. Am. Chem. Soc.* **86**, 3895 (1964).

149. Gillespie, R. J., and Schrobilgen, G. J., *J. Chem. Soc., Chem. Commun.* 90 (1974).

150. Sladky, F., Bulliner, P. A., and Bartlett, N., *J. Chem. Soc. A* p. 2179 (1969).

151. Claassen, H. H., Goodman, G. L., and Kim, H., *J. Chem. Phys.* **56**, 5042 (1972).

152. Beattie, I., Crocombe, R., German, A., Jones, P., Marsden, C., Van Schelkwyk, G., and Bukovszky, A., *J. Chem. Soc. Dalton*, p. 1380 (1976).

THE MÖSSBAUER EFFECT IN SUPPORTED MICROCRYSTALLITES

FRANK J. BERRY

Universitv of Cambridge Chemical Laboratory, Cambridge, England

I. Introduction 255
II. Iron and Iron Oxides 259
   A. Iron Oxides 259
   B. Hydrogen Reduction of Iron Oxides 269
   C. Heating of Iron Oxides in Vacuo. 272
   D. Catalysts and the Adsorption of Gases 276
III. Tin 280
IV. Gold, Europium, and Ruthenium 281
   References 282

I. Introduction

In 1958, Rudolph Mössbauer (1) discovered that γ-ray emission without loss of energy from recoil of the nucleus could be achieved by incorporating the emitting nucleus in a crystal lattice. The phenomenon is known as the "Mössbauer effect."

Mössbauer also demonstrated the inverse process of recoil-free absorption of γ-rays over a narrowly defined energy spectrum when the absorbing nucleus is similarly bound in a crystal lattice. This resonance process occurs only when the nuclear energy levels of both absorber and emitter are identical. When the absorber nucleus is in a different electronic environment from the source nucleus, the nuclear energy levels no longer coincide and absorption can only occur when the energy of the photons emitted by the source is modified using the Doppler effect. This modification is produced by oscillating the absorber relative to the emitter, or vice versa, and a range of velocities are scanned until maximum absorption occurs. A Mössbauer spectrum usually consists of a plot of γ-ray counts against the relative velocity in millimeters per second of the source with respect to the absorber (Fig. 1). The magnitude of the required applied velocity is known as the chemical isomer shift, δ, of the absorber relative to the source employed. It is a measure of the difference in nuclear excitation energies between the nuclei in the source and the absorber that results

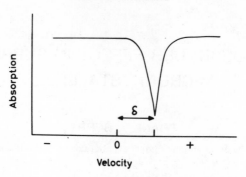

Fɪɢ. 1. Mössbauer spectrum showing the chemical isomer shift.

from their different electronic environments. Isomer shifts arise because the nucleus occupies a finite volume and during a nuclear gamma transition it is usual for the effective nuclear size to alter, thereby changing the nucleus–electronic field interaction energy. The nuclear excitation energy is sensitive to changes in the electron density at the nucleus, and the chemical isomer shift, which is an indication of the variation of this electron density in different compounds, is a function of the s-electron density at the nucleus. By making certain assumptions, it can be shown that the chemical isomer shift is given by

$$\delta = \text{constant} \times \frac{\Delta R}{R}\left(\left|\psi s(0)\right|^2_a - \left|\psi s(0)\right|^2_s\right) \tag{1}$$

where $\Delta R/R$ is the fractional change in the nuclear charge radius on excitation and $\left|\psi s(0)\right|^2_a$ and $\left|\psi s(0)\right|^2_s$ are, respectively, the total s-electron densities at the nuclei of the absorber and source. In compounds containing iron-57, higher s-electron density results in a decreased, i.e. more negative, chemical isomer shift. The chemical isomer shift is, however, also sensitive to the d-electron density since the 3d electrons have a finite probability of penetrating the s-electron shell and thereby shielding s-electrons from the nucleus. The removal of d electrons, therefore, effectively increases the s-electron density at the nucleus and consequently ferric species have lower, i.e. more negative, chemical isomer shifts than ferrous species. Additionally, for iron atoms with the same oxidation state and electronic configuration and with identical ligands, the chemical isomer shift is dependent on the number and symmetry of the coordinating ligands (2, 3). For this reason tetrahedral iron compounds have a lower chemical isomer shift than octahedral compounds with the same ligands. It has also been suggested that the chemical isomer shift is affected in a small way by changes in ligands a number of atoms away from the iron atom (4–6).

Any nuclear state with a spin $I > \frac{1}{2}$ has a nuclear quadrupole moment, i.e., the nuclear charge distribution may be elongated along the intrinsic axis of symmetry labeled the z-axis in which case the nuclear quadrupole moment, Q, is positive or it may be compressed along this axis in which case Q is negative. The interaction of the nuclear charge density with asymmetric extranuclear electric fields, i.e., situations in which the principal components of the electric field gradient are such that $V_{zz} \neq V_{xx} \neq V_{yy}$, results in a splitting of the nuclear energy levels. The axes are labeled so that $V_{zz} > V_{xx} \geq V_{yy}$ and the electric field gradient may, therefore, be expressed in terms of V_{zz}, usually written as eq and an asymmetry parameter η which is described as:

$$\eta = \frac{V_{xx} - V_{yy}}{V_{zz}} \qquad (2)$$

For the iron-57 isotope, the electric field gradient arises from (a) the charge distribution of the 3d electrons and (b) the charge distribution or· crystal field of the neighboring ions in a crystal structure. The first term usually dominates the second. The coupling of Q to eq splits the excited $I = \frac{3}{2}$ level into two, whereas the $I = \frac{1}{2}$ ground level remains degenerate. Transitions from the ground state to these two excited levels can be observed as a two-peak spectrum as shown in Fig. 2 which is a reflection of the distortion of the crystal structure. The centroid of the two peaks relative to the source is equivalent to the chemical isomer shift, and the velocity difference between the two peaks in millimeters per second is called the quadrupole splitting, Δ, and is related to the quadrupole coupling by

$$\Delta = \tfrac{1}{2}e^2qQ\left(1 + \frac{\eta^2}{3}\right)^{1/2} \qquad (3)$$

High-spin ferric compounds, which possess the iron nucleus in a spherically symmetric electronic configuration, usually have small quadrupole splittings (0–1 mm sec^{-1}), whereas high-spin ferrous species frequently have large quadrupole splittings arising from the sixth d electron. The quadrupole splitting has been found to be sensitive to changes in ligands a number of atoms away from iron (5, 6).

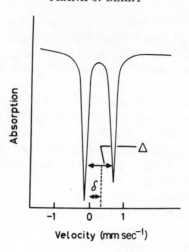

FIG. 2. Mössbauer spectrum showing quadrupole splitting (Δ).

If the nuclear dipole moment interacts with a magnetic field, all the degeneracy of the magnetic sublevels is lifted and each spin state splits into $2I + 1$ levels, where I is the nuclear spin quantum number. The Mössbauer spectrum of metallic iron, therefore, gives a six-peak pattern of equal spacing as a result of the magnetic field generated by ferromagnetic exchange interaction (Fig. 3). The spectrum is complicated by any contribution of quadrupole splitting to the magnetic splitting as in antiferromagnetic α-Fe_2O_3, which has an axially symmetric electric field gradient. In such a situation the shift in each of the levels of the excited state is $\epsilon = \frac{1}{4}e^2qQ$. The difference between the spacings of the pairs of outer peaks gives the quadrupole coupling constant. The magnetic field can also arise in paramagnetic

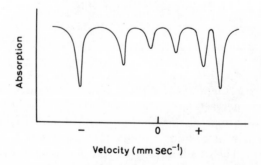

FIG. 3. Mössbauer spectrum showing magnetic hyperfine splitting.

species with long spin- lattice or spin-spin relaxation times that are temperature- and concentration-dependent, respectively.

Surface chemistry and catalytic activity are frequently dependent on surface area or crystallite size, and a knowledge of the electronic environment of surface nuclei is important in understanding the nature of catalysts and reactions at solid surfaces. Mössbauer spectroscopy, which examines directly the electronic environment at the nucleus, is a particularly favorable technique for the investigation of such matters, and several short discussions of the preliminary studies have now appeared (7–15). The examination of microcrystallites adsorbed onto high-area inert supports has been found to be informative in such investigations. The support disperses and maintains thin layers of microcrystallites over its high surface area and prevents sintering of the deposit during oxidation, reduction, and outgassing treatments.

The usual preparation of supported micrycrystalline samples by the incipient wetness technique involves the impregnation of a support, e.g., silica gel or alumina, with a solution of a metal salt to form a thick slurry that is subsequently dried and sometimes heat-treated.

II. Iron and Iron Oxides

A. Iron Oxides

Microcrystallites of iron oxides have been supported on silica gel and alumina by the incipient wetness technique and examined by Mössbauer spectroscopy. Bulk α-Fe_2O_3 has a corundum (α-Al_2O_3) crystal structure that involves a close-packed oxygen lattice containing ferric ions in octahedral sites. Above the Morin transition temperature ($T_M = 263$ K), bulk α-Fe_2O_3 shows a weak ferromagnetism due to the spins aligning with one of the vertical planes of symmetry making a small angle with the basal plane. Below T_M the spins are aligned along the [0001] axis and the oxide is pure antiferromagnetic. The magnetic ordering is reflected in the six-line Mössbauer spectrum of α-Fe_2O_3 which shows (16) the material to have a chemical isomer shift, δ (relative to iron) of 0.38 mm sec^{-1} and a quadrupole splitting of 0.12 mm sec^{-1}. The first Mössbauer investigation of microcrystalline Fe_2O_3 supported on alumina by Flinn et al. showed (17) a large quadruple split absorption, Δ 1.06 mm sec^{-1}, at 300 K, which remained constant at 77 K. The spectrum was interpreted in terms of the occupation by ferric ions of asymmetric octahedral environments on the surface of the alumina support that had one oxygen nearest neighbor missing. An expected quadrupole splitting, based on data pertaining to such a model, of ca. 1.6 mm sec^{-1} was calculated. A chemical isomer shift, corrected (as are all chemical isomer shift data reported in this work) relative to natural iron of 0.30 mm sec^{-1} was reported. The different intensities of the two peaks that disappeared at 77 K were attributed to the Goldanskii–Karyagin effect which has its origins in the anisotropy of the recoil-free fraction parallel and normal to the surface (18–20). The relative intensities of the transitions to the $\frac{3}{2}$ and $\frac{1}{2}$ levels of the excited state designated by $I_{3/2}$ and $I_{1/2}$, respectively, are given by

$$I_{3/2}(\theta) = I_0 \langle \exp - i(K \cdot X) \rangle^2 (1 + \cos^2 \theta) \qquad (4)$$

$$I_{1/2}(\theta) = I_0 \langle \exp - i(K \cdot X) \rangle^2 (\tfrac{5}{3} + \cos^2 \theta) \qquad (5)$$

where K is the wave vector of the γ-ray, X is the displacement vector of the active atom, and θ is the angle between the principal axis of the electric field gradient tensor and the direction of observation.

The ratio of the intensities can be reduced by arranging a random orientation of particles because of the differing values of $1 + \cos^2 \theta$ and $\frac{5}{3} + \cos^2 \theta$ to

$$R = \frac{I_{3/2}}{I_{1/2}} = \frac{\int_0^1 (1 + u^2) \exp(-\epsilon u^2) \, du}{\int_0^1 (\tfrac{5}{3} - u^2) \exp(-\epsilon u^2) \, du} \qquad (6)$$

where $u = \cos \theta$ and $\epsilon = K^2 \langle Z^2 \rangle - \langle X^2 \rangle$. Here $\langle Z^2 \rangle$ and $\langle X^2 \rangle$ are the mean square amplitudes of the vibrations parallel and perpendicular, respectively, to the electric field gradient tensor. A measurement of the absolute Debye–Waller factor would give $\langle X^2 \rangle$ and $\langle Z^2 \rangle$ but this is difficult to make.

A later study (21) showed that the quadrupole splittings obtained from finely divided α-Fe$_2$O$_3$ particles absorbed on alumina and silica were identical. The absence of octahedral sites in silica rendered the earlier explanation inadequate and the new observation was interpreted in terms of the adsorption of identically small superparamagnetic ferric oxide particles. The importance of Mössbauer spectroscopy in subsequent investigations of the nature of this phenomenon has been the subject of a separate review (22). The six-line spectrum of bulk α-Fe$_2$O$_3$ arises from room-temperature weak antiferromagnetism. The magnetic ordering is a cooperative property and is volume and temperature dependent. When the microcrystallites are sufficiently small, thermal energy overcomes the cooperative forces aligning the magnetic moments of the ferric ions thereby allowing them to change rapidly from one direction to another as in paramagnetic compounds to give an averaged zero effect during the time of measurement (23–25). When the relaxation time is shorter than the period for precession of the nuclear spin about the direction of the effective magnetic field, the substance is said to be superparamagnetic and the six-line Mössbauer spectrum collapses to two lines.

Subsequent work (26) by Kundig et al. used the changes in the peak areas that accompany the transition from the magnetically ordered to the superparamagnetic state to determine the crystallite sizes. A Larmor frequency for the ^{57}Fe nucleus in α-Fe_2O_3 was calculated (26, 27) to be $4 \times 10^7 \; sec^{-1}$, which corresponds to an observer relaxation time of 2.5×10^{-8} sec. The relaxation time, τ, for the spontaneous change in direction of the magnetic moment in single domain crystallites with uniaxial anisotropy may be written:

$$\tau = \frac{I}{f}\exp\left(\frac{Kv}{kT}\right) \tag{7}$$

where Kv is the energy barrier containing the magnetocrystalline anisotropy constant K and the volume v, T is the temperature, k is the Boltzmann constant, and f is set equal to the gyromagnetic precessional frequency of the magnetization vector about the effective field.

In a single-domain particle of α-Fe_2O_3 the magnetization vector is held in the c-plane perpendicular to the c-axis by the magnetocrystalline field. Mössbauer studies use the ^{57}Fe nucleus as the "observer" to record when the relaxation time τ becomes shorter than the period for precession of the nuclear spin about the direction of the effective field. Substitution into the equation for the Larmor frequency, or observer relaxation time, with an expression for the frequency factor proportional to the specific volume and anisotropy constant of the oxide gave (26, 27) the relationship:

$$\ln(2 \times 10^{-4}K) = \frac{Kv}{kT} \tag{8}$$

The value of K was obtained either from spectra of a series of samples having known average particle sizes at constant temperature or from spectra recorded as a function of temperature of a sample of known particle size. The determination of K was made at the point where half the total area under the spectrum resulted from the Zeeman pattern and the other half from the superparamagnetic fraction. Spectra used for such calculations are exemplified in Fig. 4. These spectra (28) were obtained from microcrystalline α-Fe_2O_3 produced by thermal decomposition of ferric nitrate on silica gel and subsequent

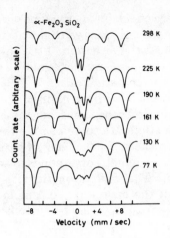

FIG. 4. Mössbauer spectra of α-Fe$_2$O$_3$ supported on silica gel as a function of temperature of the sample. [From M. C. Hobson and H. M. Gager, *J. Catal.* **16**, 254 (1970).]

calcination. Crystallite size measurements and the microcrystallite distribution was expressed in terms of mixtures of large antiferromagnetic and small superparamagnetic particles and were supported by X-ray diffraction data. The size and distribution was found to depend on the extent of calcination. Other results (*26*) showed that the quadrupole splitting increased from 0.44 mm sec^{-1} for microcrystals, which were reported to be of 180 Å diameter, to 0.57 mm sec^{-1} for crystals of 135 Å diameter. Smaller particles gave larger quadrupole splittings (0.98 mm sec^{-1}), whereas larger particles gave six-line spectra characteristic of magnetically ordered, bulk α-Fe$_2$O$_3$. The finely divided, supported α-Fe$_2$O$_3$ was not observed to undergo a Morin transition.

Further studies of this phenomenon (*29*) were reported as indicating that supported nickel oxide microcrystals doped with ^{57}Co on silica gel gave spectra consistent with the presence of ferric ions whereas the bulk material gave spectra corresponding to a mixture of both ferrous and ferric ions. An estimate of the nickel oxide particle size was made, and a later report (*30*) described the reduction of the nickel oxide to nickel of 60 Å diameter. Mössbauer studies of relaxation times and the transition to the superparamagnetic state with and without the presence of a magnetic field were described. It should be noted here that such source experiments are often complicated, and the consequences of nuclear transformations that involve both oxidation and reduction are not fully understood. Subsequent studies of small-particle relaxation times of Co$_3$O$_4$ on silica gel were reported

(31), and the calculation of microcrystallite size were supported by values obtained from X-ray line-broadening experiments.

The correlation of Mössbauer parameters with particle size involves the use of the "shell model" that describes the environment of surface nuclei as being of a lower symmetry than those within the particle. The model has been used to rationalize superimposed quadrupole-split spectra in terms of interior and surface iron nuclei.

The first of these reports (27) described the spectrum of a supported, iron oxide, microcrystallite sample as being composed of two super-imposed doublets, the most prominent having a quadrupole splitting, Δ, of 1.38 mm sec^{-1}. The spectrum was attributed to the presence of different sized particles arising from lattice expansion or chemical modification of the surface shell. Similar conclusions were reported in the following year by Hobson from an independent study of the variation of quadrupole splitting with particle size (32). Other studies (33, 34) suggested that the increase in quadrupole splitting with decreasing particle size was the result of homogeneous lattice expansion throughout the whole crystal lattice.

Application of the shell model to the study of catalysts implies that a decrease in particle size occurs with an increase in dispersion and that the resolution of separate quadrupole splittings for surface and bulk nuclei should be possible. It would appear important, when using this model, that consideration be given to the dependence of the quadrupole splitting for a highly dispersed system on the gaseous environment of the surface atoms (35) and also the lattice distortion of the particle due to internal pressure effects (8, 32, 34). It is also important that the factors giving rise to the quadrupole splittings are fully appreciated. Mössbauer studies of minerals (36) have clearly demonstrated the sensitivity of Mössbauer parameters to changes in the geometry of the ligand environment when the nearest ligands to iron remain the same. In the supported, iron oxide microcrystallites, as in many minerals, the immediate ligand to an iron atom is oxygen. It would, therefore, be expected that the Mössbauer parameters of supported, iron oxide microcrystallites would not only vary with oxidation state, electronic configuration, and coordination number, but also show marked dependence on the oxygen ligand environment. The Mössbauer investigations of minerals have demonstrated an overall correlation between the distortion of the oxygen octahedron and the quadrupole splittings. It has been shown that quadrupole splittings decrease with increased distortion according to the calculations (37) of the dependence of the quadrupole splittings in ferrous species on covalency effects and the distortion of the ligand environment from cubic symmetry.

An appreciation of these observations requires further consideration of the two major contributions to the quadrupole splitting which may be described as (a) the electric field gradient that arises from the electronic environment about the iron atom—the valence term; and (b) the electric field gradient originating from surrounding charged entities—the lattice term. The quadrupole splitting is proportional to V_{zz} the electric field gradient if both the valence and lattice charge distributions are considered as being symmetric about the z-axis:

$$V_{zz}/e = (I - R)q_{val} + (I - \gamma\alpha)q_{lat} \tag{9}$$

where q_{val} = valence contribution, q_{lat} = lattice contribution, and $(I - R)$ and $(I - \gamma\alpha)$ are Sternheimer antishielding factors.

Although these terms can in principle be calculated, their determination in practice is complicated by many factors (37). However, a cursory examination of small-particle iron oxide might well suggest that the interior iron nuclei would be expected to experience a reduced lattice contribution. The pure lattice contribution is, however, usually small and consequently any major modification and effect on the quadrupole splitting will only be observed if an appreciable number of iron nuclei occupy surface sites. Presumably the effect of microcrystallite size on the quadrupole splitting arises as a result of outer-layer ligand vacancies. It would seem to be inevitable that the electric field gradient at the superficial iron nuclei will be different from that at the nuclei in the bulk, and, consequently, it appears important that future investigations should concentrate on monolayers of supported iron oxide microcrystallites. Such investigations using samples enriched with ^{57}Fe, which would enhance the sensitivity of the Mössbauer technique, are feasible and the examination of 20 μg of ^{57}Fe, which occupies an area of 300 cm^2, should be possible.

Another method of correlating the superparamagnetic nature of ferric oxide with particle size has been described (38–40). It involves the computerized comparison of theoretically generated Mössbauer spectra for various relaxation times with the observed spectra of the highly dispersed systems.

The dependence of the isomer shift (41) and recoil-free fraction (9, 42–48) on particle size has also been suggested, but such relationships may be somewhat tenuous. It is clear that caution must be exercised in the use of methods hitherto described in the interpretation and correlation of microcrystallite size. Recent work has suggested (12) that ferric oxide may react with the support when calcined at high temperatures, e.g. for 2 hr at 500°C. The presence and contribution of such uncharacterized species to Mössbauer spectra of samples containing both large antiferromagnetic and small superparamagnetic particles renders the allocation of areas under specific peaks difficult, and it follows that the determination and interpretation of changes in peak areas that accompany the transition from the magnetically ordered state to the superparamagnetic state lack considerable precision. X-ray diffraction studies have failed to show the presence in these samples of any species other than α-Fe_2O_3, an observation suggesting that this physical method of characterization of solid surfaces may be unable, by itself, to establish the identity of all the important species on the surface. Subsequent studies of supported iron and iron oxide microcrystallites (12, 49, 50) and supported platinum and palladium iron clusters (51–54) have also indicated significant interaction between the metal and the support, illustrating that the inert supports are not as inert as they were once thought. Additionally, it is by no means clear that the size of particles that give Mössbauer parameters characteristic of surface nuclei are identical to those required for manifestation of the magnetic transitions used for making microcrystallite size determinations and that attempted correlations and interpretations must be prudently applied. Further, it is now appreciated how critically the quadrupole splitting depends on the chemical environment at the surface and that some supported microcrystallites at least are sensitive to the presence of atmospheric gases (35).

More recent studies of supported iron oxides have been directed toward an understanding of the nature of the adsorbed material. Table I illustrates the diversity of results that have been recorded for species giving quadrupole-split absorptions. Confusion still remains as to whether larger crystallites containing more defects tend to be found on silica gel or alumina (55, 56). Some studies of the thermal decomposition of adsorbed iron salts on alumina, silica gel, magnesium oxide, chromium oxide, and zinc oxide have reported a higher dispersion of microcrystallites on alumina than on silica gel (57), and others (49) suggest that iron dispersion is influenced by both surface area and pore volume, both of which are reported to increase in silica gel samples with increasing iron concentration but to decrease in alumina samples. Some recent work has suggested that both europium (58) and ruthenium (59) form stronger interactions with alumina than with silica gel, and an iron oxide species that was initially uncharacterized (28) has subsequently been attributed to a reaction product with the support (60). Studies (28) of iron oxide samples reduced with hydrogen have shown that those supported on silica gel

TABLE I

IRON-57 MÖSSBAUER PARAMETERS FOR QUADRUPOLE-SPLIT
SPECTRA OF SUPPORTED IRON OXIDE MICROCRYSTALLITES

Support	δ (mm sec^{-1})	δ(Fe) (mm sec^{-1})	Δ (mm sec^{-1})	Ref.
Alumina	0.45[a]	0.30	1.06	17
Silica	—	0.38	0.44	26
	—	0.32	0.57	26
	—	0.32	0.98	26
	—	—	1.38	27
	0.56[b]	0.31	0.69	28
	0.65[b]	0.40	0.74	28
	0.60[b]	0.35	0.77	28
	0.56[b]	0.31	0.71	28
	0.60[b]	0.35	1.17	35
	0.62[b]	0.37	0.84	49
	0.63[b]	0.38	0.73	49
α-Alumina	0.58[b]	0.33	0.71	49
	0.63[b]	0.38	0.72	49
η-Alumina	0.58[b]	0.33	0.95	49
	0.64[b]	0.39	0.82	49
γ-Alumina	0.59[b]	0.34	0.99	49
	0.60[b]	0.35	0.87	49
Alumina	—	—	0.97	56
	—	—	0.87	56
Silica	—	—	0.75	56
	—	—	1.82	56
	0.60[b]	0.35	0.75	62
	0.13[c]	0.31	0.60	63
	0.16[c]	0.34	0.75	63
	0.54[b]	0.29	1.87	65

[a] Values of δ relative to Fe/Cr.

[b] Values of δ relative to sodium nitroprusside.

[c] Values of δ relative to Co/Pd.

reproduced the original spectrum when reoxidized. Such behavior was not observed with the alumina-supported samples and was attributed to the reaction of iron in its reduced state with the support.

A six-line Mössbauer spectrum containing a superimposed doublet attributed to $ZnFe_2O_4$ was recorded (50) after $^{57}Fe_2O_3$ supported on zinc oxide was calcined at 400°C. The intensity of the doublet increased with higher calcining temperatures until, at 850°C, it was exclusive, demonstrating that reaction of Fe_2O_3 with the ZnO support had occurred. The controlled thermal degradation of ferricinium nitrate

(61) was reported to produce an iron oxide (mainly magnetite) in an inert carbonaceous matrix. The small particle size caused the material to be superparamagnetic down to and below 77 K, whereas at 4.2 K the material showed magnetic ordering with local disorder around the magnetic ions manifesting itself in a broadening of the spectrum.

Samples containing more than 10% by weight of supported iron oxide on alumina or silica have been shown to give a six-line Mössbauer spectrum corresponding to the antiferromagnetic character of large-particle ferric oxide (56, 62). A group of Russian workers (63) investigated a sample of Fe_2O_3, prepared by the impregnation of silica gel with ferric nitrate, which gave a quadrupole-split Mössbauer spectrum similar to that previously reported (64) and which remained unchanged after outgassing at 500°C. The chemical isomer shift δ of 0.31 mm sec^{-1} was characteristic of a ferric species, and the quadrupole splitting Δ of 0.60 mm sec^{-1} was smaller than that reported elsewhere for small superparamagnetic Fe_2O_3 crystals, indicating a slightly larger crystallite size. The magnetic susceptibility, which was independent of the field strength, also suggested that the Fe_2O_3 particles were superparamagnetic. A similar two-line spectrum δ of 0.34 mm sec^{-1} was obtained from a sample prepared from $Fe(CO)_5$ in absolute ether, but the large quadrupole splitting of 0.75 mm sec^{-1}, a value also reported (62) for a sample containing 2% by weight of Fe_2O_3, suggested the presence of smaller particles. A third sample prepared from $FeCl_3$ gave a six-line spectrum corresponding to coarse α-Fe_2O_3 and possessed a magnetic susceptibility that varied with magnetic field strength. Many of the early Mössbauer spectroscopic studies of α-Fe_2O_3 microcrystallites supported on silica gel and alumina were repeated (49) in 1970 and the spectra classified according to whether they were six-line, two-line, or amalgams. For the two-line spectra, the chemical isomer shift, $\delta = 0.33$–0.37 mm sec^{-1}, fell within a narrow range characteristic of trivalent iron. The decrease in quadrupole splitting Δ from 0.99 to 0.71 mm sec^{-1} was attributed to variation in particle size from 40 to 130 Å. The transition from the superparamagnetic state to the antiferromagnetic state that occurred with increasing loadings of iron oxide was attributed to the increase in particle size.

Hobson and Campbell reported that the Mössbauer spectrum of a sample of iron oxide on silica gel, which has been calcined at 500°C for 16 hr, showed (65) a small chemical isomer shift of 0.29 mm sec^{-1} and an unusually large quadrupole splitting of 1.87 mm sec^{-1}. This splitting was noted to be larger than 1.60 mm sec^{-1} calculated by Flinn et al. (17) for a ferric ion in an octahedral environment with 1 oxygen atom missing but less than 2.26 mm sec^{-1} calculated for the same model by using a more recent value of the Sternheimer anti-shielding factor (66). Both parameters were explained in terms of the presence of a very small particle, estimated by Kundig's methods (26) as being ca. 20 Å in diameter. The differences in chemical isomer shifts and quadrupole splittings produced by particles of different sizes were temperature-dependent and were tentatively associated with the number of defects in the crystal structure. Similar variations in Mössbauer parameters, although not so large, have been observed in ferric ions when located in octahedral and tetrahedral sites (66, 67). The asymmetry of the two peaks was independent of temperature, unlike the initial (17) report, and the matter has recently been considered (68) and explained in terms of asymmetrical surface bonding producing a large electric field gradient, even though other workers have claimed a complete absence of asymmetry (63) when the crystallites are large.

It would be expected that the Goldanskii effect, which is the result of anisotropy of the recoil-free fraction in the γ-ray emission, would be observed in highly dispersed systems such as supported iron or iron oxide microcrystallites since the vibrational amplitude of the iron nucleus normal to the surface would not be likely to be the same as that parallel to the surface. If taken as being normal to the surface, V_{zz} would be expected to be large compared with the other two components and, consequently, the asymmetry parameter [Eq. (2)] would be small unless the difference between V_{xx} and V_{yy} was very large. Superimposed on this may be an asymmetry produced by relaxation effects (65). It has been shown (69) that for ferric ions the $\frac{1}{2} \rightarrow \frac{3}{2}$ transition begins to broaden before the $\frac{1}{2} \rightarrow \frac{1}{2}$ transition as the spin-lattice relaxation time of the 3d electrons approaches the Larmor precessional frequency of the nucleus when the temperature is lowered. The asymmetry thereby produced in the peaks is the reverse of the temperature dependence of the Goldanskii effect. It has been suggested (65) that temperature-independent asymmetry in supported microcrystalline materials may be interpreted in terms of the two effects being cooperative and canceling each other out over the temperature range concerned.

The linewidths of the Mössbauer spectra of supported iron oxide microcrystallites are larger than those recorded for iron in well-crystallized compounds and may suggest the superimposition of different quadrupole-split doublets arising from a number of ferric ions in a heterogeneity of sites.

It is possible that samples of alleged composition Fe_2O_3 may have been confused with hydrolyzed ferric species that give similar Mössbauer parameters (70, 71). Authentic ferric oxide is known to give a magnetically split, room temperature, Mössbauer spectrum when the particles are larger than 135 Å in diameter (26). Furthermore, it is reported (72) that only superparamagnetic particles smaller than 70 Å in diameter fail to show some magnetic hyperfine structure when cooled to 78 K. The species $Fe(OH)_3 \cdot O.9H_2O$ of ca. 39 Å diameter has been reported (70) to be superparamagnetic down to 10 K, and qualitatively similar spectra have been observed for samples with less water in the empirical formula, although their quadrupole splittings at 298 K were smaller, and magnetic hyperfine splitting at 4 K greater. The possibility that the particles might be γ-FeOOH, which is paramagnetic (71) at 77 K, cannot be ignored. It must also be remembered that the preparations of samples reported in the literature have varied considerably, and comparison of results must be made with caution.

B. Hydrogen Reduction of Iron Oxides

Initial studies (55, 64, 73) of the hydrogen reduction of supported Fe_2O_3 reported spectra, as depicted in Fig. 5, which were interpreted in terms of the superimposition of a ferrous doublet on a six-line metallic iron pattern. The reduction by hydrogen of large-particle Fe_2O_3, prepared by the impregnation of silica gel or alumina with an oxide loading in excess of 10% by weight, has been reported to give a surface mixture of ferrous oxide and metallic iron, whereas lower loadings were reduced to ferrous oxide only (56, 57). The samples containing larger amounts of zero-valent iron were found to be more effective as catalysts for the hydrogenation of butene (56).

A somewhat contradictory study (49) reported that reduction at 450°C for 8 hr transformed magnetically ordered Fe_2O_3 to iron, whereas superparamagnetic specimens of supported Fe_2O_3 were reduced to metallic iron–ferrous iron mixtures. The degree of transformation of

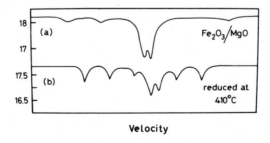

Velocity

FIG. 5. Mössbauer spectra of supported Fe_2O_3 before and after reduction with hydrogen. [From H. Koelbel and B. Kuespert, Z. Phys. Chem. (Frankfurt) **69**, 313 (1970).]

the superparamagnetic state to the ferromagnetic state by heat treatment at 780°C for 16 hr showed good agreement with the degrees of reducibility. The metal-support interaction was reported to decrease in the order γ-alumina > silica > η-alumina > α-alumina, but the nature of the interaction was not investigated.

Hobson reported that the reduction of small-particle α-Fe_2O_3 on silica gel by hydrogen followed by evacuated heating at 450°C resulted in products that gave three-peaked Mössbauer spectra (74). A doublet, thought to represent a high-spin ferrous species, $\delta = 1.04$ mm sec^{-1} and $\Delta = 1.64$ mm sec^{-1}, containing a superimposed central peak, was distinguished and the composite spectrum attributed to the presence of iron atoms in two different states. The spectrum was interpreted in terms of a surface mixture of small-particle ferrous and ferric oxide, the second peak of the latter being superimposed on one of the peaks of the ferrous doublet. Other investigations have been performed, and the results are given in Table II.

Similar spectra have subsequently been reported (65) and are shown in Fig. 6. The variation in the intensities of the three peaks with temperature and length of reduction were interpreted in terms of the dependence of the Fe^{3+}/Fe^{2+} ratio on the conditions of hydrogen

TABLE II

IRON-57 MÖSSBAUER PARAMETERS FOR QUADRUPOLE-SPLIT SPECTRA OF
SUPPORTED IRON OXIDE MICROCRYSTALLITES REDUCED IN HYDROGEN

Temperature (°C)	Peak	δ (mm sec^{-1})[a]	δ(Fe) (mm sec^{-1})	Δ (mm sec^{-1})	Ref.
200	1–2	1.02	0.77	0.98	65
	1–3	1.36	1.11	1.65	65
	2	1.51	1.26	—	65
300	1–2	1.04	0.79	0.99	65
	1–3	1.36	1.11	0.63	65
	2	1.54	1.29	—	65
450	1–2	1.07	0.82	1.01	65
	1–3	1.43	1.18	1.72	65
	2	1.58	1.33	—	65
450	1–3	1.29	1.04	1.64	74
	1–2	0.96	0.71	0.99	74
	2	1.46	1.21	—	74

[a] Values of δ relative to sodium nitroprusside.

reduction. Two surface iron sites were tentatively ascribed to a ferric ion, $\delta = 0.77$ mm sec^{-1} and $\Delta = 0.98$ mm sec^{-1}, and a high-spin ferrous ion, $\delta = 1.11$ mm sec^{-1} and $\Delta = 1.65$ mm sec^{-1}. These data were reinterpreted (75) in 1970 on the assumption that high-spin ferric ions in oxide environments normally have chemical isomer shifts of ca. 0.6 mm sec^{-1} and that the data could be interpreted in terms of a model (76) in which small surface cubo-octahedral crystals approximate spheres. Hobson and Gager (75) suggested that the samples were composed of microcrystalline ferrous oxide in which the ferrous ions occupied octahedral positions in a cubic lattice of oxide ions containing two distinct environments. The doublet characterized by the smaller chemical isomer shift and quadrupole splitting was attributed to a surface ferrous ion, i.e., a ferrous ion in an environment described by four oxide ions in the same plane and one below, whereas the other doublet was assigned to a ferrous ion within the interior of the deposit with six nearest oxide neighbors equally distributed above and below. The failure of the reduction process to go further than the ferrous state has been attributed to reaction of the adsorbed material with the support (77).

The quadrupole splitting for a ferrous high-spin compound arises from the valence, q_{val}, and lattice, q_{lat}, contributions to the electric field gradient. The q_{val} term has its origins in the field gradient resulting from the sixth d electron in ferrous compounds, but the total quadrupole splitting is smaller than the maximum value due to Boltzmann population of the upper d levels because of crystalline

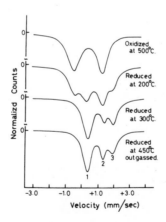

FIG. 6. Mössbauer spectra of 3% by weight iron on silica gel sample undergoing progressive reduction with hydrogen. [From M. C. Hobson and A. D. Campbell, *J. Catal.* 8, 294 (1967).]

field and covalency effects that are difficult to calculate in complex systems. The lattice term, which originates from the surrounding charged entities, may be expressed as:

$$q_{lat} = \sum \frac{z_i \, 3 \cos_3{}^2 \theta_i - 1}{r_i} \tag{10}$$

where z_i is the charge on each atom, r_i is the distance of each atom from the iron nucleus, and θ_i is the angle that the atom subtends from the chosen z-axis.

Provided that the crystal structure is known, the q_{lat} term can, in principle, be calculated (78), although the correct assignment of charges to the different kinds of oxygen atoms in the structure and the convergence of the lattice term with increasing r and its sensitivity to small errors in crystallographic parameters are frequent sources of error. The prediction of quadrupole splitting by application of a ligand field method (37) that is independent of structural details has been shown to be successful in overcoming some of these difficulties. Although later results, obtained by a direct lattice summation technique (79), have contradicted some of the findings of the ligand field method (37), it does appear that a cautious application of the ligand field method may give useful information about surface structure. The method considers (80, 81) that a cubic ligand environment about iron in a ferrous compound gives little, or no, quadrupole splitting and that $q_{val} = q_{lat} = 0$. If, however, the ligand environment distorts slightly from cubic symmetry, then a very large (82) quadrupole splitting of ca. 3.70 mm sec^{-1} or higher may be observed and attributed almost exclusively to q_{val}. As the distortion of the ligand environment from cubic symmetry increases, the q_{lat} term increases and the quadrupole splitting decreases. It might, therefore, be expected that, in supported iron oxide microcrystallites, there would exist an overall correlation between distortion of the oxygen octahedra and the magnitude of the quadrupole splitting and which, in ferrous species, would result in smaller quadrupole splittings implying increased distortion.

C. Heating of Iron Oxides *in Vacuo*

The heating of a sample containing 2% by weight of α-Fe$_2$O$_3$ on silica gel at 475°C for 3 hr *in vacuo* (62) gave a product with Mössbauer parameters (see Table III) similar to those reported by Hobson *et al.* (65, 74) for the hydrogen-reduced specimens. Although one of the two superimposed doublets, $\delta = 0.80$ mm sec^{-1} and $\Delta = 1.14$ mm sec^{-1},

TABLE III

IRON-57 MÖSSBAUER PARAMETERS FOR QUADRUPOLE-SPLIT SPECTRA OF
SUPPORTED IRON OXIDE MICROCRYSTALLITES HEATED IN VACUO⁻

Temperature (°C)	Peak	δ (mm sec^{-1})^a	δ(Fe) (mm sec^{-1})	Δ (mm sec^{-1})	Ref.
100	—	0.53	0.28	1.80	35
500	—	0.53	0.28	2.23	35
600	—	0.52	0.27	2.25	35
475	1–2	1.05	0.80	1.14	62
	1–3	1.37	1.12	1.77	62

^{a} Values of δ relative to sodium nitroprusside.

was not identified, the authors concluded that heating *in vacuo* caused at least partial reduction to a ferrous species, as in the partial reduction of some powdered metal oxides, such as TiO_2 and MoO_3, by evacuated heating (*83, 84*). Infrared spectroscopic evidence for the presence of ferrous ions in the surface of α-Fe_2O_3 that had been subjected to prolonged heating *in vacuo* (*85*) supports this hypothesis. The vacuum-reduced sample reproduced the original Mössbauer spectrum of supported α-Fe_2O_3 on silica gel after it had been exposed to oxygen at 1 atm pressure at 200°C. Reduction of the oxidized sample was reported to be impossible after threefold reduction and reoxidation, but reducibility was again achieved by exposing the sample to water vapor at room temperature. The water vapor appears to restore hydroxyl groups to the surface of the ferric oxide, and it is significant that such hydroxyl groups have been identified as likely active sites for surface chemical reactions on ferric oxide (*86*). Other investigations (*63*) have reported that coarse Fe_2O_3 samples heated at 500°C and at a residual pressure of 3×10^{-3} mm Hg give Mössbauer spectra consistent with partial reduction to ferromagnetic Fe_3O_4. These samples were found to have a greater magnetic susceptibility than the prereduced samples. These results (*62, 63*) are contrary to those obtained by Gager *et al.* (*35*) who reported α-Fe_2O_3 supported on silica gel and calcined in air to give Mössbauer parameters of $\delta = 0.35$ mm sec^{-1} and $\Delta = 1.17$ mm sec^{-1}, which are consistent with high-spin ferric species (Fig. 7). When heated *in vacuo* at 100°, 500°, and 600°C, the quadrupole splitting increased to 2.25 mm sec^{-1}, a value that correlates well with Hobson's quadrupole splitting of 2.26 mm sec^{-1} calculated for an iron atom with one oxygen atom removed from its coordination sphere, and the chemical isomer shift decreased to 0.27 mm sec^{-1}. The exposure of the outgassed sample to dry oxygen produced no change in the Mössbauer spectrum implying

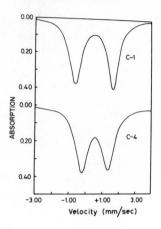

FIG. 7. Mössbauer spectra of α-Fe$_2$O$_3$. (C-1) Oxidized and outgassed at 600°C; (C-4) 205 × 10^{-3} mmole of water added. [From H. M. Gager, M. C. Hobson, and J. F. Lefelhocz, *Chem. Phys. Lett.* **15**, 124 (1972).]

physical, rather than chemical, adsorption of oxygen in accordance with the infrared evidence (*85, 87*).

The reduction of microcrystalline iron oxide is clearly another area that requires further attention. Although α-Fe$_2$O$_3$ is reduced to iron metal at 673 K in hydrogen, it appears that the degree of reduction of supported iron oxide depends on particle size and the nature of the support and that, whereas antiferromagnetic large-particle α-Fe$_2$O$_3$ may be reduced to metallic iron, the paramagnetic or superparamagnetic phase is not. It may be possible at low surface concentrations for ferric ions to occupy sites in the surface that allow strong attraction between the iron and the support such that a complex similar to a surface silicate or aluminate would be formed. Reduction of this species to a ferrous complex may be possible, and it may be envisaged that such an entity would enable interior and superficial contributions to the Mössbauer spectra to be distinguished as has been achieved in the Mössbauer studies of minerals (*36*). Higher concentrations of ferric oxide may result in the iron atoms being less intimately bonded to the support, thereby making them less susceptible to its influence and more readily reducible to metallic iron. It is also possible that the superimposed doublet, which has frequently been observed, may be related to the iron oxide called wüstite (*88*), Fe$_{1-x}$O.

The principles involved in the investigations described in the foregoing may be important in future studies of the nature of catalysts. Weak interactions between a support and the catalytically active species may lead to sintering and a poorly dispersed catalyst, whereas strong interaction may result in the stabilization of an undesired oxidation state of the catalyst. The nature of such interactions is amenable to investigation by Mössbauer spectroscopy if the catalytically active component contains a Mössbauer nucleus. In addition to the silica gel and alumina that have been discussed, carbon has been found to give too weak an interaction with iron (89) for catalytic usefulness, but magnesium oxide has demonstrated a potential utility for the preparation of highly dispersed metallic iron (90). It is still clear, however, that the synthesis of small, supported iron microcrystallites may necessitate the use of alternative methods of preparation. It is also clear that the support, although frequently considered as an inert carrier, may possess intrinsic activity due to surface acidity or through its interactions with small particles, and it is, therefore, not unreasonable to expect the support to be less inert as previously supposed.

Bartholemew and Boudart reported that the Mössbauer spectra of supported Pt—Fe clusters (91) permitted the resolution of both bulk and surface contributions to the spectra. The latter have been shown to be sensitive to the gaseous surface environment (92). Garten and Ollis (93) have investigated the particle size, composition, and component distribution in Fe—Pd clusters. The calculation of the size of metal atom clusters in which the atomic properties give way to the metallic properties (51–53) suggested that clusters of up to fifty atoms did not have the exact electronic structure of the bulk and that perturbation of the electronic structure of the metal particle by bonding with the support was probably significant. Such further evidence of support–crystallite interaction reinforces the previous cautionary comments on this matter and is again emphasized by some very recent work (94) comparing the chemical states of iron on silica gel and alumina with the chemical states of iron in platinum- or palladium-related catalysts. In spite of reports that ferric ions supported on SiO_2 or Al_2O_3 can only be reduced to ferrous ions at 500°C, the Mössbauer spectra show that reduction in the presence of platinum or palladium under similar conditions gives Pt—Fe and Pd—Fe clusters. The oxidation–reduction behavior of the iron and the spectral changes that occur on agglomerating the metals confirm this observation. It was shown that results obtained for the reduction of $PdFe/SiO_2$ catalysts at 700°C were different from those obtained for $PdFe/Al_2O_3$ catalysts and were attributed to the incorporation of silicon of the support into the PdFe clusters. Evidence for the formation of bimetallic clusters of PdFe and PtFe on Al_2O_3 and SiO_2 supports has also been reported and may be relevant to the recent application of supported bimetallic catalysts in refining petroleum.

D. Catalysts and the Adsorption of Gases

The application of the Mössbauer effect to the study of supported microcrystalline catalysts is currently receiving significant attention. Although the Mössbauer effect is primarily a solid-state bulk effect, the relatively high penetrating power of γ-radiation facilitates the investigation of catalytic pores, and, when a significant proportion of the Mössbauer nuclei are on the surface as supported microcrystallites, the active components of a catalyst may be readily investigated. The feasibility of monitoring the modification of the electric field gradient at the surface nuclei by Mössbauer spectroscopic investigation of monolayers of supported ^{57}Fe before and after the adsorption of gaseous molecules has been referred to earlier in this work. In principle, such investigations may be applied to the nature of the quadrupole splitting before and after use of the sample as a catalyst, while the sample is exposed to an adsorbing gas, or while the sample is actually operating as a catalyst.

Initial studies have involved the investigation of the effect on the Mössbauer spectrum of supported microcrystallites when exposed to polar molecules. Suzdalev et al. reported (95) microcrystalline ferric oxide to give a quadrupole-split Mössbauer spectrum characteristic of a ferrous species when carbon dioxide was adsorbed and the observation was attributed to the complex formation of a surface carbonate that could be reversed by heating in vacuo. Gager et al. (35) reported that α-Fe_2O_3 supported on silica gel that had been heated in vacuo at 600°C showed, when exposed to measured amounts of water vapor, a decrease in quadrupole splitting from 2.25 to 1.56 mm sec^{-1} and an increase in chemical isomer shift from 0.27 to 0.32 mm sec^{-1}. A subsequent report (96) gave parameters of $\delta = 0.35$ mm sec^{-1} and $\Delta = 1.17$ mm sec^{-1}. The quantitative addition of methanol and ammonia has also been shown to produce similar modifications of the Mössbauer spectrum (35), which were reversed by heating in vacuo and attributed to the hydration and subsequent dehydration at the surface. The dehydration process was presumed to decrease the number of nearest neighbors of the surface ferric ions thereby increasing the distortion of the electric field gradient to give an increased quadrupole splitting. Such changes in coordination number of surface cations following hydration and dehydration of oxides have been identified by other spectroscopic techniques (97, 98). Infrared spectroscopic evidence for the removal by outgassing of hydroxyl groups formed on activated α-Fe_2O_3 by the adsorption of water has also been reported (83, 87). The dependence of the quadrupole splitting on the chemical environment at the surface has been cited (35) as evidence for the inadequacy of the shell (27) and lattice expansion (33, 34) models for the determination of microcrystallite size. The attainment of reproducible and comparable results would require the calibration under some set of pretreatment conditions, such as rehydration to monolayer coverage. It must also be noted that aggregation by surface particles may result in a supported microcrystalline monolayer becoming multilayered.

Gager et al. (96) reported that treatment of an outgassed sample of supported α-Fe_2O_3 with hydrogen sulfide caused a decrease in the quadrupole splitting from 2.24 to 1.18 mm sec^{-1}. The chemical isomer shift, $\delta = 0.41$ mm sec^{-1}, confirmed that the iron remained in the ferric state. An additional Mössbauer absorption seen as a shoulder to the doublet at $\delta = 0.65$ mm sec^{-1} was regarded as an indication of either hydrogen sulfide occupying more than one type of adsorption site or the adsorption of a different species. Subsequent outgassing at room temperature failed to regenerate the original spectrum, indicating that chemisorption of hydrogen sulfide had occurred. Infrared spectroscopic data suggested that H_2S is dissociatively adsorbed to give HS$^-$ and HO$^-$ surface groups (85, 87), and the additional peak in the Mössbauer spectrum was, therefore, attributed to one-half of a quadrupole doublet resulting from coordination of the HS surface group with the iron. The major doublet was considered to arise from the Fe—OH linkages, thereby giving a spectrum similar to that found with water.

Skalkina et al. (99) have considered the effect of adsorbed molecules on quadrupole splitting and have correlated values of Δ for iron in mixed oxides with catalytic activity for the ammoxidation of propylene.

Some work has also been attempted on the adsorption of polar molecules onto the surfaces of the reduced iron oxides. Chemisorption of ammonia was reported by Hobson (74) to produce a sample in which the extraneous center peak of the Mössbauer spectrum had disappeared (Fig. 8). It was suggested that amine radicals formed by chemisorption of ammonia transferred electrons to the adsorption site causing easy reduction of ferric to ferrous ions at the surface. The original spectrum was recovered by outgassing at elevated temperatures. Further work by Hobson and Gager (75) showed that the initial adsorption of ammonia had no effect on the spectrum of the reduced and outgassed sample, but that further addition of ammonia decreased the area of the relative center peak (attributed to half a doublet representing the surface ferrous species) of the three-peak spectrum. Heat treatment caused the desorption of ammonia until the original spectrum of the reduced and outgassed sample was recovered at 300°C. Complete desorption of the ammonia was not achieved, suggesting that the initial easy adsorption of ammonia was to the free hydroxyl

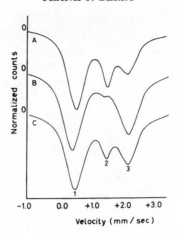

FIG. 8. Mössbauer spectra of adsorption and desorption of ammonia on an iron-on-silica gel catalyst at 25°C. [From M. C. Hobson, *Nature (London)* **214**, 79 (1967).]

groups of the silica [previously shown by infrared techniques to adsorb ammonia very strongly (*100, 101*)] and the ammonia was shown to form a 1:1 surface complex with the surface ferrous ions. Another report by Hobson (*102*) showed that addition of ammonia to silica gel-supported iron caused a decrease in the quadrupole splitting of the outgassed ferric state and an increase in the surface ferrous state. The difference in behavior was explained by the factors that produce the electric field gradient. The five 3d electrons of a high-spin ferric ion are symmetrically distributed about the nucleus, and the ligands surrounding the ion are, therefore, primarily responsible for producing the electric field gradient. Addition of ammonia would be expected to increase the symmetry of the electric field gradient thereby decreasing the quadrupole splitting. The electric field gradient at the ferrous ion is a result of the lattice, the surrounding ligands, and also the extra 3d electrons. The last is reported to be usually quite large compared with the lattice and ligand contributions and to be of opposite sign (*37*). However, the sum of the contributions gives a relatively small quadrupole splitting for the surface ferrous ion, and, although the addition of ammonia increases the symmetry at the surface site, the net contribution to the electric field gradient of the ligands decreases. The sum of the 3d-electron and ligand contributions increases and, consequently, the quadrupole splitting increases.

Qualitatively the adsorption of methanol (*75*) gave a similar change in the Mössbauer spectrum to that observed for the adsorption of ammonia, but only half as much methanol was required to change the spectrum from its initial to final state. In accordance with the infrared evidence (100), these observations were explained in terms of both strong chemisorption of methanol on the surface ferrous ion sites followed by dissociation and also by weak adsorption on the surface hydroxyl groups of the silica. Reaction on desorption to new products was also postulated. The adsorption of water vapor by the reduced and outgassed samples has also been investigated by Hobson et al. (96). The quadrupole splitting and chemical isomer shift were both found to increase to 1.90 and 1.07 mm sec^{-1}, respectively, which are parameters typical of high-spin ferrous iron. Outgassing at 500°C returned the spectrum to its original state.

The Mössbauer spectrum of ferrous Y-zeolite is somewhat similar to that of the reduced silica gel samples (103). The spectrum consists of two overlapping and partially resolved doublets with the inner doublet, $\delta = 0.89$ mm sec^{-1} and $\Delta = 0.62$ mm sec^{-1}, being attributed to the ferrous ion on the surface. In both the Y-zeolite and the reduced iron oxide on silica samples, the inner doublets representing surface ferrous states are the first to be affected by adsorption of polar molecules, but in the case of Y-zeolite the addition of excess amounts of water or ammonia causes the disappearance of the spectrum, and this has been interpreted in terms of "solvation" of the ferrous ions by absorbate causing weakening of the bonding to the crystalline lattice. It is also possible that the spectrum is a composite representing a multiplicity of parameters.

The spectral changes from three peaks to two were not observed when hydrogen sulfide was adsorbed on supported iron, although the quadrupole splitting did increase by 0.18 mm sec^{-1}. Pumping on the sample did not change the spectrum, suggesting that the H_2S is chemisorbed, and the difficulty in returning the sample to its original form was taken to indicate that either H_2S reacts directly with iron or, alternatively, dissociates on active sites of the silica gel and is followed by reaction of the resultant products with the iron surface sites to form iron sulfides. The latter explanation, supported by work on the adsorption of thiols on nickel (104) and Mössbauer studies of the reaction of H_2S with Raney iron (105) to give Fe_2S_3, was favored.

The use of Mössbauer spectroscopy to investigate the state of heterogeneous catalysts during catalysis (106) showed that when small particles of iron (ca. 5 nm) supported on magnesium oxide were used as catalysts for ammonia synthesis, the superparamagnetic portion of the spectrum was sensitive to both chemisorption of hydrogen and also to treatment of the particles by methods known to change their catalytic properties. It was suggested that surface iron atoms with seven nearest neighbors were particularly active for the ammonia synthesis.

A study of the oxidizing ammonolysis of propylene (*107*) by use of $Fe_2O_3 \cdot P_2O_5$ supported on silica gel showed that a small quadrupole splitting, corresponding to slight distortion of the lattice, gave CO_2, whereas samples with a larger quadrupole-split Mössbauer spectrum, indicating increased lattice distortion, were found to yield acrylic acid cyanide. It was suggested that catalytic selectivity and rate were determined by the degree of symmetry in the neighborhood of the iron. Another Mössbauer study (*108*) of mixed iron and antimony oxide catalysis for the ammoxidation of propene showed that Fe_2O_3 formed a compound with Sb_2O_5 or Sb_2O_3 when mixed with 25% by weight of SiO_2, which was believed to give $FeSbO_4$ as the active part of the catalyst. The isomer shift and quadrupole splitting for ^{57}Fe were correlated with the acrylonitrile yield, and a medium strength of lattice distortion or optimal change in the M—O bond was held responsible for a good selective oxidation catalyst. An electron-transfer mechanism was suggested as a necessity for activating the reacting molecules.

Accurate and reproducible Mössbauer spectra of supported iron catalysts require the prevention of the adsorption of impurities including oxygen and water onto the highly reactive sample surface. Cells designed to allow chemical reactions and *in vacuo* pretreatments at temperatures up to 673 K while the Mössbauer spectrum is being recorded have recently been reported and represent a significant and important development in the application of Mössbauer spectroscopy to catalytic and surface studies (*109–111*).

III. Tin

A limited number of Mössbauer studies of supported tin micro-crystallites have been performed. Karasev *et al.* investigated (*112*) tetraphenyltin on silica gel under different conditions of temperature and pressure in an attempt to follow the nature of the adsorption process and to determine the structure of the adsorbed layers. Studies of chemisorbed tetraalkyltin on γ-Al_2O_3 reported the surface species to be SnR_4, $SnR_{4-n}(OMe)_n$ ($n = 1,2,3$), and $Sn(OMe)_4$, where OMe represents a surface oxygen–metal group (*113, 114*). Investigations of the dynamic motion of tin atoms on silica gel have shown the presence of both tetravalent and divalent surface tin species (*115–120*). The temperature dependence of the intensity data has been interpreted in terms of physically adsorbed tetravalent tin and chemisorbed divalent tin (*115*). The electric field gradient at the ^{119}Sn nucleus in the adsorbed species increased with increasing temperature and was larger than that recorded for bulk crystalline SnO(*115*). The doublet peaks for the adsorbed SnO showed considerable asymmetry (115) and a quantitative interpretation of the Goldanskii–Karyagin effect in tin was subsequently reported (116). Attempts to estimate the zero-point vibrations and the bond energies of the tin-support species (112) were subsequently shown to be of questionable significance (121). The dynamical properties of divalent tin ions and SnO_2nH_2O molecules adsorbed on silica gel have been reported to depend on the pore diameter of the silica gel (117). These studies on supported tin species (115–120) illustrate how recoil-free fraction and linewidth data obtained from the Mössbauer spectra may be used in the determination of the mobility of catalytic components. Such information is important because surface and bulk mobility frequently govern the degree of dispersion of the catalytically active phases.

Firsova et al. (122) reported that the room temperature Mössbauer spectrum of supported tin molybdate, which had been aged in vacuo at 723 K, showed the presence of tetravalent tin. Only after exposure to oxygen at 473 K did the sample act as an adsorbent for propylene. It then gave a Mössbauer spectrum that showed the reduction of the tetravalent tin to the divalent state. Reduction without exposure to oxygen was achieved at 673 K but supported tin in the absence of molybdenum was not reduced. The results were interpreted in terms of the proposals (123) for the synergistic oxidation–reduction during catalysis.

A recent in situ Mössbauer study (124) of a mixed tin–platinum oxide catalyst supported on zinc aluminate at 500°–600°C indicated the presence of tin(IV), tin(II), and an alloy of tin and platinum in the active catalyst. Changes in the nature of the tin species with time and temperature were correlated with the catalytic activity of the material.

IV. Gold, Europium, and Ruthenium

Mössbauer investigations (125) of magnesium oxide and alumina impregnated with solutions of $KAu(CN)_2$ and $HAuCl_4$ showed that thermal decomposition above 140°C occurred without complete decomposition to metallic gold. Mössbauer chemical isomer shift, electron diffraction, and X-ray broadening data indicated a distribution of gold crystals between 100 and 1000 Å. An uncharacterized electron-deficient gold species attributed to the interaction between the gold and the support was also identified.

Alumina, impregnated with $Eu(NO_3)_3$ solution and dried at 413 K, has been reported (126) to give a Mössbauer spectrum indicating the presence of a hydroxy oxide. Mössbauer studies (127) of the reduction of trivalent europium supported on alumina showed that the fraction actually reduced was dependent on the europium-support interaction. This observation reinforces the findings of work involving supported iron microcrystallites and strongly suggests that the oxidation state of a supported catalyst depends on the interaction between the support and the active phase.

Mössbauer studies of the impregnation of silica with ruthenium chloride solution and subsequently dried at 383 K have reported (59) the presence of a ruthenium surface complex resembling $RuCl_3 \cdot xH_2O$. Recent work (128) has shown that Mössbauer spectra of ^{99}Ru supported on alumina, silica, activated charcoal, and X- and Y-zeolite are sensitive to the nature of the preparation and treatment of the samples.

Application of the Mössbauer effect, which is essentially a bulk phenomenon, to the study of surfaces has received significant attention in recent years. The usefulness of this technique lies in its ability to determine the electronic environment and symmetry of the surface nucleus, and it offers a method of investigation that is clearly complementary to other physical methods for the characterization of solid surfaces. Mössbauer spectroscopy has the attractive advantage that it may be used at a variety of pressures and can be applied to the study of heterogeneous catalysis and adsorption processes to probe the nature of the solid surface and its electronic modification when holding adsorbed species.

The first and most studied Mössbauer nucleus, iron-57, displays specific catalytic behavior. Mössbauer investigations of supported microcrystallites of iron and its oxide have demonstrated the importance of the techniques in the investigation of surface structure and chemistry. The application to other nuclei that have important catalytic qualities indicates the potential importance of the study of supported microcrystallites by Mössbauer spectroscopy in future investigations of catalysts. Developments in experimental techniques enabling *in situ* investigations are enhancing the scope of the technique.

ACKNOWLEDGMENTS

The author is indebted to Dr. A. G. Maddock for stimulating discussions during the writing of this work. The award of a Fellowship by I. C. I. Ltd. to the author is also acknowledged.

REFERENCES

1. Mössbauer, R. L., *Z. Phys.* **151**, 124 (1958).
2. Bancroft, G. M., Maddock, A. G., Ong, W. K., and Prince, R. H., *J. Chem. Soc.* p. 723 (1966).

3. Gibb, T. C., and Greenwood, N. N., *J. Chem. Soc.* p. 6989 (1966).

4. Epstein, L. M., *J. Chem. Phys.* **40**, 435 (1964).

5. Ablov, A. V., Belozerskii, G. N., Goldanskii, V. I., Makarov, E. F., Trukhtanov V. A., and Khrapov V. V., *Proc. Acad. Sci. USSR, Phys. Chem. Sect.* **151**, 712 (1963).

6. Lang, G., and Marshall, W. *Proc. Phys. Soc.* **87**, 3 (1966).

7. Herber, R. H., ed., *Am. Chem. Soc. Adv. Chem. Ser.* **68**, 30 (1966).

8. Schroeer, D., *in* "Mössbauer Effect Methodology" (I. J. Gruverman, ed.), Vol. 5, p. 141. Plenum, New York, 1969.

9. Goldanskii, V. I., and Suzdalev, I. P., *Russ. Chem. Rev.* **39**, 609 (1970).

10. Fluck, E., and Taube R., *in* "Development in Applied Spectroscopy" (E. L. Grove, ed.), Vol. 8, p. 244. Plenum, New York, 1970.

11. Hobson, M. C., *Progr. Surface Membrane Sci.* **5**, 1 (1972).

12. Hobson, M. C., *in* "Characterization of Solid Surfaces" (P. F. Kane, and G. B. Larrabee, eds.), p. 379. Plenum, New York, 1974.

13. Gager, H. M., and Hobson, M. C., *Catal. Rev.–Sci. Eng.* **11**, 117 (1975).

14. Delgass, W. N., *in* "Mössbauer Effect Methodology (I. J. Gruverman, and C. W. Seidel, eds.), Vol. 10. Plenum, New York, 1976.

15. Dumesic, J. A., *J. Phys. (Paris) Colloq. C6*, **37**, 279 (1976).

16. Kistner, O. C., and Sunyar, A. W., *Phys. Rev. Lett.* **4**, 412 (1960).

17. Flinn, P. A., Ruby, S. L., and Kehl, W. L., *Science* **143**, 1434 (1964).

18. Karyagin, S. V., *Dokl. Akad. Nauk SSSR* **148**, 1102 (1963).

19. Goldanskii, V. I., Gorodinskii, G. M., Karyagin, S. V., Korytko, L. A., Kriszhanski, L. M., Makarov, E. F., Suzdalev, I. P., and Khrapov, V. V., *Dokl. Akad. Nauk SSSR* **147**, 127 (1962).

20. Goldanskii, V. I., Makarov, E. F., and Khrapov, V. V., *Phys. Lett.* **3**, 344 (1963).

21. Constaboris, G., Lindquist, R. H., and Kundig, W., *Appl. Phys. Lett.* **7**, 59 (1965).

22. Collins, D. W., Dehn, J. T., and Malay, L. N., *in* "Mössbauer Effect Methodology" (I. J. Gruverman, ed.), Vol. 3, p. 103. Plenum, New York, 1967.

23. Neel, L., *C. R. Acad. Sci. Paris* **228**, 664 (1949).

24. Neel, L., *J. Phys. Soc. Jpn.* **17**, Suppl. B-1, 676 (1962).

25. Bean, C. P., and Livingston, J. D., *J. Appl. Phys.* **30**, 1205 (1959).

26. Kundig, W., Bömmel, H., Constaboris, G., and Lindquist, R. H., *Phys. Rev.* **142**, 327 (1966).

27. Kundig, W., Ando, K. J., Lindquist, R. H., and Constaboris, G., *Czech. J. Phys.* **17**, 467 (1967).

28. Hobson, M. C., and Gager, H. M., *J. Catal* **16**, 254 (1970).

29. Lindquist, R. H., Ando, K. J., Kundig, W., and Constaboris, G., *J. Phys. Chem. Solids* **28**, 2291 (1967).

30. Lindquist, R. H., Constaboris, G., Kundig, W., and Portis, A. M., *J. Appl. Phys.* **39**, 1001 (1968).

31. Constaboris, G., and Lindquist, R. H., *J. Phys. Chem. Solids* **30**, 819 (1969).

32. Hobson, M. C., *J. Electrochem. Soc.* **175c**, 115 (1968).

33. Schroeer, D., and Nininger, R. C., *Phys. Rev. Lett.* **19**, 632 (1967).

34. Schroeer, D., *Phys. Lett.* **27a**, 507 (1968).

35. Gager, H. M., Hobson, M. C., and Lefelhocz, J. F., *Chem. Phys. Lett.* **15**, 124 (1972).

36. Bancroft, G. M., Burns, R. G., and Maddock, A. G., *Geochim. Cosmochim. Acta* **31**, 2219 (1967).

37. Ingalls, R., *Phys. Rev.* **133**, A787 (1964).

38. Wickman, H. H., Klein, M. P., and Shirley, D. A., *Phys. Rev.* **152**, 345 (1966).

39. Blume, M., *Phys. Rev. Lett.* **18**, 305 (1967).

40. Blume, M., and Tjon, J. A., *Phys. Rev.* **165**, 446 (1968).

41. Schroeer, D., Marzke, R. F., Erickson, D. J., Marshall, S. W., and Wilenzick, R. M., *Phys. Rev. B* **2**, 4414 (1970).
42. Marshall, S. W., and Wilenzick, R. M., *Phys. Rev. Lett.* **16**, 219 (1966).
43. Suzdalev, I. P., Ya Gen, M., Goldanskii, V. I., and Makarov, E. F., *Soviet Phys.—JETP* **51**, 118 (1966).
44. Akselrod, S., Pasternak, M., and Bukshpan, S., *Phys. Rev. B* **11**, 1040 (1975).
45. Roth, S., and Hörl, E. M., *Phys. Lett. A* **25**, 299 (1967).
46. Viegers, M. P. A., Van Eijkeran, J. C. H., Van Deventer, M. M., and Trooster, J. M., *Proc. Int. Conf. Mössbauer Spectrosc., 1975.*
47. Van Wieringer, J. S., *Phys. Lett. A* **26**, 370 (1968).
48. Ruppin, R., *Phys. Rev. B* **2**, 1229 (1970).
49. Yoshioka, T., Koezuka, J., and Ikoma, H., *J. Catal.* **16**, 264 (1970).
50. Winzer, A., Vogt, F., Schödel, R, Bremmer, H., and Wiesner, E., *Z. Chem.* **10**, 312 (1970).
51. Baetzold, A. C., and Mack, L. E., *J. Chem. Phys.* **6**, 1513 (1975).
52. Johnson, K., and Messmer, R. P., *J. Vac. Sci. Technol.* **11**, 236 (1974).
53. Fripiat J. G., Chow K. T., Boudart, M., Diamond, J. G., and Johnson, K. H., *J. Mol. Catal.* **1**, 59 (1975).
54. Garten, R. L. *in* "Mössbauer Effect Methodology" I. J. Gruverman, and C. W. Seidel, (eds.), Vol. 10. Plenum, New York, 1976.
55. Dunken, H., and Hobart, H., *Z. Chem.* **6**, 276 (1966).
56. Hobson, M. C., and Gager, H. M., *Proc. Int. Congr. Catal., 4th, 1968.* p. 28
57. Hobart, H., and Arnold, D., *Proc. Conf. Application Mössbauer Effect, 1969*, p. 325.
58. Ross, P. N., and Delgass, W. N., *J. Catal.* **33**, 219 (1974).
59. Clausen, C. A., and Good, M. C., *J. Catal.* **38**, 92 (1975).
60. See Hobson (*12*), p. 397.
61. Aharoni, S. M., and Litt, M. H., *J. Appl. Phys.* **42**, 352 (1971).
62. Tachibama, T., and Ohya, T., *Bull. Chem. Soc. Jpn.* **42**, 2180 (1969).
63. Rabashov, A. M., Fabrichnyi, P. B., Strakhov, B. V., and Babeshkin, A. M., *Russ. J. Phys. Chem.* **46**, 765 (1972).
64. Arnold, D., and Hobart, H., *Z. Chem.* **8**, 197 (1968).
65. Hobson, M. C., and Campbell, A. D., *J. Catal.* **8**, 294 (1967).
66. Nicholson, W. J., and Burns, G., *Phys. Rev.* **133**, A1568 (1964).
67. Armstrong, R. J., Morrish, A. H., and Sawatzky, G. A., *Phys. Lett.* **23**, 414 (1966).
68. Suzdalev, I. P., and Makarov, E. F., *Proc. Conf. Applications Mössbauer Effect, 1969*, p. 201.
69. Blume, M., *Phys. Rev. Lett.* **14**, 96 (1965).
70. Coey, J. M. D., and Readman, P. W., *Earth Planet Sc. Lett.* **21**, 45 (1973).
71. Johnson, C. E., *J. Phys. Chem.* [2] **2**, 1996 (1969).
72. Shinjo, T., *J. Phys. Soc. Jpn.* **21**, 917 (1966).
73. Koelbel, H., and Kuespert, B., *Z. Phys. Chem. (Frankfurt)* **69**, 313 (1970).
74. Hobson, M. C., *Nature (London)* **214**, 79 (1967).
75. Hobson, M. C., and Gager, H. M., *J. Colloid Interface Sci.* **34**, 357 (1970).
76. Van Hardeveld, R., and Hartog, F., *Surface Sci.* **15**, 189 (1969).
77. See Hobson (*12*), p. 390.
78. Burns, G., *Phys. Rev.* **124**, 524 (1961).
79. Nozik, A. J., and Kaplan, M., *Phys. Rev.* **159**, 273 (1967).
80. Shirane, G., Cox, D. E., Ruby, S. L., *Phys. Rev.* **125**, 1158 (1962).
81. Coston, C. J., Ingalls, R., and Drickamer, H. G., *Phys. Rev.* **145**, 409 (1966).
82. Johnson, C. E., Marshall, W., and Perlow, G. J., *Phys. Rev.* **126**, 1503 (1962).
83. Gebhardt, J., and Herrington, K., *J. Phys. Chem.* **63**, 120 (1958).

84. Ouchi, M., and Kusunoki, I., *J. Chem. Soc. Jpn.* **85**, 612 (1964).
85. Blyholder, G., and Richardson, E. A., *J. Phys. Chem.* **68**, 3882 (1964).
86. Okamoto, G., Furnichi, R., and Sato, N., *Electrochim. Acta* **12**, 1287 (1967).
87. Blyholder, G., and Richardson, E. A., *J. Phys. Chem.* **66**, 2597 (1962).
88. Greenwood, N. N., and Gibb, T. C., "Mössbauer Spectroscopy," p. 248. Chapman and Hall, London, 1971.
89. Kalvius, M., *in* "Mössbauer Effect Methodology" (I. J. Gruverman, ed.), Vol. 1, p. 163. Plenum, New York, 1965.
90. Boudart, M., Delbouille, A., Dumesic, J. A., Khammoma, S., and Topsoe, H., *J. Catal.* **37**, 486 (1975).
91. Bartholemew, C. H., and Boudart, M., *J. Catal.* **29**, 278 (1973).
92. Williams, I. L., and Nason, D., *Surface Sci.* **45**, 377 (1974).
93. Garten, R. L., and Ollis, D. F., *J. Catal.* **35**, 2 (1974).
94. Garten, R. L., "Mössbauer Effect Methodology" (I. J. Gruverman, and C. W. Seidel eds.), Vol. 10. Plenum, New York, 1976.
95. Suzdalev, I. P., Shkarin, A. V., and Zhabrova, G. M., *Kinet. Catal. USSR* **10**, 179 (1969).
96. Gager, H. M., Lefelhocz, J. F., and Hobson, M. C., *Chem. Phys. Lett.* **23**, 386 (1973).
97. Anderson, J. H., *J. Catal.* **28**, 76 (1973).
98. Dominique, P. H., and Danon, J., *Chem. Phys. Lett.* **13**, 365 (1972).
99. Skalkina, L. V., Suzdalev, I. P., Kolchin, I. K., and Margolis, L. Y., *Kinet. Katal.* **10**, 456 (1969).
100. Hair, M. C., and Hertl, W. J., *J. Phys. Chem.* **73**, 4269 (1969).
101. Folman, M., and Yates, D. J. C., *J. Phys. Chem.* **63**, 183 (1959).
102. See Hobson (*12*), p. 393.
103. Delgass, W. N., Garten, R. L., and Boudart, M., *J. Phys. Chem.* **73**, 2970 (1969).
104. Blyholder, G., and Bowen, D. O., *J. Phys. Chem.* **66**, 1288 (1972).
105. Arnold, D., Kuchnel, S., and Hobart, H., *Z. Anorg. Allg. Chem.* **379**, 35 (1970).
106. Dumesic, J. A., Maksimov, Y., and Suzdalev, I. P., *in* "Mössbauer Effect Methodology" (I. J. Gruverman, and C. W. Seidel, eds.), Vol. 10. Plenum, New York, 1976.
107. Shalkina, L. V., Suzdalev, I. P., Kolehim, I. K., and Ya Margolis, L., *Kinet. Catal.* **10**, 378 (1969).
108. Kriegsmann, H., Ohlmann, G., Scheve, J., and Ulrich, F. J., *Int. Congr. Catal. 6th 1976.*
109. See Dumesic, J. A., Maksimov, Y., and Suzdalev, I. P., *in* "Mössbauer Effect Methodology," (I. V. Gruverman and C. W. Seidel, eds.), Vol. 10. Plenum, New York, 1976.
110. Delgass, W. N., Chen, L. Y., and Vogel, G., *Rev. Sci. Instrum.* **47**, 136 (1976).
111. Dumesic, J. A., and Topsoe, A. H., *Adv. Catal.* **26**, 122 (1977).
112. Karasev, A. N., Polak, L. S., Shlikhter, E. B., and Shpinel, V. S., *Kinet. Katal.* **6**, 710 (1965).
113. Karasev, A. N., Polak, L. S., Shlikhter, E. B., and Shpinel, V. S., *Zh. Fiz. Khim.* **39**, 3117 (1965).
114. Karasev, A. N., Kolbanovskii, Yu. A., Polak, L. A., and Shlikhter, E. B., *Kinet. Katal.* **8**, 232 (1967).
115. Suzdalev, I. P., Goldanskii, V. I., Makarov, E. F., Plachinda, A. S., and Korytko, L. A., *Sov. Phys.—JETP* **22**, 979 (1966).
116. Suzdalev, I. P., Plachinda, A. S., and Makarov, E. F., *Sov. Phys.—JETP* **26**, 897 (1968).
117. Goldanskii, V. I., Neimark, I. E., Plachinda, A. S., and Suzdalev, I. P., *Teor. Eksp. Khim.* **6**, 347 (1970).

118. Goldanskii, V. I., Suzdalev, I. P., Plachinda, A. S., and Shtyrkov, L. G., *Dokl. Akad. Nauk SSSR* **169**, 872 (1966).

119. Suzdalev, I. P., Plachinda, A. S., Makarov, E. F., and Dolgopolov, V. A., *Russ. J. Phys. Chem.* **41**, 1522 (1967).

120. Kordynk, S. L., Lisichenko, V. I., and Suzdalev, I. P., *Kolloid Zh.* **33**, 374 (1971).

121. See Hobson (*12*), p. 389.

122. Firsova, A. A., Khovanskaya, N. N., Tsygranov, A. D., Suzdalev, I. P., and Margolis, L. Ya., *Kinet. Katal.* **12**, 792 (1971).

123. Margolis, L. Ya, *J. Catal.* **21**, 93 (1971).

124. Gray, P. R., *in* "Mössbauer Effect Methodology" (I. J. Gruverman, and C. W. Seidel, eds.), Vol. 10. Plenum, New York, 1976.

125. Delgass, W. N., Boudart, M., and Parravano, G., *J. Phys. Chem.* **72**, 3563 (1968).

126. Ross, P. N., and Delgass, W. N., *in* "Catalysis" (J. W. Hightower, ed.), Vol. 1, p. 597. North-Holland Publ., Amsterdam, 1973.

127. Ross, P. N., and Delgass, W. N., *J. Catal.* **33**, 219 (1974).

128. Clausen, C. A., and Good, M. C., "Mössbauer Effect Methodology," (I. V. Gruveman, and C. W. Seidel, eds.), Vol. 10. Plenum, New York, 1976.

Subject Index

A

Alkoxy(aryloxy)cyclophosphazene, 59
Alumina support in Mössbauer studies, 265–267
Aluminum-cyclophosphazene complex, 67
Aluminum fluoride, 234, 235, 238, 242, 244, 247, 249
Amine hydrochloride, 55
o-Aminobenzaldehyde, self-condensation, 7, 8
Aminochlorocyclophosphazene, 56
Aminodichlorophosphorane, dehydrohalogenation, 46
Aminohalogenocyclophosphazene, ^{1}H NMR, 75, 76
Ammonocyclophosphazene, 72, 109
Angular overlap method, 114–143
    d orbital, 114–117
    energy-level diagram, 115, 123, 124
    interaction energy, 116, 122
    overlap, 116
    overlap integral, 115, 116
        angular dependence, 116
    σ orbital, 114–117
    stabilization energy, 118–120
        d orbital, 114, 115
Anilinocyclophosphazene, 72
Antibiotics, 36
Antiferromagnetism, 260
Antimony complex, binuclear, 195
Antimony-cyclophosphazene complex, 67
Antimony fluoride, 234, 235, 238, 240, 242, 244, 247, 249
Arsenic fluoride, 234, 235, 238, 240, 244
Azidophosphine, thermal decomposition, 44, 45

B

Barium fluoride, 234, 235, 237, 249
Beryllium fluoride, 234–236, 247, 249
Binuclear complex, 180, see also specific element complex
    bridged, 195–202
        charge transfer spectra, 195–197
            energy parameters, 196

inert chromophores, 198–200
labile, of unknown structure, 197, 198
symmetrical inert, 200–202
energy level diagram, 183
mixed-valence, 183
valency states, 182–184
Bismuth fluoride, 235
Borinic acid ester, macrocycle synthesis, 19, 20
Boron fluoride, 234, 235, 238, 242, 247
Boron–halogen bond, 148
Boron trifluoride adduct, 147, 148
Boron trihalide, 147–172, see also specific compounds
    chemistry, 147
    halogen-exchange reactions, 148, 149
    mixed, 147, 148
    NMR, 148
Boron trihalide (mixed) adducts, 147–172
    bridging mechanism, 160, 161
    comparison with BH_2X and BHX_2 adducts, 157, 158
    dissociation of halide ion, 159, 160
    donor–acceptor bond, 148, 150, 151
        dissociation of, 158, 159
    halogen-exchange reactions
        difluoroboron cations as donor, 166, 167
        donor, 166, 167
        equilibria, 162–166
        mechanisms, 158–162
    infrared spectroscopy, 151
    isotope substitution studies, 160, 161
    mass spectroscopy, 151, 152
    NMR, 151–157, 167–172
        complexation shifts, 168–171
        donor–acceptor bond strength, 168–171
        pairwise interaction, 171, 172
    preparation, 149–151
    properties, 152–157
    Raman spectroscopy, 151
Boron trihalide-amine adducts, 153, 154
Boron trihalide-dimethylsulfide adducts, 155
Boron trihalide-ester adducts, 156
Boron trihalide-ether adducts, 152, 153
Boron trihalide-ketone adducts, 157

Boron trihalide-phosphine adducts, 155, 156

Boron trihalide-tetrahaloborate anion adducts, 154, 155

Boron trihalide-tetramethylthiourea adducts, 157

Boron trihalide-tetramethylurea adducts, 156

Boron trihalide-thioester adducts, 156

Boron trihalide-trimethylphosphine oxide adducts, 157

Boron trihalide-trimethylphosphine sulfide adducts, 157

Bromine fluoride, 235, 236, 238, 240, 242, 244, 247, 249

Bromocyclophosphazene
alkyl and aryl derivatives, 43
synthesis, 43, 44

Bromocyclotriphosphazene, aminolysis, 52, 53

C

Cadmium complex, bond length, 139

Calcium fluoride, 234–236, 249

Calcium ion
in crown ether synthesis, 24
in template synthesis, 24

Carbon fluoride, 234–236, 238, 239, 242, 247, 249

Catalyst, Mössbauer effect, 276–280

Cesium complex
bond length, 136
geometry, 129

Cesium fluoride, 234, 235, 237, 247

Cesium ion, in crown ether synthesis, 24

Charge-transfer process
converse, 190
energy parameters, contact ion pairs, 192
ligand-to-metal, 203–206
reactions, 179
single-ion, 208, 209
spectrum, 190

Chlorine fluoride, 235, 236, 238, 240, 242, 244, 246, 247, 249
oxygen-containing, 245

Chloroaminocyclotetraphosphazene, isomerization, 56

Chlorobromocyclophosphazene, synthesis, 43

Chlorocyclophosphazene, 42
alkyl and aryl derivatives, 43
cis-trans isomerization, 55
^{35}Cl NQR spectrum, 93
hydrolysis, 58, 59
metathetical exchange reactions, 63
peptide coupling agents, 75
reaction with thiolate, 61
ring degradation, 73
synthesis, 43
thermal polymerization, 71

Chlorodimethylaminocyclophosphazene, 74

Chlorophyll, 2, 14

Chlorothioalkoxy(aryloxy)cyclophosphazene, zene, 61

Chromatography of cyclophosphazenes, 46, 59

Chromium-carbonyl complex, 131
^{13}C-labeled, 140
geometry, 130
octahedral, 113
photochemical rearrangement, 141–143
Walsh diagram, 134

Chromium complex
binuclear
energy parameter, 196
inert chromophore, 198–200
cis isomer, 140

Chromium-cyclophosphazene complex, 68, 69

Chromium-iridium complex, binuclear, inert chromophore, 198

Chromophore, 198–200

Clathrate, 66, 70

Cobalt-carbonyl complex, 129

Cobalt complex
bond length, 139
charge-transfer process, 192, 193
cis isomer, 140
directly bonded, energy parameters, 205
ligand-to-metal charge transfer, 203, 206
square planar, 128
symmetrical bridged mixed-valence, 197
Walsh diagram, 135

Cobalt-cyclophosphazene complex, 67, 68

Cobalt ion, in template synthesis, 30
Cobalt–iron complex
  ion pair, charge-transfer process, 193
  linked pair, 194
Cobalt-macrocycle complex, 8–11, 13, 15,
    16, 18–20
  hexadentate, 15, 16
  octahedral, 8, 9
  quadridentate, 10, 15
Cobalt–ruthenium complex, linked pairs,
    194
Complex, see specific type
Copper complex
  binuclear, 195, 197, 198
  bond lengths, 136, 139
  d⁸ system, 137
Copper-cyclophosphazene complex, 67
Copper ion, in crown ether synthesis, 24
Copper-macrocycle complex, 8, 10–13, 15,
    19, 20, 29
  quadridentate, 10, 15
  square planar, 8
Copper tetrachloride, 132
  square planar, 128
Coproporphyrin, 36
Corrin, 36
Crown ether, 17, 18, 23–25, 36, 37
  metathetical exchange reactions of cy-
    clophosphazenes, 63
Crystal field splitting, 118, 119, 121, 122
Cyclam, 16, 17, 36
Cyclophosphazadiene, 59
Cyclophosphazene, 41–96, 108–111, see
    also specific compounds
  adducts, 66, 69, 70
    infrared spectroscopy, 84, 85
    X-ray crystallography, 69
  alcoholysis, 46
  aminolysis, 46–57
    reaction products, 48, 49
    S_N2 mechanism, 49
  applications, 96
  basicity measurements, 69, 91–93
  bonding, 94–96
  chromatography, 46, 59
  cis-trans isomerizations, 55, 56, 109
  clathrates, 66, 70
  crystal structure, 88–91
  cyclic tetramer, 42
  cyclic trimer, 42

dipole moment, 93, 94
electronic structure, 94–96
Friedel–Crafts reactions, 65, 66
halogen replacement reactions, 46–66
hydrolysis, 57–59
infrared spectroscopy, 82–85
kinetic studies, 56
mass spectra, 94
metathetical exchange reactions, 46,
    61–63, 109
mixed amino derivatives, 54, 55
nomenclature, 43
NMR spectroscopy, 69, 75–82
    ¹⁹F NMR, 81, 82
    ¹H NMR, 75–78
    ³¹P NMR, 78–81
nuclear quadrupole resonance spec-
    troscopy, 93
phenolysis, 46
Raman spectroscopy, 88
reaction at side chains, 70, 71
    with alkoxides, 59–61
    with aryloxides, 59–61
    with organometallic reagents, 63–65
    with thiolates, 59–61
ring degradation, 72, 73
ring-forming reactions, 44–46
salts, 66, 69, 70
solvent effects, 56, 57
spectrophotometry, 69
synthesis, 43–46
    from ammonium halides and
        halophosphoranes, 43, 44
    cyclization of linear phosphazenes, 45
    miscellaneous, 46
    thermal decomposition of azidophos-
        phines, 44, 45
thermal polymerization, 71, 72
thioalcoholysis, 46
X-ray crystallography, 86–91, 110
Cyclophosphazene–metal complex,
    66–69
X-ray crystallography, 67
Cyclotetraphosphazene, 43
  halogen replacement, 48
  infrared spectroscopy, 83
  mixed amino derivatives, 54, 55
  ³¹P NMR, 78–81
Cyclotriphosphazene, 43
  clathrates, 70

Cyclotriphosphazene (cont'd)
 infrared spectroscopy, 83
 kinetic studies, 56
 mixed amino derivatives, 54, 55
 synthesis, 45, 46

D

2,6-Diacetylpyridine, template synthesis,
 14
Diaminodialdehyde, condensation, 10–12
Diboron tetrafluoride adduct, reactions of,
 151
Difluoroboron cation, halogen-exchange
 reactions, 166, 167
Dimethylaminochlorocyclophosphazene,
 fluorination, 62, 63
Dimethylaminocyclotriphosphazene
 salts, 70
Directly bonded complexes, 202–207
Donor–acceptor pairs, 148, 150, 151,
 168–171
 charge-transfer process, 189, 190
Donor–acceptor processes, 221–225
Doppler effect, 255
d orbital, see Angular overlap method

E

Electron-transfer reaction, 179
 continuum theory, 213–217
 optical process, see Optical electron
  transfer process
 reverse, 181, 182
 thermal process, 179, 180, 191
Etioporphyrin, 36
Europium, microcrystallites, Mössbauer
 spectrum, 281, 282

F

Ferrocene, 201, 202
Ferrocyanide ion
 Franck–Condon energy, 188, 189
 photoemission spectra, 186

Fluoride, see also specific compounds
 binary
  matrix isolation studies, 246–250
  stretching force constants, 233, 234
  symmetry point groups, 233, 234
  vibrational spectra, 231–250
Fluoroalkoxycyclophosphazene, exchange reactions, 63, 109
Fluorocyclodiphosphazane, dehydrohalogenation, 46
Fluorocyclophosphazene, 44
 ^{19}F NMR, 81, 82
 metathetical exchange reactions, 61–63
 synthesis, 46
Fluorocyclotriphosphazene, 110
 aminolysis, 52, 53
Franck–Condon energy, 180, 188, 189
Franck–Condon principle, 179, 181
Friedel–Crafts reaction, cyclophosphazene, 65, 66

G

Gadolinium fluoride, 235, 238, 247, 249
Germanium fluoride, 234–236, 239, 244,
 247, 249
Gold, microcrystallites, Mössbauer spec-
 trum, 281, 282
Goldanskii–Karyagin effect, 260, 268

H

Halogen complex, 206, 207
 directly bonded, energy parameters,
  205
Halogenocyclophosphazene, 46, see also
 specific compounds
 adducts, 69, 70
 aminolysis, 46–57
  reaction pattern and mechanism,
   47–54
 halogen replacement, 47
 hydrolysis, 57–59
 kinetic studies, 56
 mass spectra, 94
 metal complexes, 67–69
 polymerization, 71, 72

reaction with organometallic reagents, 63–65
salts, 69
solvent effects, 56, 57
Halogenocyclotetraphosphazene, aminolysis, 53, 54
Halophosphorane, reaction with ammonium halides, 43, 44
Hemoglobin, 2, 135
Hexachlorocyclophosphazene aminolysis
proton abstraction mechanism, 51
reaction products, 49
Hexachlorocyclotriphosphazene
aminolysis, 46–49
fluorination, 61, 62
halogen replacement pathways, 47, 48
hydrolysis, 57, 58
reaction with catechol, 60
with diols, 60
with organometallic reagents, 64, 65
ring degradation, 73
Hydridocyclophosphazene, 45, 110
metalation, 74
Hydrogen ion, thermodynamic parameters, 186, 187
Hydroxyoxophosphazane, 57, 58

I

Indium fluoride, 235
Infrared spectroscopy,
of cyclophosphazenes, 69, 82–85
of mixed boron trihalide adducts, 151
Iodine fluoride, 234, 235, 238, 240, 242, 244, 247, 249
Iodophosphazene, 44
Ion pair, charge-transfer process, 190–194
Iridium complex, photoassociation, 138
Iron-carbonyl complex
bond length, 139
geometry, 131
trigonal bipyramidal, 113, 130
Iron–chromium complex, ion pair, 191
Iron–cobalt complex, binuclear, inert chromophore, 198
Iron complex
binuclear, 195
energy parameters, 196

inert chromophores, 198, 199
cis isomer, 140
directly bonded, 202
energy parameters, 205
geometry, 135
ligand-to-metal charge transfer, 204, 206
linked pairs, 194
symmetrical bridged mixed-valence, 197
Iron-57 compounds,
chemical isomer shift, 256
electric field gradient, 257
Iron ions, in template synthesis, 30
Iron–iron complex, ion pair, 191
Iron–macrocycle complex, 12–14, 16, 19, 20, 29
hexadentate, 15, 16
Iron–molybdenum complex
binuclear, inert chromophore, 199
energy parameter, 196
Iron oxide, Mössbauer spectra
hydrogen reduction, 269–272
microcrystallites, 259–275
heating in vacuo, 272–275
Isothiocyanatocyclophosphazene, mass spectra, 94

J

Jahn–Teller theorem, 113, 132–134, 136, 137

K

Kirsanov reaction, 46
Krypton fluoride, 236, 247

L

Lead fluoride, 235, 237, 239, 249
Lifschitz's salt, 137
Ligand, 114–143
closed chain, see Macrocycle
in electron transfer process, 184, 185, 191
multidentate, 1
open chain, 1

Ligand (*cont'd*)
  σ orbital, 116–121
    overlap integrals, 118
    site preferences, 138–141
    template effect, 13
Lithium fluoride, 234–236, 247, 249
Lithium ions, in template synthesis,
  24–26

M

Macrocycle, 1, 2, *see also* specific types
  geometry, 8–10
  mixed donor, 27–33
    nitrogen and oxygen, 27–29
    nitrogen, oxygen, and sulfur, 33
    nitrogen and phosphorus, 31,32
    nitrogen and sulfur, 29–31
    oxygen and sulfur, 32
  nitrogen-donor, 7–21
    benzenoid, 7–16
    nonbenzenoid, 16–19
  oxygen-donor, 22–26
  stability constants, 35, 36
  sulfur-donor, 26, 27
  template synthesis, 3, 4, 7–33
    choice, 33, 34
    equilibrium, 6, 7
    kinetic, 4, 5
    thermodynamic, 5, 6
Macrocycle–metal complex, 2, 4, 6, 7, 34,
  see also specific metal
  properties, 4
Magnesium complex, d^8 system, 137
Magnesium fluoride, 234–236, 247, 249
Magnesium ions, in template synthesis,
  24
Magnesium-macrocycle complex, 14
Manganese-carbonyl complex,
  geometry, 130
  Walsh diagram, 135
Manganese complex
  ^{13}C-labeled, 140
  symmetrical bridged mixed-valence,
    197
Manganese-macrocycle complex, 29
Mass spectroscopy
  of cyclophosphazenes, 94
  of mixed boron trihalide adducts, 151,
    152

Matrix isolation of transient species,
  246–250
Mercury complex, bond length, 139, 140
Metal, *see also* specific element; Transi-
  tion metal complex
  as template, 33, 34
Metal-cyclophosphazene complex, 67–69
Metal-hexamine complex, 193
Metalloporphyrin, 11
Metal-phthalocyanine complex, 3
Metal-salicylaldehyde complex, 27
Metal-trisethylenediamine complex, 193
Methylcyclophosphazene
  salts, 70
  synthesis, 109
Microcrystallite
  of iron oxide, 259–269
  Mössbauer effect, 255–282
  supported, 255–282
Mössbauer effect, 255
  in microcrystallites, 255–282
Mössbauer spectrum, 255, 256
  catalysts and gas adsorption, 276–280
  chemical isomer shift, 256
  europium microcrystallites, 281, 282
  gold microcrystallites, 281, 282
  iron oxide, 259–275
  magnetic hyperfine splitting, 258
  quadrupole splitting, 258
  ruthenium microcrystallites, 281, 282
  tin microcrystallites, 280, 281
Molecular orbital method, *see* Angular
  overlap method; specific ion
Molybdenum-carbonyl complex,
  geometry, 130
Molybdenum complex, cis isomer, 140
Molybdenum-cyclophosphazene complex, 67, 68

N

Nickel-carbonyl complex, tetrahedral,
  113, 128
Nickel complex, 134
  bond length, 139
  d^8 system, 137
  photochemical rearrangement, 141, 143
  square planar, 128, 130
  trigonal bipyramidal, 130
  trigonal planar, 131

Nickel ions, in template synthesis, 25, 28, 30
Nickel-macrocycle complex, 4, 5, 27, 29, 31
  hexadentate, 15, 16
  pseudo-octahedral, 8
  quadridentate, 10, 15
  square planar, 8
  square pyramidal, 9
Nickel oxide, microcrystallites, 262
Nitrogen fluoride, 235, 236, 238, 242, 247, 249
  oxygen-containing, 245
Nuclear magnetic resonance spectroscopy
  of boron trihalides, 148
  of cyclophosphazenes, 69, 75–82
  of mixed boron trihalide adducts, 151–157, 167–172
  template reactions, 35
Nuclear quadrupole resonance spectroscopy of cyclophosphazenes, 93

O

Optical electron transfer process, 179–225
  single-ion, *see* Photoemission
Osmium complex
  cis isomer, 140
  directly bonded, 202
Oxidant-reductant pair
  charge transfer process, 189
  linked, 194, 195
Oxocyclophosphazene, 60
Oxygen fluoride, 235, 236, 247, 249
Oxyhemoglobin, 135

P

Palladium-macrocycle complex, 20
Pentakisaminomonochlorocyclotriphosphazene, 52
Pentaphenylchlorocyclotriphosphazene
  hydrolysis, 58
  synthesis, 66
o-Phenylenediamine, template condensation, 11, 12
Phospham, 72, 109
Phosphazene
  bicyclic, 57

adduct formation, 69
crystal structure, 90, 110
linear cyclization, 45
Phosphorine, 73
Phosphorotrithioite, 61
Phosphorus fluoride, 234–236, 238, 240, 244, 247, 249
Phosphorus–nitrogen ring system, 43
Phosphorus pentachloride, ammonolysis, 43, 44
Photoemission, 185–189, 217–221
  electron transfer from solvent, 210, 211
    to solvent, 207–210
  free-energy parameters, 188
  spectra, 186
Phthalocyanin, 35
Poly(bisorganophosphazene), 72
Poly(dichlorophosphazene), 71, 72
Polyether macrocycle, 22
Poly(organophosphazene)
  applications, 96
  high molecular weight, 71
Polyphosphazene, 44
Polythioether macrocycle, 26, 27
Porphyrin, 14, 36, 135
Potassium fluoride, 234–236, 247, 249
Potassium ions, in template synthesis, 23, 24

R

Raman spectroscopy
  of cyclophosphazenes, 88
  of mixed boron trihalide adducts, 151
Redox potential, 191
Reorganization energy of optical electron transfer process, 179–225
  inner-sphere component, 184, 185
    bonding, 184
    ligand, 184
  outer sphere component, 184, 185
    solvent, 184
  using ellipsoid model, 223
  using separate-spheres model, 218
Rhenium-carbonyl complex, geometry, 130
Rhenium complex
  binuclear, directly bonded, 202
  ^{13}C-labeled, 140
Rhodium complex, binuclear, directly bonded, 202

Rubidium fluoride, 234, 235, 237, 247
Rubidium ions, in crown ether synthesis, 24
Ruthenium, microcrystallites, Mössbauer spectrum, 281, 282
Ruthenium complex
    binuclear
        directly bonded, 202
        symmetrical inert, 200–202
    charge-transfer process, 192, 193
    cis isomer, 140
    directly bonded, energy parameters, 205
    ligand-to-metal charge transfer, 203, 204, 206
    linked pairs, 194
    symmetrical bridged mixed-valence, 197

S

Selenium fluoride, 234, 239, 242, 244, 247
σ orbital, see Angular overlap method
Silica as support in Mössbauer spectra, 261, 262, 265, 267, 278, 280–282
Silicon fluoride, 234–236, 238, 239, 242, 244, 247, 249
Silver-manganese complex
    binuclear, inert chromophore, 199
    energy parameter, 196
Sodium fluoride, 234–236, 247, 249
Sodium ions, in template synthesis, 22, 25, 32
Solvation, ionic, 211–213
Spectrophotometry, see also specific types
    template reactions, 35
Spirobicyclotriphosphazene, 110
Spirocyclophosphazene, 72, 109
Spirophosphorane, 72
Strontium fluoride, 234, 235, 237, 249
Sulfur fluoride, 231, 234, 238–240, 242, 244, 247, 249
Support material, see specific substance

T

Tellurium fluoride, 234, 239, 242, 244, 247
Template reaction, 1–37, see also Macrocycle

applications, 36, 37
kinetic, 4, 5
physical studies, 34–36
thermodynamic, 5, 6
Tetraethoxycarbonylporphyrin, 36
Tetrahaloborate
    adducts, 154, 155
    exchange mechanism, 161, 162
Thallium fluoride, 235, 247, 249
Tin, microcrystallites, Mössbauer spectrum, 280, 281
Tin complex, binuclear, 195, 198
Tin fluoride, 235, 237, 239, 244, 249
Titanium complex, binuclear, 195
Titanium tetrachloride, 132
Transition metal complex, see also specific complex
    angular overlap method, 114–143, see also Angular overlap method
    bonding, 113–143
    cis-divacant, 128
    crystal forces, 114
    electronic spectra, 114
    geometry, 127–134
        five-coordinate, 130, 131
        four-coordinate, 127–130
        three-coordinate, 131, 132
    ligand site preferences, 138–141
    octahedral, 127
    photochemical rearrangements, intramolecular, 141–143
    polymeric, 113, 114
    square planar, 128
    structure, 113–143
        d^8 system, 135–138
        d^9 system, 135–138
        gas-phrase, 113
    tetrahedral, 127, 128
    trigonal bipyramidal, 127
    trigonal planar, 127
    Walsh diagrams, 113, 134, 135
Transition metal hexafluoride, 113, 132
Transition metal ion
    cis-divacant, 124
    hexaquo
        molecular orbital rationalization, 125, 126
        octahedral, 125
        rates of reaction, 125–127
        square pyramidal, 125
    octahedral, 117–122

crystal field method, 119–121
    ligand σ-orbital labeling, 117
    molecular orbital method, 119–121
    π bonding, 121, 122
    polar coordinates, 117
    σ-type interactions, 117–121
    stabilization energy, 118–121
  square planar, 124
  tetrahedral coordination, 123, 124
  trigonal bipyramidal, 124
  trigonal planar, 124
Transition metal-macrocycle complex, 17
Transition metal tetrachloride, 113
Triethyltetramine, condensation, 19
Tungsten-carbonyl complex, geometry,
    130
Tungsten complex, binuclear, 195
Tungsten-cyclophosphazene complex, 67,
    68

U

Uranium complex, binuclear, 195

V

Vanadium complex
  binuclear, 195
  symmetrical bridged mixed-valence,
    197

Vanadium tetrachloride, 132
Vibrational spectroscopy
  of binary fluorides, 231–250
  molecular force constants, 232
Vitamin B$_{12}$, 2, 13
VSEPR scheme, 113, 117, 127, 133, 135

W

Walsh diagram, 113
  derivation of, 134, 135

X

Xenon fluoride, 237, 239, 247
X-ray crystallography of cyclophos-
    phazenes, 67, 69, 86–91, 110

Z

Zeolite, 279
Zinc ions
  in crown ether synthesis, 24
  in template synthesis, 24, 28, 30
Zinc-macrocycle complex, 12, 16, 20, 29
  hexadentate, 15, 16
Zinc tetrachloride, 132
  tetrahedral, 128

CONTENTS OF PREVIOUS VOLUMES

VOLUME 1

Mechanisms of Redox Reactions of
Simple Chemistry
H. Taube

Compounds of Aromatic Ring Systems
and Metals
E. O. Fischer and H. P. Fritz

Recent Studies of the Boron Hydrides
William N. Lipscomb

Lattice Energies and Their Significance
in Inorganic Chemistry
T. C. Waddington

Graphite Intercalation Compounds
W. Rüdorff

The Szilard-Chalmers Reactions in Solids
Garman Harbottle and Norman Sutin

Activation Analysis
D. N. F. Atkins and A. A. Smales

The Phosphonitrilic Halides and Their
Derivatives
N. L. Paddock and H. T. Searle

The Sulfuric Acid Solvent System
R. J. Gillespie and E. A. Robinson

AUTHOR INDEX—SUBJECT INDEX

VOLUME 2

Stereochemistry of Ionic Solids
J. D. Dunitz and L. E. Orgel

Organometallic Compounds
John Eisch and Henry Gilman

Fluorine-Containing Compounds of
Sulfur
George H. Cady

Amides and Imides of the Oxyacids of
Sulfur
Margot Becke-Goehring

Halides of the Actinide Elements
Joseph J. Katz and Irving Sheft

Structures of Compounds Containing
Chains of Sulfur Atoms
Olav Foss

Chemical Reactivity of the Boron
Hydrides and Related Compounds
F. G. A. Stone

Mass Spectrometry in Nuclear Chemistry
*H. G. Thode, C. C. McMullen, and
K. Fritze*

AUTHOR INDEX—SUBJECT INDEX

VOLUME 3

Mechanisms of Substitution Reactions of
Metal Complexes
Fred Basolo and Ralph G. Pearson

Molecular Complexes of Halogens
L. J. Andrews and R. M. Keefer

Structures of Interhalogen Compounds
and Polyhalides
*E. H. Wiebenga, E. E. Havinga, and
K. H. Boswijk*

Kinetic Behavior of the Radiolysis
Products of Water
Christiane Ferradini

The General, Selective, and Specific
Formation of Complexes by Metallic
Cations
G. Schwarzenbach

Atmosphere Activities and Dating
Procedures
A. G. Maddock and E. H. Willis

Polyfluoroalkyl Derivatives of Metalloids
and Nonmetals
R. E. Banks and R. N. Haszeldine

AUTHOR INDEX—SUBJECT INDEX

VOLUME 4

Condensed Phosphates and Arsenates
Erich Thilo

Olefin, Acetylene, and π-Allylic
Complexes of Transition Metals
R. G. Guy and B. L. Shaw

Recent Advances in the Stereochemistry
of Nickel, Palladium, and Platinum
J. R. Miller

The Chemistry of Polonium
K. W. Bagnall

The Use of Nuclear Magnetic Resonance
in Inorganic Chemistry
E. L. Muetterties and W. D. Phillips

Oxide Melts
   J. D. Mackenzie
AUTHOR INDEX—SUBJECT INDEX

VOLUME 5

The Stabilization of Oxidation States of
   the Transition Metals
   R. S. Nyholm and M. L. Tobe

Oxides and Oxyfluorides of the Halogens
   M. Schmeisser and K. Brandle

The Chemistry of Gallium
   N. N. Greenwood

Chemical Effects of Nuclear Activation in
   Gases and Liquids
   I. G. Campbell

Gaseous Hydroxides
   O. Glemser and H. G. Wendlandt

The Borazines
   E. K. Mellon, Jr., and J. J. Lagowski

Decaborane-14 and Its Derivatives
   M. Frederick Hawthorne

The Structure and Reactivity of
   Organophosphorus Compounds
   R. F. Hudson

AUTHOR INDEX—SUBJECT INDEX

VOLUME 6

Complexes of the Transition Metals with
   Phosphines, Arsines, and Stibines
   G. Booth

Anhydrous Metal Nitrates
   C. C. Addison and N. Logan

Chemical Reactions in Electric
   Discharges
   Adli S. Kana'an and John L. Margrave

The Chemistry of Astatine
   A. H. W. Aten, Jr.

The Chemistry of Silicon–Nitrogen
   Compounds
   U. Wannagat

Peroxy Compounds of Transition Metals
   J. A. Connor and E. A. V. Ebsworth

The Direct Synthesis of Organosilicon
   Compounds
   J. J. Zuckerman

The Mössbauer Effect and Its Application
   in Chemistry
   E. Fluck

AUTHOR INDEX—SUBJECT INDEX

VOLUME 7

Halides of Phosphorous, Arsenic,
   Antimony, and Bismuth
   L. Kolditz

The Phthalocyanines
   A. B. P. Lever

Hydride Complexes of the Transition
   Metals
   M. L. H. Green and D. L. Jones

Reactions of Chelated Organic Ligands
   Quintus Fernando

Organoaluminum Compounds
   Roland Köster and Paul Binger

Carbosilanes
   G. Fritz, J. Grobe, and D. Kummer

AUTHOR INDEX—SUBJECT INDEX

VOLUME 8

Substitution Products of the Group VIB
   Metal Carbonyls
   Gerard R. Dobson, Ingo W. Stolz, and
   Raymond K. Sheline

Transition Metal Cyanides and Their
   Complexes
   B. M. Chadwick and A. G. Sharpe

Perchloric Acid
   G. S. Pearson

Neutron Diffraction and Its Application in
   Inorganic Chemistry
   G. E. Bacon

Nuclear Quadrupole Resonance and Its
   Application in Inorganic Chemistry
   Masaji Kubo and Daiyu Nakamura

The Chemistry of Complex
   Aluminohydrides
   E. C. Ashby

AUTHOR INDEX—SUBJECT INDEX

VOLUME 9

Liquid–Liquid Extraction of Metal Ions
   D. F. Peppard

Nitrides of Metals of the First Transition
Series
R. Juza

Pseudohalides of Group IIIB and IVB
Elements
M. F. Lappert and H. Pyszora

Stereoselectivity in Coordination
Compounds
J. H. Dunlop and R. D. Gillard

Heterocations
A. A. Woolf

The Inorganic Chemistry of Tungsten
R. V. Parish

AUTHOR INDEX—SUBJECT INDEX

VOLUME 10

The Halides of Boron
A. G. Massey

Further Advances in the Study of
Mechanisms of Redox Reactions
A. G. Sykes

Mixed Valence Chemistry—A Survey and
Classification
Melvin B. Robin and Peter Day

AUTHOR INDEX—SUBJECT INDEX—
VOLUMES 1–10

VOLUME 11

Technetium
K. V. Kotegov, O. N. Pavlov, and
V. P. Shvedov

Transition Metal Complexes with Group
IVB Elements
J. F. Young

Metal Carbides
William A. Frad

Silicon Hydrides and Their Derivatives
B. J. Aylett

Some General Aspects of Mercury
Chemistry
H. L. Roberts

Alkyl Derivaties of the Group II Metals
B. J. Wakefield

AUTHOR INDEX—SUBJECT INDEX

VOLUME 12

Some Recent Preparative Chemistry of
Protactinium
D. Brown

Vibrational Spectra of Transition Metal
Carbonyl Complexes
Linda M. Haines and M. H. Stiddard

The Chemistry of Complexes Containing
2,2'-Bipyridyl, 1,10-Phenanthroline,
or 2,2', 6',2"-Terpyridyl as Ligands
W. R. McWhinnie and J. D. Miller

Olefin Complexes of the Transition
Metals
H. W. Quinn and J. H. Tsai

Cis and Trans Effects in Cobalt(III)
Complexes
J. M. Pratt and R. G. Thorp

AUTHOR INDEX—SUBJECT INDEX

VOLUME 13

Zirconium and Hafnium Chemistry
E. M. Larsen

Electron Spin Reasonance of Transition
Metal Complexes
B. A. Goodman and J. B. Raynor

Recent Progress in the Chemistry of
Fluorophosphines
John F. Nixon

Transition Metal Clusters with π-Acid
Ligands
R. D. Johnston

AUTHOR INDEX—SUBJECT INDEX

VOLUME 14

The Phosphazotrihalides
M. Bermann

Low Temperature Condensation of High
Temperature Species as a Synthetic
Method
P. L. Timms

Transition Metal Complexes Containing
Bidentate Phosphine Ligands
W. Levason and C. A. McAuliffe

Beryllium Halides and Pseudohalides
N. A. Bell

Sulfur–Nitrogen–Fluorine Compounds
  O. Glemser and R. Mews
AUTHOR INDEX—SUBJECT INDEX

VOLUME 15

Secondary Bonding to Nonmetallic
  Elements
  N. W. Alcock

Mössbauer Spectra of Inorganic
  Compounds: Bonding and Structure
  G. M. Bancroft and R. H. Platt

Metal Alkoxides and Dialkylamides
  D. C. Bradley

Fluoroalicyclic Derivatives of Metals and
  Metalloids
  W. R. Cullen

The Sulfur Nitrides
  H. G. Heal

AUTHOR INDEX—SUBJECT INDEX

VOLUME 16

The Chemistry of
  Bis(trifluoromethyl)amino
  Compounds
  H. G. Ang and Y. C. Syn

Vacuum Ultraviolet Photoelectron
  Spectroscopy of Inorganic Molecules
  R. L. DeKock and D. R. Lloyd

Fluorinated Peroxides
  Ronald A. De Marco and
  Jean'ne M. Shreeve

Fluorosulfuric Acid, Its Salts, and
  Derivatives
  Albert W. Jache

The Reaction Chemistry of Diborane
  L. H. Long

Lower Sulfur Fluorides
  F. Seel

AUTHOR INDEX—SUBJECT INDEX

VOLUME 17

Inorganic Compounds Containing the
  Trifluoroacetate Group
  C. D. Garner and B. Hughes

Homopolyatomic Cations of the
  Elements
  R. J. Gillespie and J. Passmore

Use of Radio-Frequency Plasma in
  Chemical Synthesis
  S. M. L. Hamblyn and B. G. Reuben

Copper(I) Complexes
  F. H. Jardine

Complexes of Open-Chain Tetradenate
  Ligands Containing Heavy Donor
  Atoms
  C. A. McAuliffe

The Functional Approach to Ionization
  Phenomena in Solutions
  U. Mayer and V. Gutmann

Coordination Chemistry of the Cyanate,
  Thiocyanate, and Selenocyanate Ions
  A. H. Norbury

SUBJECT INDEX

VOLUME 18

Structural and Bonding Patterns in
  Cluster Chemistry
  K. Wade

Coordination Number Pattern
  Recognition Theory of Carborane
  Structures
  Robert E. Williams

Preparation and Reactions of
  Perfluorohalogenoorganosulfenyl
  Halides
  A. Haas and U. Niemann

Correlations in Nuclear Magnetic
  Shielding. Part I
  Joan Mason

Some Applications of Mass Spectroscopy
  in Inorganic and Organometallic
  Chemistry
  Jack M. Miller and Gary L. Wilson

The Structures of Elemental Sulfur
  Beat Meyer

Chlorine Oxyfluorides
  K. O. Christe and C. J. Schack

SUBJECT INDEX

VOLUME 19

Recent Chemistry and Structure
    Investigation of Nitrogen Triiodide,
    Tribromide, Trichloride, and Related
    Compounds
    Jochen Jander

Aspects of Organo-Transition-Metal
    Photochemistry and Their Biological
    Implications
    Ernst A. Koerner von Gustorf,
    Luc H. G. Leenders, Ingrid Fischler,
    and Robin N. Perutz

Nitrogen-Sulfur-Fluorine Ions
    R. Mews

Isopolymolybdates and Isopolytungstates
    Karl-Heinz Tytko and Oskar Glemser

SUBJECT INDEX

VOLUME 20

Recent Advances in the Chemistry of the
    Less-Common Oxidation States of the
    Lanthanide Elements
    D. A. Johnson

Ferrimagnetic Fluorides
    Alain Tressaud and Jean Michel Dance

Hydride Complexes of Ruthenium
    Rhodium, and Iridium
    G. L. Geoffroy and J. R. Lehman

Structures and Physical Properties of
    Polynuclear Carboxylates
    Janet Catterick and Peter Thornton

SUBJECT INDEX

A
B
C 8
D 9
E 0
F 1
G 2
H 3
I 4
J 5